A4 1$30.15

Introduction to Queueing Theory

Second edition

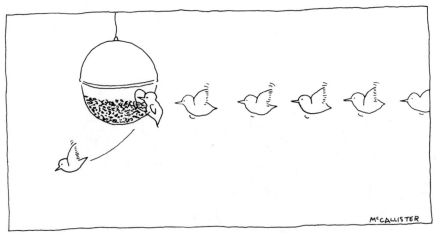

Drawing by McCallister; © 1977 *The New Yorker* Magazine, Inc.

I hope you find this book useful.

Robert B Cooper

Introduction to Queueing Theory

Second Edition

Robert B. Cooper
Computer Systems and Management Science
Florida Atlantic University
Boca Raton, Florida

North Holland
New York • Oxford

Elsevier North Holland, Inc.
52 Vanderbilt Avenue, New York, New York 10017

Distributors outside the United States and Canada:
Edward Arnold (Publishers) Limited
41 Bedford Square
London WC1B 3DQ, England

© 1981 by Elsevier North Holland, Inc.

Library of Congress Cataloging in Publication Data

Cooper, Robert B
    Introduction to queueing theory.

    Bibliography: p.
    Includes index.
    1. Queueing theory. I. Title.
T57.9.C66   1981   519.8'2   80-16481
ISBN 0-444-00379-7

Copy Editor   Joe Fineman
Desk Editor   Louise Calabro Schreiber
Design   Edmée Froment
Art Editor   Glen Burris
Cover Design   Paul Agule Design
Production Manager   Joanne Jay
Compositor   Science Typographers, Inc.
Printer   Haddon Craftsmen

Manufactured in the United States of America

For
   my son
      Bill

Contents

List of Exercises		xi
Preface to the Second Edition		xiii
1	**Scope and Nature of Queueing Theory**	1
2	**Review of Topics from the Theory of Probability and Stochastic Processes**	9
	2.1 Random Variables	9
	2.2 Birth-and-Death Processes	14
	2.3 Statistical Equilibrium	19
	2.4 Probability-Generating Functions	26
	2.5 Some Important Probability Distributions	34
	Bernoulli Distribution	34
	Binomial Distribution	34
	Multinomial Distribution	35
	Geometric Distribution	36
	Negative-Binomial (Pascal) Distribution	37
	Uniform Distribution	38
	Negative-Exponential Distribution	42
	Erlangian (Gamma) Distribution	50
	Poisson Distribution	64
	2.6 Remarks	71
3	**Birth-and-Death Queueing Models**	73
	3.1 Introduction	73
	3.2 Relationship Between the Outside Observer's Distribution and the Arriving Customer's Distribution	77
	3.3 Poisson Input, s Servers, Blocked Customers Cleared: The Erlang Loss System	79

3.4	Poisson Input, s Servers with Exponential Service Times, Blocked Customers Delayed: The Erlang Delay System	90
3.5	Quasirandom Input	102
3.6	Equality of the Arriving Customer's n-Source Distribution and the Outside Observer's $(n-1)$-Source Distribution for Birth-and-Death Systems with Quasirandom Input	105
3.7	Quasirandom Input, s Servers, Blocked Customers Cleared: The Engset Formula	108
3.8	Quasirandom Input, s Servers with Exponential Service Times, Blocked Customers Delayed	111
3.9	Summary	116

4 Multidimensional Birth-and-Death Queueing Models   123

4.1	Introduction	123
4.2	Product Solutions (Mixed Traffic, Queues in Tandem)	126
	Infinite-Server Group with Two Types of Customers	126
	Finite-Server Group with Two Types of Customers and Blocked Customers Cleared	128
	Queues in Tandem	132
4.3	Generating Functions (Overflow Traffic)	139
4.4	Macrostates (Priority Reservation)	150
4.5	Indirect Solution of Equations (Load Carried by Each Server of an Ordered Group)	153
4.6	Numerical Solution of State Equations by Iteration (Gauss-Seidel and Overrelaxation Methods)	158
4.7	The Equivalent Random Method	165
4.8	The Method of Phases	171

5 Imbedded-Markov-Chain Queueing Models   176

5.1	Introduction	176
5.2	The Equation $L = \lambda W$ (Little's Theorem)	178
5.3	Equality of State Distributions at Arrival and Departure Epochs	185
5.4	Mean Queue Length and Mean Waiting Time in the $M/G/1$ Queue	189
5.5	Riemann-Stieltjes Integrals	192
5.6	Laplace-Stieltjes Transforms	197
5.7	Some Results from Renewal Theory	200
5.8	The $M/G/1$ Queue	208
	The Mean Waiting Time	209
	The Imbedded Markov Chain	210
	The Pollaczek-Khintchine Formula	216
	The Busy Period	228
5.9	The $M/G/1$ Queue with Finite Waiting Room	235
5.10	The $M/G/1$ Queue with Batch Arrivals	241
5.11	Optimal Design and Control of Queues: The N-Policy and the T-Policy	243
5.12	The $M/G/1$ Queue with Service in Random Order	253
5.13	Queues Served in Cyclic Order	261
5.14	The $GI/M/s$ Queue	267
5.15	The $GI/M/s$ Queue with Service in Random Order	275

Contents ix

6 Simulation of Queueing Models   **281**

    6.1 Introduction   281
    6.2 Generation of Stochastic Variables   284
    6.3 Simulation Programming Languages   287
    6.4 Statistical Questions   288
    6.5 Examples   298
        Waiting Times in a Queue with Quasirandom Input and Constant Service Times   298
        Response Times in a Processor-Sharing Operating System   301

7 Annotated Bibliography   **308**

Appendix   **315**

References   **325**

Index   **341**

List of Exercises

Chapter 1	
1	6
2	6
3	6
4	6
5	7
6	8

Chapter 2	
1	14
2	17
3	24
4	30
5	31
6	34
7	35
8	38
9	38
10	41
11	49
12	49
13	49

Chapter 2	
(continued)	
14	49
15	49
16	49
17	51
18	56
19	56
20	59
21	62
22	63
23	63
24	63
25	70
26	71

Chapter 3	
1	78
2	79
3	81
4	82
5	82

Chapter 3	
(continued)	
6	82
7	82
8	82
9	83
10	85
11	87
12	87
13	89
14	90
15	90
16	92
17	92
18	93
19	95
20	95
21	95
22	98
23	98
24	98
25	98
26	98

Chapter 3 (continued)		Chapter 4 (continued)		Chapter 5 (continued)	
27	99	8	135	16	233
28	99	9	136	17	233
29	99	10	138	18	233
30	99	11	138	19	233
31	99	12	148	20	234
32	101	13	153	21	238
33	102	14	157	22	240
34	110	15	170	23	240
35	111	16	170	24	243
36	114	17	171	25	250
37	116	18	175	26	252
38	116	19	175	27	258
39	116			28	259
40	116			29	260
41	118	**Chapter 5**		30	267
42	118	1	188	31	267
43	118	2	188	32	273
44	119	3	196	33	273
45	119	4	199	34	273
46	120	5	199	35	273
		6	199	36	279
		7	203		
Chapter 4		8	203		
1	125	9	206	**Chapter 6**	
2	129	10	206	1	285
3	130	11	207	2	285
4	130	12	222	3	298
5	130	13	223	4	298
6	131	14	224		
7	131	15	227		

Preface to the Second Edition

This is a revised, expanded, and improved version of my textbook, *Introduction to Queueing Theory*. As before, it is written primarily for seniors and graduate students in operations research, computer science, and industrial engineering; and as before, the emphasis is on insight and understanding rather than either cookbook application or fine mathematical detail. Although it has been structured for use primarily as a textbook, it should be a useful reference for researchers and practitioners as well. The second edition reflects the feedback of students and instructors who have used the first edition (which began as a set of notes for an in-house course I taught at Bell Laboratories), as well as my own experience teaching from the book at the Georgia Institute of Technology (in the School of Industrial and Systems Engineering and the School of Information and Computer Science), the University of Michigan (in the Department of Industrial and Operations Engineering), the New Mexico Institute of Mining and Technology (in the Department of Mathematics), and Florida Atlantic University (in the Department of Computer Systems and Management Science). The objective of the second edition is to improve on the first edition with respect to clarity, comprehensiveness, currency, and, especially, its utility both for teaching and self-study.

In particular, improvements over the first edition include

1. the availability from the publisher of an instructor's manual containing detailed solutions, by Børge Tilt, to essentially all of the many exercises;
2. a new chapter, on simulation of queueing models;
3. an expanded and more complete treatment of background mathematical topics, such as Laplace-Stieltjes transforms and renewal theory;
4. an annotated bibliography of almost all of the English language books on queueing theory in existence;
5. the placement of exercises in the text just after the discussion they are

designed to illuminate, rather than all together at the end of each chapter, and the addition of new exercises, including numerical exercises that reflect engineering and economic concepts;
6. the presentation in the text of some of the more difficult and important material that was left to the exercises in the first edition;
7. the integration of the material of Chapter 6 of the first edition *("Waiting Times")* into the other chapters;
8. the inclusion of new material and references published after the first edition;
9. revision throughout to improve clarity and correct errors;
10. arrangement of the material so that the book can be used either for a course that emphasizes practical applications, or for a course that emphasizes more advanced mathematical concepts with an eye toward more sophisticated applications or research.

The first chapter illustrates the nature of the subject, its potential for application, and the interplay between intuitive and mathematical reasoning that makes it so interesting. Chapter 2 contains a review of topics from applied probability and stochastic processes. Since the reader is presumed to have had a course in applied probability, some of the material (such as the discussion of common probability distributions) may well be familiar, but other topics (such as elementary renewal theory) may be less familiar. The instructor can decide which topics to emphasize and which to assign for review or self-study. Chapter 3 covers the basic one-dimensional birth-and-death queueing models, including the well-known Erlang B, Erlang C, and finite-source models. Care is taken to distinguish between the viewpoint of the arriving customer and that of the outside observer. Chapter 4 discusses multidimensional birth-and-death models. This chapter is organized by model, each of which exemplifies a different method of solution of the multidimensional birth-and-death equations. The methods include product solutions, probability-generating functions, inspection, the method of phases, and numerical analysis. The models include networks of queues (with many examples and references), systems with alternate routing, and systems with multiple types or classes of traffic. Little's theorem, Kendall's notation, and the Pollaczek-Khintchine formula are mentioned, but detailed discussion and derivations are left to Chapter 5; the instructor can refer ahead or not as he sees fit. Relatively "advanced" mathematical concepts, such as Laplace-Stieltjes transforms and imbedded Markov chains, are not used until Chapter 5. The intention is that an elementary course, requiring no knowledge of "advanced" applied mathematics but nevertheless conveying the main results and the essence of the subject, can be constructed from topics covered in the first four chapters. For the very practical-minded, a course can be constructed using only Chapters 1 and 3.

Chapter 5, which is the longest chapter, uses more advanced mathematics and is more rigorous in its arguments than the preceding chapters. It covers much of the material of classical queueing theory, as well as some more specialized topics, such as queues with service in random order, which can be skipped without

Preface

loss of continuity. Also, it includes a summary treatment of Riemann-Stieltjes integration and Laplace-Stieltjes transforms, tools that are often used but rarely explained in textbooks on queueing theory or applied stochastic processes. An intermediate-advanced course for students with prior knowledge of elementary queueing theory can be built around Chapter 5. Chapter 6, which is completely new, surveys the field of simulation as applied to queueing models. It contains, among other things, one of the first textbook treatments of the regenerative method in the context of simulation of queueing models, and discussions of two structurally simple but nevertheless sophisticated simulation studies. Chapter 7 concludes with an annotated bibliography of (almost) all the books on queueing theory and its application that have been published in the English language. (The reader may well turn to Chapter 7 first.)

Colleagues who have contributed to this book, either directly by their reading and commenting on the manuscript or indirectly by their personal and professional influence, include Carl Axness, Paul J. Burke, Grace M. Carter, Ralph L. Disney, Hamilton Emmons, Carl M. Harris, Philip Heidelberger, Daniel P. Heyman, Marcel F. Neuts, Stephen St. John, Donald G. Sanford, Bruce W. Schmeiser, Samuel S. Stephenson, Shaler Stidham, Jr., Ryszard Syski, and Eric Wolman. Special thanks go to Børge Tilt who read the manuscript in its entirety (through several versions), made many excellent suggestions, and wrote the solutions manual (whose existence is surely one of the major attractions of this book). Finally, I want to thank Elizabeth Fraissinet and Dora Yates, who typed the manuscript, and Kenneth J. Bowman and Louise Calabro Schreiber of North Holland Publishing Company, who oversaw the editing and production.

Robert B. Cooper

Introduction to Queueing Theory

Second edition

[1]
Scope and Nature of Queueing Theory

This text is concerned with the mathematical analysis of systems subject to demands whose occurrences and lengths can, in general, be specified only probabilistically.

For example, consider a telephone system, whose function is to provide communication paths between pairs of telephone sets (customers) on demand. The provision of a permanent communications path between each pair of telephone sets would be astronomically expensive, to say the least, and perhaps impossible. In response to this problem, the facilities needed to establish and maintain a talking path between a pair of telephone sets are provided in a common pool, to be used by a call when required and returned to the pool when no longer needed. This introduces the possibility that the system will be unable to set up a call on demand because of a lack of available equipment at that time. Thus the question immediately arises: How much equipment must be provided so that the proportion of calls experiencing delays will be below a specified acceptable level?

Questions similar to that just posed arise in the design of many systems quite different in detail from a telephone system. How many taxicabs should be on the streets of New York City? How many beds should a hospital provide? How many teletypewriter stations can a time-shared computer serve?

The answers to these questions will be based in part on such diverse considerations as politics, economics, and technical knowledge. But they share a common characteristic: In each case the times at which requests for service will occur and the lengths of times that these requests will occupy facilities cannot be predicted except in a statistical sense.

The purpose of this text is to develop and explicate a mathematical theory that has application to the problems of design and analysis of such systems. Although these systems are usually very complex, it is often possible to abstract from the system description a mathematical model whose analysis yields useful information.

This mathematical theory is a branch of applied probability theory and is known variously under the names traffic theory, queueing theory, congestion theory, the theory of mass service, and the theory of stochastic service systems. The term traffic theory is often applied to theories of telephone and communications traffic, as well as to theories of vehicular traffic flow. These two areas share some common ground, and the material we shall develop will be useful in both fields. The term queueing theory is often used to describe the more specialized mathematical theory of waiting lines (queues). But some of the most interesting and most useful models are based on systems in which queues are not allowed to form, so that the term queueing theory does not seem completely appropriate. The subject of this text is perhaps better described by the broader terms congestion theory and stochastic service system theory. However, the name queueing theory has become most widely used to describe the kind of material presented in this text, and therefore I chose to entitle the book *Introduction to Queueing Theory*.

Historically, the subject of queueing theory has been developed largely in the context of telephone traffic engineering. Some of our examples will be drawn from this area, but it should be kept in mind that the theory is widely applicable in engineering, operations research, and computer science. Also, it is worth noting that the mathematics underlying queueing theory is quite similar to that underlying such seemingly unrelated subjects as inventories, dams, and insurance.

Consider the following model. Customers request the use of a particular type of equipment (server). If a server is available, the arriving customer will seize and hold it for some length of time, after which the server will be made immediately available to other incoming or waiting customers. If the incoming customer finds no available server, he then takes some specified action such as waiting or going away. Such models often can be defined in terms of three characteristics: the input process, the service mechanism, and the queue discipline.

The *input process* describes the sequence of requests for service. Often, for example, the input process is specified in terms of the distribution of the lengths of time between consecutive customer arrival instants. The *service mechanism* includes such characteristics as the number of servers and the lengths of time that the customers hold the servers. For example, customers might be processed by a single server, each customer holding the server for the same length of time. The *queue discipline* specifies the disposition of blocked customers (customers who find all servers busy). For example, it might be assumed that blocked customers leave the system

Scope and Nature of Queueing Theory

immediately or that blocked customers wait for service in a queue and are served from the queue in their arrival order.

We now use a model of this type to illustrate some important points about queueing theory: (1) The subject matter is of great practical value. (2) Heuristic and intuitive reasoning is often useful. (3) Mathematical subtleties abound.

Consider two cities interconnected by a group of s telephone trunks (servers). Suppose that arrivals finding all trunks busy do not wait, but immediately depart from the system. (Technically, no "queueing" occurs.) What proportion of incoming calls (customers) will be unable to find an idle trunk (and thus be lost)?

We wish to derive a formula that will predict the proportion of calls lost as a function of the demand; that is, we wish to derive a formula that allows estimation of the number of trunks required to meet a prespecified service criterion from an estimate of the telephone traffic load generated between the two cities. The great practical value of any model that leads to such a formula is obvious. The fact that such models have been successfully employed has spurred continuing investigations by industrial and academic researchers throughout the world.

We shall now give a heuristic derivation of the required formula, using a concept of great importance in science and engineering, that of *conservation of flow*. It is important for the reader to realize that since the following derivation is heuristic, he should not expect to understand it completely; our "derivation" is really a plausibility argument that, as will be demonstrated later, is correct in certain circumstances. With this disclaimer, let us now proceed with the argument.

When the number of customers in the system is j, the system is said to be in state E_j ($j=0,1,\ldots,s$). Let P_j be the proportion of time that the system spends in state E_j; P_j is therefore also the proportion of time that j trunks are busy. Denote by λ the call arrival rate; λ is the average number of requests for service per unit time. Consider first the case $j<s$. Since calls arrive with overall rate λ, and since the proportion of time the system spends in state E_j is P_j, the rate at which the transition $E_j \to E_{j+1}$ occurs (the average number of such transitions per unit time) is therefore λP_j. Now consider the case when $j=s$. Since the state E_{s+1} represents a physically impossible state (there are only s trunks), the transition $E_s \to E_{s+1}$ cannot occur, so the rate of transition $E_s \to E_{s+1}$ is zero. Thus the rate at which the upward transition $E_j \to E_{j+1}$ occurs is λP_j when $j=0,1,\ldots,s-1$ and is zero when $j=s$.

Let us now consider the downward transition

$$E_{j+1} \to E_j \qquad (j=0,1,\ldots,s-1).$$

Suppose that the mean holding time (the average length of time a call holds a trunk) is τ. Then, if a single trunk is busy, the average number of calls terminating during an elapsed time τ is 1; the termination rate for a single call is therefore $1/\tau$. Similarly, if two calls are in progress simultaneously and the average duration of a call is τ, the average number of calls terminating during an elapsed time τ is 2; the termination rate for two simultaneous calls is therefore $2/\tau$. By this reasoning, then, the termination rate for $j+1$ simultaneous calls is $(j+1)/\tau$. Since the system is in state E_{j+1} a proportion of time P_{j+1}, we conclude that the downward transition $E_{j+1} \to E_j$ occurs at rate $(j+1)\tau^{-1}P_{j+1}$ transitions per unit time $(j=0,1,\ldots,s-1)$.

We now apply the principle of conservation of flow: We equate, for each value of the index j, the rate of occurrence of the upward transition $E_j \to E_{j+1}$ to the rate of occurrence of the downward transition $E_{j+1} \to E_j$. Thus we have the equations of *statistical equilibrium* or *conservation of flow*:

$$\lambda P_j = (j+1)\tau^{-1}P_{j+1} \qquad (j=0,1,\ldots,s-1). \tag{1.1}$$

These equations can be solved recurrently; the result, which expresses each P_j in terms of the value P_0, is

$$P_j = \frac{(\lambda\tau)^j}{j!} P_0 \qquad (j=1,2,\ldots,s). \tag{1.2}$$

Since the numbers $\{P_j\}$ are proportions, they must sum to unity:

$$P_0 + P_1 + \cdots + P_s = 1. \tag{1.3}$$

Using the normalization equation (1.3) together with (1.2), we can determine P_0:

$$P_0 = \left(\sum_{k=0}^{s} \frac{(\lambda\tau)^k}{k!}\right)^{-1}. \tag{1.4}$$

Thus we obtain for the proportion P_j of time that j trunks are busy the formula

$$P_j = \frac{(\lambda\tau)^j/j!}{\sum_{k=0}^{s} (\lambda\tau)^k/k!} \qquad (j=0,1,\ldots,s). \tag{1.5}$$

An important observation to be made from the formula (1.5) is that the proportions $\{P_j\}$ depend on the arrival rate λ and mean holding time τ only through the product $\lambda\tau$. This product is a measure of the demand made on the system; it is often called the *offered load* and given the symbol a ($a=\lambda\tau$). The numerical values of a are expressed in units called *erlangs* (erl), after the Danish mathematician and teletraffic theorist A. K. Erlang, who first published the formula (1.5) in 1917.

When $j=s$ in (1.5), the right-hand side becomes the well-known *Erlang loss formula*, denoted in the United States by $B(s,a)$ and in Europe by $E_{1,s}(a)$:

$$B(s,a) = \frac{a^s/s!}{\sum_{k=0}^{s} a^k/k!}. \qquad (1.6)$$

We shall derive these results more carefully later in the text. The point to be made here is that some potentially useful mathematical results have been derived using only heuristic reasoning. The question we must now answer is: What, if any, are the conditions under which these results are valid?

More precisely, what assumptions about the input process and service mechanism are required for the validity of the formulas (1.5) and (1.6)? [We have described the input process by giving only the arrival rate; similarly, the service times have been specified only through their mean value. It turns out that our conclusion (1.5) is valid for a particular type of input process called a *Poisson process* and, surprisingly, for *any* distribution of service times whatever.] Can the assertion that the downward transition rate is proportional to the reciprocal of the mean holding time be justified? What is the relationship between the proportion P_j of time that j calls are in progress and the proportion Π_j, say, of arriving calls that find j other calls in progress? [It turns out that for systems with Poisson input these two distributions are equal. Thus, the Erlang loss formula (1.6) gives the proportion of customers who will be denied service, that is, lost, when the arrivals occur according to a Poisson process.] How widely applicable is the conservation-of-flow analysis? How does one handle processes for which this type of analysis is inapplicable?

Questions of this nature sometimes require highly sophisticated mathematical arguments. In this text we shall take a middle ground with regard to the use of advanced mathematics. We shall attempt to impart an understanding of the theory without recourse to the use of abstract and measure-theoretic tools; the material is presented in a mathematically informal manner with emphasis on the underlying physical ideas. Mathematical difficulties and fine points will be noted but not dwelled upon. The material should be accessible to a student who *understands* applied probability theory and those areas of mathematics traditionally included in undergraduate programs in engineering and the physical sciences.

A word about the range of applicability of these models is in order here. The example discussed above is based on a telephone traffic application, but it should be clear that the identification of the customers as "calls" and the servers as "trunks" in no way limits the generality of the model.

As mentioned previously, telephone applications have provided the context for the development of much of queueing theory (and this example illustrates why). We shall occasionally refer to telephone applications, thereby illustrating the general principles of the theory with a coherent class of authentic examples. Of course, no technical knowledge of telephony is required.

In summary, this is a text and reference in queueing theory, a subject of both practical importance and theoretical interest. Although the book is directed primarily toward the student, some of the lesser-known concepts and results from telephone traffic theory should prove of interest, apart from their value in illustrating the theory, to engineers and researchers. The material is presented with an emphasis on the underlying physical processes and the interplay between physical and mathematical reasoning. Heuristic and intuitive approaches are used where helpful. Mathematical subtleties are observed and are explored where it is profitable to do so without recourse to methods of abstract analysis.

Since the publication of Erlang's first paper in 1909 (which is regarded by many as marking the birth of queueing theory), and especially since the renewed interest in queueing theory that accompanied the formalization of the field of operations research after World War II, about 2000 papers and at least 35 books on this subject have been published. It is hoped that the reader will find the present book a useful and interesting addition to the literature.

Exercises

1. In what ways are telephone traffic theory and vehicular traffic theory similar, and in what ways are they different?

2. List some applications of the Erlang loss model.

3. Discuss ways in which queueing models might be used in the following:
   a. Highway design.
   b. City planning.
   c. Hospital management.
   d. Airport design.
   e. Reliability engineering.
   f. Computer design.

4. Extend the heuristic conservation-of-flow argument to include the case in which all customers who find all servers busy wait until served.
   a. Argue that

$$P_j = \begin{cases} \dfrac{a^j}{j!} P_0 & (j=1,2,\ldots,s-1), \\ \dfrac{a^j}{s!\,s^{j-s}} P_0 & (j=s,s+1,\ldots), \end{cases} \tag{1}$$

where

$$P_0 = \left(\sum_{k=0}^{s-1} \frac{a^k}{k!} + \frac{a^s}{s!(1-a/s)} \right)^{-1}. \qquad (2)$$

b. What restriction must be placed on the magnitude a of the offered load in order for (2) to be correct? Give a physical interpretation of this restriction.

c. Show that the proportion of time that all servers are busy (which, in this case, turns out to equal the proportion of customers who find all servers busy and therefore must wait for service) is given by

$$C(s,a) = \frac{\dfrac{a^s}{s!(1-a/s)}}{\displaystyle\sum_{k=0}^{s-1} \frac{a^k}{k!} + \frac{a^s}{s!(1-a/s)}}. \qquad (3)$$

This is the well-known *Erlang delay formula*, also denoted by $E_{2,s}(a)$, which we shall discuss in detail in Chapter 3.

d. What assumption, if any, have you made about the order in which waiting customers are selected from the queue when a server becomes idle? Is such an assumption necessary? Why?

e. Show that, if all servers are busy, the probability p_j that j customers are waiting for service is given by

$$p_j = (1-\rho)\rho^j \quad (j=0,1,\ldots), \qquad (4)$$

where $\rho = a/s$. [The probabilities defined by (4) constitute the *geometric distribution*.]

f. Suppose it is known that at least $k > 0$ customers are waiting in the queue. Show that the probability that the number of waiting customers is exactly $j+k$ is p_j, given by (4). Note that this probability is independent of the value of the index k.

g. Show that $P_j = p_j$ $(j=0,1,\ldots)$ when $s=1$. Show that $C(1,a) = a$.

5. Consider the so-called *loss-delay system*, which is one with a finite number n of waiting positions: An arrival who finds all servers busy and at least one waiting position unoccupied waits as long as necessary for service, while an arrival who finds all n waiting positions occupied departs immediately (and thus is lost).

a. Using the conservation-of-flow argument, show that the probability P_j that j customers are present simultaneously is given by the formula (1) of Exercise 4, where now the largest value of the index j is $s+n$, and

$$P_0 = \left[\sum_{k=0}^{s-1} \frac{a^k}{k!} + \frac{a^s}{s!} \sum_{i=0}^{n} \left(\frac{a}{s}\right)^i \right]^{-1}.$$

b. Use these results to obtain the Erlang loss formula (1.6) and the Erlang delay formula (3) of Exercise 4.

c. What restrictions, if any, must be placed on the magnitude a of the offered load when $n<\infty$? When $n=\infty$?

6. Consider a queueing model with two servers and one waiting position, and assume that if an arriving customer finds both servers busy and the waiting position unoccupied, then with probability p the customer will wait as long as necessary for service, and with probability $1-p$ he will depart immediately. (As usual, all customers who arrive when there is an idle server enter service immediately, and all customers who arrive when the waiting position is occupied depart immediately.)
    a. Let P_j ($j=0,1,2,3$) be the probability that j customers are present simultaneously; let λ be the arrival rate and τ be the average service time; and write the conservation-of-flow equations that determine the probabilities P_j ($j=0,1,2,3$).
    b. Solve the equations of part a.
    c. Find the fraction B of arriving customers who don't get served.
    d. Suppose $\lambda=2$ customers per hour, $\tau=1$ hour, and $p=\frac{1}{2}$. Evaluate P_0, P_1, P_2, P_3, and B.
    e. Suppose the manager receives $2.00 for every customer who gets served without waiting and $1.00 for every customer who gets served after having to wait. How many dollars per hour will she receive?
    f. If the manager pays a rental fee of $.50 per hour per server (whether the server is busy or idle), and each server when busy consumes fuel at the rate of $.25 per hour, what is the manager's total operating cost in dollars per hour? What is her profit in dollars per hour?

2

Review of Topics From the Theory of Probability and Stochastic Processes

This chapter reviews and summarizes some aspects of the theory of probability and stochastic processes that have direct application to elementary queueing theory. The material given here is intended not only as a reference and refresher in probability and stochastic processes but also as an introduction to queueing theory. It is assumed that the reader is already familiar with the basic concepts of probability theory as covered, for example, in Çinlar [1975], Feller [1968], Fisz [1963], Neuts [1973], and Ross [1972]. Specifically, we assume of the reader working knowledge of the basic notions of event, probability, statistical dependence and independence, distribution and density function, conditional probability, and moment. These concepts will be used freely where required. Familiarity with the theory of Markov chains is helpful, but not essential. Birth-and-death processes, generating functions, and the properties of some important distributions will be reviewed in some detail. Examples will be drawn largely from queueing theory. The reader who feels no need for review is advised to skim, but not skip, this chapter before proceeding to the next.

2.1. Random Variables

If an experiment could be performed repeatedly under identical conditions, then, experience seems to tell us, each of the different possible outcomes of the experiment will occur with a long-run relative frequency that remains constant. Probability theory represents an attempt to model this "observed" behavior of long-run frequencies with a formal mathematical structure. As a mathematical theory, of course, probability theory need only be internally consistent. As a model of reality, however, the theory needs also to be consistent with the observed properties of the phenomena it purports to describe. To be a useful model, moreover, the theory should be capable of exposing facts about reality that were previously hidden from view.

Probability theory is now a well-developed mathematical theory, and has proved also to be an excellent model of frequencies in repeated experiments. We will take advantage of this fact and apply probability theory to the analysis of the phenomenon of queueing. Characteristic of applied probability, and queueing theory in particular, is the interaction between the mathematical and intuitive (that is, interpretive) aspects. All mathematical results should be understandable in physical terms; and all intuitive results should be capable of rigorous mathematical proof. Unfortunately, neither of these conditions can always be met. In this text intuitive arguments and interpretations will be emphasized, with the tacit understanding that, of course, one is never safe until a rigorous argument is proffered.

We begin with a brief discussion of random variables in the context of queueing theory, and we follow with an example that illustrates the intuitive "frequency" approach to the calculation of probabilities.

The mathematical definition of event involves the notions of sample space and Borel field, but for most practical purposes, intuitive notions of event are sufficient. Examples of the kinds of events of interest in queueing theory are {an arbitrary customer finds all servers busy}, {an arbitrary customer must wait more than three seconds for service}, and {the number of waiting customers at an arbitrary instant is j}.

In each of these examples, the event can be expressed in numerical terms in the following ways. Define N to be the number of customers in an s-server system found by an arbitrary arriving customer. Then the event {an arbitrary customer finds all s servers busy} can be written $\{N \geqslant s\}$, and the probability of occurrence of this event can be represented by $P\{N \geqslant s\}$.

Similarly, let W be the waiting time of an arbitrary customer, and let Q be the number of waiting customers (queue length) at an arbitrary instant. Then these events are $\{W>3\}$ and $\{Q=j\}$, with probabilities $P\{W>3\}$ and $P\{Q=j\}$, respectively.

The quantities N, W, and Q defined above are random variables. Each of them has the property that it represents an event in terms of a numerical value about which a probability statement can be made. More formally: A *random variable* is a function defined on a given sample space, and about whose values a probability statement can be made.

The question of precisely which functions can qualify as random variables is, like that of the notion of event, a mathematical one. For our purposes, it will be sufficient to consider the use of random-variable

Random Variables

notation as simply a device for describing events without explicit concern for the underlying probability space.

A random variable is defined by assigning a numerical value to each event. For example, in a coin-tossing experiment, the event {head} can be assigned the value 1, and the event {tail} the value 0. Call the describing random variable X. Then at each toss a head will occur with probability $P\{X=1\}$ and a tail with probability $P\{X=0\}$. This use of the random-variable notation may seem artificial, since the association of the event {head} with the value $X=1$ is arbitrary, and the value $X=2.6$ might just as well have been chosen. But it will be seen that judicious choices of the random-variable definitions often lead to simplifications. For example, observe that any random variable X that takes only the values 0 and 1 has expected value $E(X)=0P\{X=0\}+1P\{X=1\}=P\{X=1\}$; that is, the mean value of a zero-one random variable is equal to the probability that the random variable takes on the value 1.

Suppose, for example, that we wish to study the random variable S_n, which we define to be the number of heads occurring in n tosses. Let the random variable X_j describe the jth toss; let $X_j=1$ if the jth toss results in a head, and $X_j=0$ for a tail. Then the random variable S_n can be interpreted as the sum $S_n = X_1 + \cdots + X_n$. The interpretation of S_n as a sum of random variables allows its study in terms of the simpler component random variables $X_1,\ldots,X_n$, without further regard to the physical meaning of S_n. We shall return to this example later.

In most of the queueing-theory problems we shall encounter, however, the assignment of random-variable values to events is quite natural from the context of the problem. Only a malcontent would choose to assign the value $Q=4.2$ to the event {six customers are waiting}.

We now consider an example that will illustrate the intuitive "frequency" approach to the calculation of probabilities, while simultaneously providing us with a result that will prove useful in queueing theory. Consider a population of urns, where an urn of type j ($j=1,2,\ldots$) contains n_j balls. Let the random variable X be the type of an urn selected at random; that is, $P\{X=j\}$ is the probability that an arbitrary urn is of type j. We interpret $P\{X=j\}$ as the frequency with which urns of type j occur in the general population of urns. Let us say that a ball is of type j if it is contained in an urn of type j, and let Y be the type of a ball selected at random; we interpret $P\{Y=j\}$ as the frequency with which balls of type j occur in the general population of balls. Our objective is to derive a formula that relates the frequencies (probabilities) of occurrence of the urns to the corresponding frequencies (probabilities) of occurrence of the balls.

Suppose first that m urns are selected "at random" from the (infinite) parent population, and let m_j be the number of urns of type j among the m urns selected. Then the relative frequency of occurrence of urns of type j

among the m urns selected is m_j/m. We now assume that this ratio has a limit as m tends to infinity, and we call this hypothesized limiting value the *probability* that an "arbitrary" urn is of type j:

$$P\{X=j\} = \lim_{m\to\infty} \frac{m_j}{m} \quad (j=1,2,\ldots). \tag{1.1}$$

Similarly, the number of balls that are of type j among all the balls in the m urns selected is $n_j m_j$; hence

$$P\{Y=j\} = \lim_{m\to\infty} \frac{n_j m_j}{\sum_i n_i m_i} \quad (j=1,2,\ldots). \tag{1.2}$$

If we divide the numerator and denominator of (1.2) by m, and use (1.1), we obtain

$$P\{Y=j\} = \frac{n_j P\{X=j\}}{\sum_i n_i P\{X=i\}} \quad (j=1,2,\ldots). \tag{1.3}$$

Equation (1.3) relates the (mathematical) probability distributions of X and Y to each other, without further regard for the (physical) frequencies these probabilities are supposed to represent. It is not our purpose here to discuss the mathematical theory of probability and its use as a model of frequencies of events in infinitely repeatable experiments, but merely to point out that our strategy as applied probabilists is both to use mathematics to provide physical insight (that is, intuition) and to use intuition to guide and check the corresponding mathematical formalism.

Equation (1.3) can be interpreted as saying that the frequency with which balls of type j occur is proportional to the product of the number of balls of type j per urn of type j and the frequency of occurrence of urns of type j; that is, $P\{Y=j\} = cn_j P\{X=j\}$, where $c = (\sum_i n_i P\{X=i\})^{-1}$ is the constant of proportionality such that $P\{Y=1\} + P\{Y=2\} + \cdots = 1$.

It is easy to verify that (1.3) yields the very obvious results that (1) if there are only a finite number r of urn types ($j=1,2,\ldots,r$) and each type occurs with equal frequency, then

$$P\{Y=j\} = \frac{n_j}{n_1+n_2+\cdots+n_r} \quad (j=1,2,\ldots,r),$$

and (2) if each type of urn contains the same number of balls, then

$$P\{Y=j\} = P\{X=j\} \quad (j=1,2,\ldots).$$

As an example of the application of (1.3) in queueing theory, we consider a single-server system. We view time as being decomposed into a sequence of *cycles*, where each cycle consists of an *idle period* (during which there are no customers present and the server is idle) and an adjacent *busy period* (during which the server is continuously busy). In many important queueing models the cycles are independent and statistically identical; that is, each cycle represents an independent trial in a sequence of infinitely repeatable experiments. Now suppose that the number of customers who arrive during a busy period is N, with mean value $E(N)$. [The calculation of $E(N)$ depends on other assumptions of the model, such as whether customers who arrive during a busy period, and thus are *blocked*, wait for service or leave immediately or take some other action.] We shall show that the proportion Π of customers who are blocked is given by

$$\Pi = \frac{E(N)}{1 + E(N)}. \tag{1.4}$$

According to (1.4) the probability of blocking equals the ratio of the mean number of customers who are blocked per cycle, $E(N)$, to the mean number of customers who arrive during a cycle, $1 + E(N)$. This result has great intuitive appeal, and will also be useful when $E(N)$ is easy to calculate.

To prove (1.4), let X be the number of customers (balls) who arrive during an arbitrary cycle (urn), and let Y be the number of customers who arrive during the cycle that contains an arbitrary customer (called the *test customer*) whose viewpoint we shall adopt. Also, let a customer be of type j if exactly j customers (including himself) arrive during his cycle. Then Equation (1.3) applies, with $n_j = j$:

$$P\{Y = j\} = \frac{jP\{X = j\}}{E(X)}. \tag{1.5}$$

Now, if the test customer is one of j arrivals during a cycle, the probability that he is not the first of those j, and therefore is blocked, is $(j-1)/j$. It follows from the *theorem of total probability** that the probability Π that the test customer is blocked is

$$\Pi = \sum_{j=1}^{\infty} \frac{j-1}{j} P\{Y = j\}. \tag{1.6}$$

---

*If the events $B_1, B_2, \ldots$ are mutually exclusive (that is, $B_i B_j = \phi$ for all $i \neq j$) and exhaustive (that is, $B_1 \cup B_2 \cup \cdots = \Omega$, where Ω is the sample space), then, for any event A, $P\{A\} = P\{A|B_1\}P\{B_1\} + P\{A|B_2\}P\{B_2\} + \cdots$. (See, for example, p. 15 of Çinlar [1975].)

Using (1.5) in (1.6), we get

$$\Pi = \frac{E(X)-1}{E(X)}. \tag{1.7}$$

But the number of arrivals during a cycle equals the number who arrive during the busy period plus the one who arrives at the end of the idle period and thus initiates the busy period; that is, $X = N + 1$. Thus, $E(X) = E(N) + 1$, and Equation (1.4) follows.

Exercise

1. In the model considered above, suppose that it costs c dollars to turn on the server at the start of each busy period. Show that

$$E\left(\frac{c}{Y}\right) = \frac{c}{E(X)};$$

that is, if the cost of a cycle is divided evenly among the customers who arrive during a cycle, then the average cost per arrival equals the cost per cycle divided by the average number of arrivals per cycle.

2.2. Birth-and-Death Processes

The theory of birth-and-death processes, developed largely by Feller, comprises part of the subject matter commonly called stochastic processes. We now sketch an outline of the theory of birth-and-death processes. For a more comprehensive treatment see Chapter XVII of Feller [1968] and Chapter I.4 of Cohen [1969].

Consider a system that for each fixed t ($0 \leqslant t < \infty$) can be described by a random variable $N(t)$ with realizations $0, 1, 2, \ldots$. Examples are (1) a telephone switchboard, where $N(t)$ is the number of calls occurring in an interval of length t; (2) a queue, where $N(t)$ is the number of customers waiting or in service at time t; (3) an epidemic, where the number of deaths that have occurred in $(0, t)$ is $N(t)$; and (4) a city whose population is $N(t)$ at time t.

We wish to study processes that can be described by such a random variable $N(t)$. We shall make some assumptions about the behavior of the process that are simple enough to avoid intractable analysis but that nevertheless lead to useful models. [Note that $N(t)$ is a (different) random variable for each fixed $t \geqslant 0$. Our interest is the (uncountable) family of random variables $\{N(t), t \geqslant 0\}$, which is called a *stochastic process*. Questions concerning the existence and properties of such uncountable sequences are addressed in books on the theory of stochastic processes. The interested reader is referred to the books referenced in the last section of this chapter.]

Birth-and-Death Processes

We say that a system is in state E_j at time t if $N(t)=j$. Then a process obeying the following postulates is called a *birth-and-death process*:

If at any time t the system is in state E_j, the conditional probability that during $(t,t+h)$ the transition $E_j \to E_{j+1}$ $(j=0,1,\ldots)$ occurs equals $\lambda_j h + o(h)$ as $h \to 0$, and the conditional probability of the transition $E_j \to E_{j-1}$ $(j=1,2,\ldots)$ equals $\mu_j h + o(h)$ as $h \to 0$. The probability that during $(t,t+h)$ the index j changes by more than one unit is $o(h)$ as $h \to 0$. [A quantity $f(h)$ is said to equal $o(h)$ as $h \to 0$ if $\lim_{h \to 0} f(h)/h = 0$. Note that if $f(h) = o(h)$ as $h \to 0$, then also $xf(h) = o(h)$ as $h \to 0$, for any finite x.]

Applying the theorem of total probability, we can write

$$P\{N(t+h)=j\} = \sum_{i=0}^{\infty} P\{N(t+h)=j | N(t)=i\} P\{N(t)=i\}. \quad (2.1)$$

Now it follows from the postulates that, as $h \to 0$,

$$P\{N(t+h)=j | N(t)=i\} = \begin{cases} \lambda_{j-1} h + o(h) & \text{when } i=j-1 \\ \mu_{j+1} h + o(h) & \text{when } i=j+1 \\ o(h) & \text{when } |i-j| \geq 2. \end{cases}$$

Since we require that

$$\sum_{k=0}^{\infty} P\{N(t+h)=k | N(t)=j\} = 1,$$

it follows that, as $h \to 0$,

$$P\{N(t+h)=j | N(t)=j\} = 1 - (\lambda_j + \mu_j)h + o(h).$$

Hence, if we set $P\{N(t)=j\} = P_j(t)$, then Equation (2.1) can be written

$$P_j(t+h) = \lambda_{j-1} h P_{j-1}(t) + \mu_{j+1} h P_{j+1}(t)$$
$$+ [1 - (\lambda_j + \mu_j)h] P_j(t) + o(h)$$
$$[h \to 0; \; j=0,1,\ldots; \; \lambda_{-1} = \mu_0 = P_{-1}(t) = 0].$$

Rearranging and dividing through by h, we have

$$\frac{P_j(t+h) - P_j(t)}{h} = \lambda_{j-1} P_{j-1}(t) + \mu_{j+1} P_{j+1}(t)$$
$$- (\lambda_j + \mu_j) P_j(t) + \frac{o(h)}{h}$$
$$[h \to 0; \; j=0,1,\ldots; \; \lambda_{-1} = \mu_0 = P_{-1}(t) = 0]. \quad (2.2)$$

We now let $h \to 0$ in Equation (2.2). The result is the following set of differential-difference equations for the birth-and-death process:

$$\frac{d}{dt} P_j(t) = \lambda_{j-1} P_{j-1}(t) + \mu_{j+1} P_{j+1}(t) - (\lambda_j + \mu_j) P_j(t)$$

$$[j = 0, 1, \ldots; \quad \lambda_{-1} = \mu_0 = P_{-1}(t) = 0]. \tag{2.3}$$

If at time $t = 0$ the system is in state E_i, the initial conditions that complement Equation (2.3) are

$$P_j(0) = \begin{cases} 1 & \text{if } j = i, \\ 0 & \text{if } j \neq i. \end{cases} \tag{2.4}$$

The coefficients $\{\lambda_j\}$ and $\{\mu_j\}$ are called the *birth* and *death rates*, respectively. When $\mu_j = 0$ for all j, the process is called a *pure birth process*; and when $\lambda_j = 0$ for all j, the process is called a *pure death process*.

In the case of either a pure birth process or a pure death process, it is easy to see that the differential-difference equations (2.3) can always be solved, at least in principle, by recurrence (successive substitution).

For example, consider the important special case of the pure birth process with constant birth rate $\lambda_j = \lambda$. (This model is often used to describe the arrival process of customers at a queue.) If we assume that the system is initially in state E_0, then for this case Equations (2.3) and (2.4) become

$$\frac{d}{dt} P_j(t) = \lambda P_{j-1}(t) - \lambda P_j(t) \quad [j = 0, 1, \ldots; \quad P_{-1}(t) = 0] \tag{2.5}$$

and

$$P_j(0) = \begin{cases} 1 & \text{if } j = 0, \\ 0 & \text{if } j \neq 0. \end{cases} \tag{2.6}$$

An easy solution by recurrence (or induction) gives, for each $t \geq 0$,

$$P_j(t) = \frac{(\lambda t)^j}{j!} e^{-\lambda t} \quad (j = 0, 1, \ldots). \tag{2.7}$$

Note that the probabilities given by (2.7) satisfy the normalization condition

$$\sum_{j=0}^{\infty} P_j(t) = 1 \quad (t \geq 0). \tag{2.8}$$

(The normalization condition (2.8) is not satisfied for every choice of the birth coefficients. For a more extensive treatment, see Chapter XVII of Feller [1968].)

According to Equation (2.7), $N(t)$ has the *Poisson* distribution with mean λt; we say that $\{N(t), t \geq 0\}$ is a *Poisson process*. Since the assumption $\lambda_j = \lambda$ is often a realistic one in the construction of queueing models, the simple formula (2.7) is important in queueing theory. The Poisson distribution possesses important theoretical properties and plays a central role in queueing theory.

Another important example is the special case of the pure death process with death rate μ_j proportional to the index of the state E_j; that is, $\mu_j = j\mu$. (A real system that might fit this model is a population in which only deaths occur and where the death rate is proportional to the population size.) If we assume that the system is in state E_n at $t=0$, then Equations (2.3) and (2.4) become

$$\frac{d}{dt}P_n(t) = -n\mu P_n(t) \quad [P_n(0)=1], \tag{2.9}$$

$$\frac{d}{dt}P_j(t) = (j+1)\mu P_{j+1}(t) - j\mu P_j(t)$$

$$[P_j(0)=0; \quad j=n-1, n-2, \ldots, 2, 1, 0]. \tag{2.10}$$

Solving these equations by recurrence, it is easy to verify that the general form of the solution $P_j(t)$ is

$$P_j(t) = \binom{n}{j}(e^{-\mu t})^j(1-e^{-\mu t})^{n-j} \quad (j=0,1,\ldots,n). \tag{2.11}$$

We remark that the probabilities defined by (2.11) constitute a *binomial* distribution, and therefore sum to unity.

Exercise

2. Consider a population modeled as a pure birth process in which the birth rates are proportional to the population size, that is, $\lambda_j = j\lambda$ ($j=0,1,\ldots$). Write the differential-difference equations that determine the probability $P_j(t)$ that $N(t) = j$ for $j=1,2,\ldots$ and all $t \geq 0$; and show that if $N(0)=1$, then

$$P_j(t) = e^{-\lambda t}(1-e^{-\lambda t})^{j-1} \quad (j=1,2,\ldots).$$

We have observed that in the case of the pure birth process ($\mu_j = 0$), the differential-difference equations (2.3) can always be solved recurrently, at least in principle. Therefore, even though the number of equations is, in general, infinite, there is no question about the existence of a solution, since the solution can be exhibited. On the other hand, it is not necessarily true that the solution $\{P_j(t)\}$ is a proper probability distribution. In any particular case, a good strategy may be to find the solution first, and then determine if the solution is a proper probability distribution, rather than vice versa.

In the case of the pure death process ($\lambda_j = 0$), the differential-difference equations (2.3) offer less theoretical difficulty, since not only can they be solved recurrently, but also they are finite in number.

In contrast, in the general case, the equations (2.3) of the birth-and-death process do not yield to solution by recurrence, as one can easily verify. Also, in general, these equations are infinite in number. Therefore, both practical and theoretical difficulties present themselves.

The questions of existence and uniqueness of solutions are difficult and will not be discussed here. Suffice it to say that in almost every case of practical interest, Equations (2.3) and (2.4) have a unique solution that satisfies

$$\sum_{j=0}^{\infty} P_j(t) = 1 \quad [0 \leqslant P_j(t) \leqslant 1; \quad 0 \leqslant t < \infty]. \tag{2.12}$$

The question is discussed in some detail in Chapter XVII of Feller [1968], who also gives several pertinent references.

As an example of the use of the birth-and-death process in a queueing-theory context, we consider a model of a queueing system with one server and no waiting positions. Specifically, we assume that the probability of a request in $(t, t+h)$ is $\lambda h + o(h)$ as $h \to 0$, and assume that if the server is busy with a customer at t, the probability that the customer's service will end in $(t, t+h)$ is $\mu h + o(h)$ as $h \to 0$. Assume further that every customer who finds the server occupied leaves the system immediately and thus has no effect upon it.

In terms of the postulates for the birth-and-death process, this queueing model corresponds to a two-state birth-and-death process. E_0 corresponds to the state {server idle}, and E_1 corresponds to the state {server busy}. Since, by assumption, an arrival that occurs when the server is busy has no effect on the system, an arrival will cause a state transition if and only if it occurs when the server is idle. Therefore, the effective arrival rates are $\lambda_0 = \lambda$ and $\lambda_j = 0$ for $j \neq 0$. Similarly, no customers can complete service when no customers are in the system, so that $\mu_j = 0$ when $j \neq 1$, and, by assumption, $\mu_1 = \mu$. The birth-and-death equations (2.3) for these particular choices of the birth-and-death coefficients are

$$\frac{d}{dt} P_0(t) = \mu P_1(t) - \lambda P_0(t) \tag{2.13}$$

and

$$\frac{d}{dt} P_1(t) = \lambda P_0(t) - \mu P_1(t). \tag{2.14}$$

Statistical Equilibrium

Standard techniques exist for the solution of sets of simultaneous linear differential equations, but we shall solve this simple set by using its special properties.

First note that when Equations (2.13) and (2.14) are added, we obtain

$$\frac{d}{dt}[P_0(t) + P_1(t)] = 0,$$

so that the sum of the probabilities is constant for all $t \geqslant 0$,

$$P_0(t) + P_1(t) = c. \qquad (2.15)$$

We require that the system initially be describable by a probability distribution, so that

$$P_0(0) + P_1(0) = 1. \qquad (2.16)$$

Then Equations (2.15) and (2.16) require $c = 1$, and hence

$$P_0(t) + P_1(t) = 1 \qquad (t \geqslant 0). \qquad (2.17)$$

Substitution of (2.17) into (2.13) yields

$$\frac{d}{dt}P_0(t) + (\lambda + \mu)P_0(t) = \mu,$$

which has the general solution

$$P_0(t) = \frac{\mu}{\lambda + \mu} + \left(P_0(0) - \frac{\mu}{\lambda + \mu}\right)e^{-(\lambda + \mu)t}. \qquad (2.18)$$

By symmetry, Equations (2.17) and (2.14) yield

$$P_1(t) = \frac{\lambda}{\lambda + \mu} + \left(P_1(0) - \frac{\lambda}{\lambda + \mu}\right)e^{-(\lambda + \mu)t}. \qquad (2.19)$$

Equations (2.18) and (2.19) comprise the *transient solution*, which describes the system as a function of time. In most applications, interest centers not on the values of these probabilities at a specific point in time, but rather on their long-run values. This topic is discussed in the next section.

2.3. Statistical Equilibrium

Suppose that we are interested in the behavior of the system just described for large values of t, that is, after it has been in operation for a long period of time. The state probabilities as functions of time are given by Equations

(2.18) and (2.19). Letting $t \to \infty$ in (2.18) and (2.19), we obtain

$$P_0 = \lim_{t \to \infty} P_0(t) = \frac{\mu}{\lambda + \mu} \tag{3.1}$$

and

$$P_1 = \lim_{t \to \infty} P_1(t) = \frac{\lambda}{\lambda + \mu}. \tag{3.2}$$

Observe that

$$P_0 + P_1 = 1, \tag{3.3}$$

so that the limiting distribution is proper. Note that the limiting values of the probabilities are independent of the initial values $P_0(0)$ and $P_1(0)$. In other words, after a sufficiently long period of time the state probabilities are independent of the initial conditions and sum to unity; the system is then said to be in *statistical equilibrium* or, more simply, *equilibrium*.

An important characteristic of the statistical-equilibrium distribution is that it is *stationary*; that is, the state probabilities do not vary with time. For example, if our system were assumed to be in equilibrium at some time t, say $t = 0$, so that $P_0(0) = \mu/(\lambda + \mu)$ and $P_1(0) = \lambda/(\lambda + \mu)$, then Equations (2.18) and (2.19) make it clear that these initial values will persist for all $t \geq 0$. In other words, when a system is in statistical equilibrium no net trends result from the statistical fluctuations.

Another important property of a system possessing a statistical equilibrium distribution is that it is *ergodic*, which means that in each realization of the process, the proportion of the time interval $(0, x)$ that the system spends in state E_j converges to the equilibrium probability P_j as $x \to \infty$.

Note carefully that the concept of statistical equilibrium relates not only to the properties of the system itself, but also to the observer's knowledge of the system. In our example, if an observer were to look at the system at any time t, then he would find the system in either state E_1 or E_0, say E_0. If he were to look again at any later time $x = t + t'$, the probability that he would then find the system in state E_1, say, is given by (2.19) with $t = t'$ and $P_1(0) = 0$; whereas the corresponding probability would be given by (3.2) if equilibrium had prevailed at time t and the observer had not looked. Thus, the value of the probability $P_j(t + t')$ depends on whether or not an observation was made at t, even though the system itself is not physically affected in either case.

Clearly, practical applications of queueing theory will be largely concerned with the statistical-equilibrium properties of a system. Therefore, it would be useful to be able to obtain the equilibrium distribution directly

Statistical Equilibrium

(when it exists), without having to find the time-dependent probabilities first. In our example, since the limiting probabilities $\lim_{t\to\infty} P_j(t) = P_j$ have been shown directly to exist, it follows from Equations (2.13) and (2.14) that $\lim_{t\to\infty} (d/dt) P_j(t) = 0$. This suggests that the equilibrium probabilities might follow directly from Equations (2.13) and (2.14) when the time derivatives are set equal to zero. That is, letting $t \to \infty$ in (2.13) and (2.14), we obtain

$$0 = \mu P_1 - \lambda P_0 \tag{3.4}$$

and

$$0 = \lambda P_0 - \mu P_1. \tag{3.5}$$

Equations (3.4) and (3.5) are identical; they yield

$$P_1 = \frac{\lambda}{\mu} P_0. \tag{3.6}$$

Since the initial conditions (2.16), which led to the normalization equation (2.17), no longer appear, we must specify that

$$P_0 + P_1 = 1. \tag{3.7}$$

We then obtain from Equations (3.6) and (3.7) that $P_0 = \mu/(\lambda + \mu)$ and $P_1 = \lambda/(\lambda + \mu)$, in agreement with the previous results (3.1) and (3.2). Thus we have obtained the statistical-equilibrium solution by solving the linear difference equations (3.4) and (3.5) instead of the more difficult linear differential-difference equations (2.13) and (2.14).

We have shown that, in this simple example at least, the statistical-equilibrium distribution can be obtained in two different ways:

1. Solve the differential-difference equations (2.3), with appropriate initial conditions, to obtain $P_j(t)$, and then calculate the limits $\lim_{t\to\infty} P_j(t) = P_j$.
2. Take limits as $t \to \infty$ throughout the basic differential-difference equations (2.3), set $\lim_{t\to\infty} (d/dt) P_j(t) = 0$ and $\lim_{t\to\infty} P_j(t) = P_j$, solve the resulting set of difference equations, and normalize so that $\sum_{j=0}^{\infty} P_j = 1$.

Method 2 is clearly the easier way to obtain the equilibrium distribution, since the problem is reduced to solving a set of difference equations instead of a set of differential-difference equations.

Let us now move from this simple motivating example to the general birth-and-death process. One might hope that method 2 applies to the general birth-and-death equations (2.3). In essentially all birth-and-death models of practical interest, it does. The following informal theorem, which we state without proof, is useful in this regard.

Theorem. *Consider a birth-and-death process with states $E_0, E_1, \ldots$ and birth and death coefficients $\lambda_j > 0$ $(j = 0, 1, \ldots)$ and $\mu_j > 0$ $(j = 1, 2, \ldots)$; and let*

$$S = 1 + \frac{\lambda_0}{\mu_1} + \frac{\lambda_0 \lambda_1}{\mu_1 \mu_2} + \cdots + \frac{\lambda_0 \lambda_1 \cdots \lambda_{j-1}}{\mu_1 \mu_2 \cdots \mu_j} + \cdots. \tag{3.8}$$

Then the limiting probabilities

$$P_j = \lim_{t \to \infty} P_j(t) \qquad (j = 0, 1, \ldots) \tag{3.9}$$

exist, and do not depend on the initial state of the process. If $S < \infty$, then

$$P_j = \begin{cases} S^{-1} & (j = 0), \\ \dfrac{\lambda_0 \lambda_1 \cdots \lambda_{j-1}}{\mu_1 \mu_2 \cdots \mu_j} P_0 & (j = 1, 2, \ldots). \end{cases} \tag{3.10}$$

Since the probabilities (3.10) sum to unity, they constitute the *statistical-equilibrium distribution*. If $S = \infty$, then $P_j = 0$ for all finite j, so that, roughly speaking, the state of the process grows without bound and no statistical-equilibrium distribution exists. If the process has only a finite number of allowed states $E_0, E_1, \ldots, E_n$, then put $\lambda_n = 0$ in (3.8); in this case we shall always have $S < \infty$, and hence (3.10) is the statistical-equilibrium distribution, where, of course, $P_j = 0$ for all $j > n$.

For example, consider again the (finite-state) system described by (3.1) and (3.2), for which the birth coefficients and death coefficients are $\lambda_0 = \lambda$, $\lambda_1 = 0$, and $\mu_1 = \mu$. In this case, $S = 1 + \lambda/\mu < \infty$, and therefore the statistical-equilibrium distribution exists and is given by (3.10), in agreement with (3.1) and (3.2).

As a second example, consider the birth-and-death process in which $\lambda_{j-1} = \mu_j > 0$ for $j = 1, 2, \ldots$. Then $S = 1 + 1 + \cdots = \infty$; hence $P_j = 0$ ($j = 0, 1, \ldots$) and no equilibrium distribution exists.

A third example is provided by the three-state birth-and-death process characterized by $\lambda_0 = 0$, $\lambda_1 = \mu_1 > 0$, and $\lambda_2 = \mu_2 = 0$. Our theorem is inapplicable, because the condition $\mu_j > 0$ $(j = 1, 2, \ldots)$ is violated [causing S, given by (3.8), to be undefined]. However, it is clear from physical considerations that a limiting distribution does exist: If the initial state is E_i and $i \neq 1$, then $P_i = 1$; and if the initial state is E_1, then $P_0 = P_2 = \frac{1}{2}$ (and, again, $P_1 = 0$). In other words, the system is not ergodic, because if it ever occupies state E_0 it will remain there forever, and similarly for E_2. Thus, the limiting distribution depends on the initial conditions and is therefore not a statistical-equilibrium distribution.

Statistical Equilibrium

It is instructive to assume the existence of $\lim_{t \to \infty} P_j(t) = P_j$, and then derive (3.10). Observe first that, by (2.3), the existence of $\lim_{t \to \infty} P_j(t) = P_j$ for all j implies the existence of $\lim_{t \to \infty} (d/dt) P_j(t)$ for all j. Consequently, $\lim_{t \to \infty} (d/dt) P_j(t) = 0$, because any other limiting value would contradict the assumed existence of $\lim_{t \to \infty} P_j(t)$. Thus, taking limits throughout the equations (2.3) and rearranging, we have:

$$(\lambda_j + \mu_j) P_j = \lambda_{j-1} P_{j-1} + \mu_{j+1} P_{j+1}$$
$$(j = 0, 1, \ldots; \quad \lambda_{-1} = \mu_0 = 0). \tag{3.11}$$

When $j = 0$, Equation (3.11) is

$$\lambda_0 P_0 = \mu_1 P_1, \tag{3.12}$$

and when $j = 1$, Equation (3.11) is

$$(\lambda_1 + \mu_1) P_1 = \lambda_0 P_0 + \mu_2 P_2. \tag{3.13}$$

Equations (3.12) and (3.13) together yield

$$\lambda_1 P_1 = \mu_2 P_2. \tag{3.14}$$

Similarly, (3.14) and (3.11) for $j = 2$ together yield $\lambda_2 P_2 = \mu_3 P_3$; continuing in this way we derive the following set of equations, which are equivalent to (3.11) but much simpler in form:

$$\lambda_j P_j = \mu_{j+1} P_{j+1} \quad (j = 0, 1, \ldots). \tag{3.15}$$

If $\mu_{j+1} > 0$ for $j = 0, 1, \ldots$, then (3.15) yields

$$P_{j+1} = \frac{\lambda_j}{\mu_{j+1}} P_j \quad (j = 0, 1, \ldots), \tag{3.16}$$

or, equivalently,

$$P_j = \frac{\lambda_0 \lambda_1 \cdots \lambda_{j-1}}{\mu_1 \mu_2 \cdots \mu_j} P_0 \quad (j = 1, 2, \ldots). \tag{3.17}$$

In the cases of practical interest for our purposes, it will always be true that $\mu_{j+1} > 0$ ($j = 0, 1, \ldots$) and either (i) $\lambda_j > 0$ for all $j = 0, 1, \ldots$ or (ii) $\lambda_j > 0$ for $j = 0, 1, \ldots, n-1$ and $\lambda_n = 0$. In both cases, the probability P_0 is determined formally from the normalization condition

$$\sum_{j=0}^{\infty} P_j = 1. \tag{3.18}$$

In case (i) the sum on the left-hand side of (3.18) may fail to converge when (3.17) is substituted into it unless $P_0 = 0$. This would imply that $P_j = 0$ for all finite j, and therefore no statistical-equilibrium distribution would exist. In case (ii) this sum will always converge, because it is composed of only a finite number of terms [because, from (3.17), $P_j = 0$ for all $j > n$]. Therefore, if the initial state is E_i and $i \leq n$, then the statistical-equilibrium distribution exists and is given by (3.17), with P_0 calculated from (3.18). Thus we see that in both cases we obtain results that agree with (3.10).

If $\lambda_j > 0$ for all j, and $\mu_j = 0$ for some $j = k$, then, as Equation (3.15) shows, $P_j = 0$ for $j = 0, 1, \ldots, k-1$. An extreme example is provided by the pure birth process, where $\mu_j = 0$ for all values of the index j. Thus $P_j = 0$ ($j = 0, 1, \ldots,$) for the pure birth process; that is, no proper equilibrium distribution exists. In the particular case of the Poisson process, for example, where $\lambda_j = \lambda$ ($j = 0, 1, \ldots$), the time-dependent probabilities $\{P_j(t)\}$ are given by Equation (2.7), from which we see that these probabilities are all positive and sum to unity for all finite $t \geq 0$, but that each probability approaches zero as $t \to \infty$.

Exercise

3. Consider a birth-and-death process with $\mu_k = 0$ and $\mu_j > 0$ when $j > k$, and $\lambda_j > 0$ for all j. Show that when $S < \infty$,

$$P_j = \begin{cases} 0 & (j = 0, 1, \ldots, k-1), \\ S^{-1} & (j = k), \\ \dfrac{\lambda_k \lambda_{k+1} \cdots \lambda_{j-1}}{\mu_{k+1} \mu_{k+2} \cdots \mu_j} P_k & (j = k+1, k+2, \ldots), \end{cases}$$

where

$$S = 1 + \sum_{j=k+1}^{\infty} \prod_{i=k+1}^{j} \frac{\lambda_{i-1}}{\mu_i};$$

and

$$P_j = 0 \ (j = 0, 1, \ldots) \quad \text{when} \quad S = \infty.$$

Equation (3.11) admits of a simple and important intuitive interpretation: To write the *statistical-equilibrium state equations* (sometimes also called the *balance equations*), simply equate the rate at which the system leaves state E_j to the rate at which the system enters state E_j. Similarly, Equation (3.15) can be interpreted as stating that the rate at which the system leaves state E_j for a higher state equals the rate at which the system leaves state E_{j+1} for a lower state.

Equations (3.11) and (3.15) are statements of conservation of flow. Recall that in Chapter 1 we gave a heuristic derivation of an important

Statistical Equilibrium

queueing formula using this concept. Specifically, we appealed to this concept to derive the statistical-equilibrium equation (1.1) of Chapter 1:

$$\lambda P_j = (j+1)\tau^{-1} P_{j+1} \qquad (j=0,1,\ldots,s-1).$$

Observe that this equation is the special case of the statistical-equilibrium state equation (3.15) with $\lambda_j = \lambda$ and $\mu_{j+1} = (j+1)\tau^{-1}$ for $j=0,1,\ldots,s-1$ and $\lambda_j = 0$ for $j \geqslant s$, and that the heuristic conservation-of-flow argument of Chapter 1 is identical with the intuitive interpretation of Equation (3.15).

Thus far our discussion has been limited to birth-and-death processes for which the ordering of states $E_0, E_1, E_2, \ldots$ arises quite naturally. As we shall see, however, this is not always the case. Quite often problems arise in which the natural definition of states requires two variables, for example, E_{ij} ($i=0,1,2,\ldots; j=0,1,2,\ldots$). We have shown that for the one-dimensional case, the "rate out equals rate in" formulation (3.11) is equivalent to the algebraically simpler "rate up equals rate down" formulation (3.15). Of course, these two-dimensional states can always be relabeled so that the problem reduces to a one-dimensional case, but in general the result is no longer a birth-and-death process. Thus, the "rate up equals rate down" formulation, as exemplified by (3.15), is in general inapplicable in the multidimensional case. However, as we will show in Chapter 4, multidimensional birth-and-death models can be described by "rate out equals rate in" equations that are analogous to (3.11).

To summarize, from a practical point of view, it is assumed that a system is in statistical equilibrium after it has been in operation long enough for the effects of the initial conditions to have worn off. When a system is in statistical equilibrium its state probabilities are constant in time, that is, the state distribution is stationary. This does not mean that the system does not fluctuate from state to state, but rather that no net trends result from the statistical fluctuations. The statistical-equilibrium state equations are obtained from the birth-and-death equations by setting the time derivatives of the state probabilities equal to zero, which reflects the idea that during statistical equilibrium the state distribution is constant in time. Equivalently, the statistical-equilibrium state probabilities are defined by the equation "rate out equals rate in," which reduces in the one-dimensional case to "rate up equals rate down." A system with a statistical-equilibrium distribution is ergodic, and hence the probability P_j can be interpreted as the proportion of time that the system will spend in state E_j ($j=0,1,2,\ldots$), taken over any long period of time throughout which statistical equilibrium prevails.

It should be apparent that a deep understanding of the theory of birth-and-death processes requires extensive mathematical preparation. We emphasize that the present treatment is informal; the interested reader should consult more advanced texts on stochastic processes and Markov processes, such as Çinlar [1975], Cohen [1972], Feller [1968], Karlin [1968], and Khintchine [1969]. (Roughly speaking, a *Markov process* is a process whose future probabilistic evolution after any time t depends only on the state of the system at time t, and is independent of the history of the system prior to time t. It should be easy for the reader to satisfy himself that a birth-and-death process is a Markov process.) Hopefully, it is also apparent that the birth-and-death process has sufficient intuitive appeal so that useful insights can be gained from the preceding cursory summary. We shall return to the birth-and-death process in Chapter 3, where we shall develop queueing models by judiciously choosing the birth-and-death coefficients.

2.4. Probability-Generating Functions

Many of the random variables of interest in queueing theory assume only the integral values $j = 0, 1, 2, \ldots$. Let K be a nonnegative integer-valued random variable with probability distribution $\{p_j\}$, where $p_j = P\{K=j\}$ ($j = 0, 1, 2, \ldots$). Consider now the power series $g(z)$,

$$g(z) = p_0 + p_1 z + p_2 z^2 + \cdots, \tag{4.1}$$

where the probability p_j is the coefficient of z^j in the expansion (4.1). Since $\{p_j\}$ is a probability distribution, therefore $g(1) = 1$. In fact, $g(z)$ is convergent for $|z| \leq 1$ [so that the function (4.1) is holomorphic at least on the unit disk—see any standard text in complex-variable or analytic-function theory]. Clearly, the right-hand side of (4.1) characterizes K, since it displays the whole distribution $\{p_j\}$. And since the power-series representation of a function is unique, the distribution $\{p_j\}$ is completely and uniquely specified by the function $g(z)$. The function $g(z)$ is called the *probability-generating function* for the random variable K. The variable z has no inherent significance, although, as we shall see, it is sometimes useful to give it a physical interpretation.

The notion of a generating function applies not only to probability distributions $\{p_j\}$, but to any sequence of real numbers. However, the generating function is a particularly powerful tool in the analysis of probability problems, and we shall restrict ourselves to probability-generating functions. The generating function transforms a discrete sequence of numbers (the probabilities) into a function of a dummy variable, much the same way the Laplace transform changes a function of a particular variable into another function of a different variable. As with all transform methods, the use of generating functions not only preserves the information while changing its form, but also presents the information in a form that often simplifies manipulations and provides insight. In this section we summarize some important facts about probability-generating functions. For a more complete treatment, see Feller [1968] and Neuts [1973].

Probability-Generating Functions

As examples of probability-generating functions, we consider those for the Bernoulli and Poisson distributions. A random variable X has the Bernoulli distribution if it has two possible realizations, say 0 and 1, occurring with probabilities $P\{X=0\}=q$ and $P\{X=1\}=p$, where $p+q=1$. As discussed in Section 2.1, this scheme can be used to describe a coin toss, with $X=1$ when a head appears and $X=0$ when a tail appears. Referring to (4.1), we see that the Bernoulli variable X has probability-generating function

$$g(z) = q + pz. \tag{4.2}$$

(If this seems somewhat less than profound, be content with the promise that this simple notion will prove extremely useful.)

In Section 2.2 we considered the random variable $N(t)$, defined as the number of customers arriving in an interval of length t, and we showed that with appropriate assumptions this random variable is described by the Poisson distribution

$$P\{N(t)=j\} = \frac{(\lambda t)^j}{j!} e^{-\lambda t} \quad (j=0,1,\ldots).$$

If we denote the probability-generating function of $N(t)$ by $g(z)$, then

$$g(z) = \sum_{j=0}^{\infty} \frac{(\lambda t)^j}{j!} e^{-\lambda t} z^j = e^{-\lambda t} \sum_{j=0}^{\infty} \frac{(\lambda t z)^j}{j!},$$

and this reduces to

$$g(z) = e^{-\lambda t(1-z)}. \tag{4.3}$$

Having defined the notion of probability-generating function and given two important examples, we now discuss the special properties of such functions that make the concept useful. Since the probability-generating function $g(z)$ of a random variable K contains the distribution $\{p_j\}$ implicitly, it therefore contains the information specifying the moments of the distribution $\{p_j\}$. Consider the mean $E(K)$,

$$E(K) = \sum_{j=1}^{\infty} j p_j. \tag{4.4}$$

It is easy to see that (4.4) can be obtained formally by evaluating the derivative $(d/dz)g(z) = \sum_{j=1}^{\infty} j p_j z^{j-1}$ at $z=1$:

$$E(K) = g'(1). \tag{4.5}$$

From (4.2) and (4.5) we see that the Bernoulli random variable X has mean

$$E(X) = p, \qquad (4.6)$$

and from (4.3) and (4.5) we see that the Poisson random variable $N(t)$ has mean

$$E(N(t)) = \lambda t. \qquad (4.7)$$

Similarly, it is easy (and we leave it as an exercise) to show that the variance $V(K)$ can be obtained from the probability-generating function as

$$V(K) = g''(1) + g'(1) - [g'(1)]^2. \qquad (4.8)$$

This formula gives for the Bernoulli variable

$$V(X) = pq \qquad (4.9)$$

and for the Poisson variable

$$V(N(t)) = \lambda t. \qquad (4.10)$$

Sometimes, as we shall see, it is easier to obtain the probability-generating function than it is to obtain the whole distribution directly. In such cases, the formulas (4.5) and (4.8) often provide the easiest way of obtaining the mean and variance. The higher moments can also be calculated in a similar manner from the probability-generating function, but the complexity of the formulas increases rapidly. If primary interest is in the moments of a distribution rather than the individual probabilities, it is often convenient to work directly with the moment-generating function or the related characteristic function, which generate moments as (4.1) generates probabilities. These topics are covered in most texts on mathematical statistics, such as Fisz [1963].

Another important use of probability-generating functions is in the analysis of problems concerning sums of independent random variables. Suppose $K = K_1 + K_2$, where K_1 and K_2 are independent, nonnegative, integer-valued random variables. Then $P\{K=k\}$ $(k=0,1,2,\ldots)$ is given by the *convolution*

$$P\{K=k\} = \sum_{j=0}^{k} P\{K_1=j\} P\{K_2=k-j\}. \qquad (4.11)$$

Let K_1 and K_2 have generating functions $g_1(z)$ and $g_2(z)$, respectively:

$$g_1(z) = \sum_{j=0}^{\infty} P\{K_1=j\} z^j$$

and

$$g_2(z) = \sum_{j=0}^{\infty} P\{K_2 = j\} z^j.$$

Then term-by-term multiplication shows that the product $g_1(z)g_2(z)$ is given by

$$g_1(z)g_2(z) = \sum_{k=0}^{\infty} \left[\sum_{j=0}^{k} P\{K_1 = j\} P\{K_2 = k-j\} \right] z^k. \quad (4.12)$$

If K has generating function $g(z) = \sum_{k=0}^{\infty} P\{K=k\} z^k$, then (4.11) and (4.12) show that

$$g(z) = g_1(z)g_2(z). \quad (4.13)$$

Thus we have the important result: The generating function of a sum of mutually independent random variables is equal to the product of their respective generating functions.

For a more concise proof of (4.13), observe that for any random variable K with probability-generating function $g(z)$, the random variable z^K has expected value $E(z^K)$ given by

$$E(z^K) = g(z). \quad (4.14)$$

Then if $K = K_1 + K_2$, it follows that $E(z^K) = E(z^{K_1+K_2}) = E(z^{K_1} z^{K_2})$. If K_1 and K_2 are independent, then the random variables z^{K_1} and z^{K_2} are also independent, and thus $E(z^{K_1} z^{K_2}) = E(z^{K_1})E(z^{K_2})$, from which (4.13) follows.

Consider again the coin-tossing experiment described in Section 2.1. A coin is tossed n times, with $X_j = 1$ if a head appears and $X_j = 0$ if a tail appears on the jth toss. We wish to determine the probability that k heads appear in n tosses. Let

$$S_n = X_1 + \cdots + X_n, \quad (4.15)$$

so that the value of the random variable S_n is the number of heads appearing in n tosses. S_n is the sum of n mutually independent, identically distributed Bernoulli variables $\{X_j\}$, each with generating function $g(z)$ given by (4.2). Therefore, S_n has generating function $[g(z)]^n$; hence

$$(q + pz)^n = \sum_{k=0}^{\infty} P\{S_n = k\} z^k. \quad (4.16)$$

Expanding the left-hand side of (4.16) by the binomial theorem, we obtain

$$(q+pz)^n = \sum_{k=0}^{n} \binom{n}{k}(pz)^k q^{n-k}. \tag{4.17}$$

Equating coefficients of z^k in (4.16) and (4.17) yields

$$P\{S_n = k\} = \begin{cases} \binom{n}{k} p^k q^{n-k} & (k=0,1,\ldots,n), \\ 0 & (k>n), \end{cases} \tag{4.18}$$

which is, of course, the binomial distribution.

The result (4.18) could have been obtained by direct probabilistic reasoning. (Any particular sequence of k heads and $n-k$ tails has probability $p^k q^{n-k}$, and there are $\binom{n}{k}$ such sequences.) In this example, the direct probabilistic reasoning concerning the possible outcomes of n tosses is replaced by the simpler probabilistic reasoning concerning the possible outcomes of one toss and the observation (4.15) that leads to the use of probability-generating functions. In a sense, probabilistic or intuitive reasoning has been traded for more formal mathematical manipulation. In the present case both approaches are simple, but this is not always true. Roughly speaking, the difficulties encountered in the generating-function approach, when applicable, are not extremely sensitive to the underlying probabilistic structure. Thus the use of generating functions tends to simplify hard problems and complicate easy ones.

Another example is provided by the Poisson distribution. Let $N = N_1 + N_2$, where N_i is a Poisson random variable with mean λ_i; that is, $P\{N_i = j\} = (\lambda_i^j/j!)e^{-\lambda_i}$. Then N has probability-generating function

$$g(z) = e^{-\lambda_1(1-z)} e^{-\lambda_2(1-z)} = e^{-(\lambda_1+\lambda_2)(1-z)}. \tag{4.19}$$

Equation (4.19) shows (with no effort) that the sum of two independent Poisson variables with means λ_1 and λ_2 is itself a Poisson variable with mean $\lambda = \lambda_1 + \lambda_2$.

Exercise

4. *Compound distributions.* Let $X_1, X_2, \ldots$ be a sequence of independent, identically distributed, nonnegative, integer-valued random variables with probability-generating function $f(z)$; and let N be a nonnegative, integer-valued random variable, independent of $X_1, X_2, \ldots$, with probability-generating function $g(z)$. Let S_N denote the sum of a random number of random variables: $S_N = X_1 + \cdots + X_N$. (In general, the distribution of the sum of a random number of independent random variables is called a compound distribution.)
   a. Show that S_N has probability-generating function $g(f(z))$.
   b. Show that $E(S_N) = E(N)E(X)$ and $V(S_N) = E(N)V(X) + V(N)E^2(X)$.

Probability-Generating Functions

It was mentioned previously that the variable z in the generating function has no inherent significance, but that it is sometimes useful to give it a probabilistic interpretation. This may allow us to obtain the generating function directly from probabilistic considerations and, in so doing, to replace tedious arguments by elegant ones. This probabilistic interpretation, suggested by D. van Dantzig and called by him the *method of collective marks*, is: Imagine that the random variable whose probability-generating function we seek has been observed to have realization $K=j$, say. Now suppose we perform j Bernoulli trials, where at each trial we generate a "mark" with probability $1-z$ (and no mark with probability z). Then z^j is the probability that none of the j trials resulted in a mark. Hence, $\sum_{j=0}^{\infty} P\{K=j\} z^j = g(z)$ is the probability that no marks are generated by the realization of K.

For example, consider Exercise 4a, where we wish to determine the probability-generating function of the sum S_N of a random number of random variables. The probability that the realization of X_1, say, generates no marks is equal to its probability-generating function $f(z)$. Thus, the probability that the realization of $X_1 + \cdots + X_n$ generates no marks is $[f(z)]^n$. Hence, the probability that the realization of S_N generates no marks is $\sum_{n=1}^{\infty} P\{N=n\} [f(z)]^n = g(f(z))$.

We shall appeal to the method of collective marks to eliminate messy calculation at various points in the text. For further discussion see Neuts [1973] and Runnenburg [1965] (as well as the accompanying discussion of Runnenburg's paper by van der Vaart).

Exercise

5. Let N_1 and N_2 be independent, identically distributed, nonnegative integer-valued random variables. We consider two different procedures for distributing $N_1 + N_2$ balls into two cells. In procedure (a), we place N_1 balls in the first cell and N_2 balls in the second cell. In procedure (b), we consider the $N_1 + N_2 = N$ balls together, and independently place each of the N balls in a cell, with each ball having probability $\frac{1}{2}$ of being placed in the first cell and probability $\frac{1}{2}$ of being placed in the second cell. Define the generating functions

$$g(z) = \sum_{j=0}^{\infty} P\{N_\nu = j\} z^j \qquad (\nu = 1, 2),$$

and show that if procedures (a) and (b) are equivalent, then

$$g(x)g(y) = g^2\left(\frac{x+y}{2}\right).$$

[*Hint*: Define the generating function $g(x,y)$ of the joint distribution of the number of balls in each cell, and use the method of collective marks to interpret $g(x,y)$ as the probability that there are no marked balls in either cell.]

Perhaps the most important use of probability-generating functions in queueing theory is the solution of probability state equations. Consider, for example, the birth-and-death equations (2.3) with $\lambda_j = \lambda$, $\mu_j = j\mu$ ($j = 0,1,2,\ldots$), and initial condition $P_0(0) = 1$. The equations are

$$\frac{d}{dt} P_j(t) = \lambda P_{j-1}(t) + (j+1)\mu P_{j+1}(t) - (\lambda + j\mu) P_j(t)$$

$$[P_{-1}(t) = 0; \quad j = 0, 1, \ldots]. \tag{4.20}$$

These equations describe the following important queueing model. Customers request service from an infinite-server group. The probability that exactly one customer arrives in $(t, t+h)$ is $\lambda h + o(h)$ as $h \to 0$. If there are j customers in service at time t, the probability that exactly one customer will complete service in $(t, t+h)$ is $j\mu h + o(h)$ as $h \to 0$, and the probability that more than one change (arrivals and/or service completions) occurs in $(t, t+h)$ is $o(h)$ as $h \to 0$. (As we shall see, these assumptions are true if the customers arrive according to a Poisson process with rate λ, and the service times are mutually independent, identically distributed exponential random variables, each with mean μ^{-1}.) $P_j(t)$ is the probability that j customers are in service at time t.

Although this model is an important one in queueing theory, we shall defer discussion of it to later chapters. Our present concern is simply the application of the theory of probability-generating functions to the solution of the probability state equations (4.20).

We emphasize that (4.20) is an infinite set of simultaneous linear differential-difference equations, an apparently hopeless case. Undaunted, we define the probability-generating function $P(z,t)$,

$$P(z,t) = \sum_{j=0}^{\infty} P_j(t) z^j. \tag{4.21}$$

Now multiply each equation in $(d/dt) P_j(t)$ of the set (4.20) by z^j and add all the equations. The result is

$$\sum_{j=0}^{\infty} \frac{d}{dt} P_j(t) z^j = \lambda \sum_{j=1}^{\infty} P_{j-1}(t) z^j + \mu \sum_{j=0}^{\infty} (j+1) P_{j+1}(t) z^j$$

$$- \lambda \sum_{j=0}^{\infty} P_j(t) z^j - \mu \sum_{j=1}^{\infty} j P_j(t) z^j, \tag{4.22}$$

which in view of (4.21) can be written

$$\frac{\partial}{\partial t} P(z,t) = \lambda z P(z,t) + \mu \frac{\partial}{\partial z} P(z,t) - \lambda P(z,t) - \mu z \frac{\partial}{\partial z} P(z,t). \tag{4.23}$$

Probability-Generating Functions

When rearranged into standard form, Equation (4.23) reads

$$\frac{\partial}{\partial t}P(z,t) - (1-z)\mu\frac{\partial}{\partial z}P(z,t) = -\lambda(1-z)P(z,t). \tag{4.24}$$

Thus we have transformed the infinite set of simultaneous linear differential-difference equations (4.20) for $P_j(t)$ into the single linear first-order partial differential equation (4.24) for the generating function $P(z,t)$. It remains to solve (4.24) for $P(z,t)$ and invert.

Equation (4.24) can be solved in general by well-known techniques (see any standard text, such as Garabedian [1964]). The solution that satisfies $P(z,0) = 1$, corresponding to the initial condition $P_0(0) = 1$, is

$$P(z,t) = \exp\left[-\frac{\lambda}{\mu}(1 - e^{-\mu t})(1-z)\right], \tag{4.25}$$

as is readily verified by substitution into (4.24). The state probabilities $P_j(t)$ are easily found by recognizing (4.25) as the probability-generating function for the Poisson distribution with mean $(\lambda/\mu)(1 - e^{-\mu t})$:

$$P_j(t) = \frac{\left[(\lambda/\mu)(1 - e^{-\mu t})\right]^j}{j!} \exp\left[-\frac{\lambda}{\mu}(1 - e^{-\mu t})\right] \quad (j = 0, 1, \ldots). \tag{4.26}$$

Several observations are in order. As in Section 2.2, we have solved the birth-and-death equations (2.3) completely for a particular choice of birth and death rates and a particular initial condition. The normalization requirement (2.12) is satisfied, since for any t, $P_j(t)$ is a Poisson probability. Equivalently, $P(1,t) = 1$, as (4.25) shows. The statistical-equilibrium distribution can be obtained directly by taking limits in (4.26):

$$P_j = \lim_{t \to \infty} P_j(t) = \frac{(\lambda/\mu)^j}{j!} e^{-\lambda/\mu} \quad (j = 0, 1, \ldots). \tag{4.27}$$

The limiting distribution is the Poisson distribution with mean λ/μ.

The limiting distribution (4.27) can also be obtained by taking limits in the probability-generating function (4.25). In general, it is true that a sequence of probability distributions converges to a limiting distribution if and only if the corresponding generating functions converge (see Chapter XI.6 of Feller [1968]).

Finally, note that (4.27) can be obtained directly from the statistical-equilibrium state equations and the normalization condition, that is, from (3.17) and (3.18) with $\lambda_j = \lambda$, $\mu_j = j\mu$ $(j = 0, 1, 2, \ldots)$.

Exercise

6. For the model of Exercise 2, Section 2.2, define the generating function $P(z,t) = \sum_{j=0}^{\infty} P_j(t) z^j$.

  a. Show directly from the birth-and-death equations that $P(z,t)$ satisfies the partial differential equation

  $$\frac{\partial}{\partial t} P(z,t) = \lambda z (z-1) \frac{\partial}{\partial z} P(z,t).$$

  b. Verify that this equation has solution

  $$P(z,t) = \frac{z e^{-\lambda t}}{1 - z + z e^{-\lambda t}}$$

  satisfying the initial condition $N(0) = 1$ [and, of course, the normalization condition $\sum_{j=0}^{\infty} P_j(t) = 1$], and verify that $P(z,t)$ generates the distribution found in Exercise 2.

  c. Verify that the solution $P_n(z,t)$ of the partial differential equation corresponding to the initial condition $N(0) = n \geqslant 1$ is given by $P_n(z,t) = P^n(z,t)$. Give a physical interpretation of this result. [It can be shown by expansion of $P^n(z,t)$ that for $N(0) = n \geqslant 1$,

  $$P_j(t) = \binom{j-1}{j-n} e^{-n\lambda t} (1 - e^{-\lambda t})^{j-n} \quad (j = n, n+1, \ldots)$$

  and $P_j(t) = 0$ when $j < n$.]

2.5. Some Important Probability Distributions

We consider several distributions that are important in queueing theory.

Bernoulli Distribution

A random variable X has the *Bernoulli distribution* if it has two possible outcomes, say $X = 1$ and $X = 0$ (often called success and failure, respectively) occurring with probabilities $P\{X = 1\} = p$ and $P\{X = 0\} = q$, where $p + q = 1$. As mentioned previously, this scheme is often used to describe a coin toss, with $X = 1$ when a head occurs. The Bernoulli variable has probability-generating function $q + pz$, mean p, and variance pq.

Binomial Distribution

The binomial distribution describes the number of successes in n Bernoulli trials, which, by definition, are a sequence of n independent experiments, each of which results in either "success," with probability p, or "failure," with probability $1 - p = q$. If S_n is the number of successes in n Bernoulli trials, then S_n can be represented as the sum $S_n = X_1 + \cdots + X_n$, where X_j

Some Important Probability Distributions

is the Bernoulli variable with value $X_j = 1$ if the jth trial results in success and $X_j = 0$ for failure. Since the $\{X_j\}$ are mutually independent, identically distributed random variables, each with probability-generating function $q + pz$, then S_n has generating function $(q + pz)^n$. Expansion of the generating function shows that S_n has the *binomial distribution*

$$P\{S_n = k\} = \binom{n}{k} p^k (1-p)^{n-k} \quad (k = 0, 1, \ldots, n). \tag{5.1}$$

As mentioned previously, Equation (5.1) can be derived directly by probabilistic reasoning: Since the trials are mutually independent, each trial having probability p of success, any sequence of k successes and $n - k$ failures has probability $p^k(1-p)^{n-k}$. There are

$$\binom{n}{k} = \frac{n!}{k!(n-k)!}$$

ways to choose the k locations of the successes in a sequence of k successes and $n - k$ failures. Thus there are altogether $\binom{n}{k}$ different sequences of k successes and $n - k$ failures, and (5.1) follows.

The binomial distribution is often used to describe the number of busy telephone lines in a line group. This use of (5.1) is based on the implicit assumptions that each line has equal probability p of being busy, and that the lines become busy and idle independently of each other. These implicit assumptions are often not true, in which case (5.1) provides only a rough approximation.

The random variable S_n with distribution (5.1) has mean np and variance npq, which follows directly from the interpretation of S_n as a sum of n identical, mutually independent Bernoulli variables.

Exercise

7. Suppose S_{n_1} has the binomial distribution (5.1) with $n = n_1$, and S_{n_2} has the binomial distribution (5.1) with $n = n_2$. Show that if S_{n_1} and S_{n_2} are independent, then $S_{n_1} + S_{n_2}$ has the binomial distribution (5.1) with $n = n_1 + n_2$.

Multinomial Distribution

Suppose that we have a sequence of n independent trials, but at each trial the number of possible outcomes is $r \geq 2$. That is, at the jth trial there are r possible outcomes with respective probabilities $p_1, \ldots, p_r$ (subject, of course, to $p_1 + \cdots + p_r = 1$). Let N_i be the number of times the ith outcome occurs in n trials. The joint probability that in n trials the ith outcome occurs k_i times, where $k_1 + \cdots + k_r = n$, is

$$P\{N_1 = k_1, \ldots, N_r = k_r\} = \frac{n!}{k_1! \cdots k_r!} p_1^{k_1} \cdots p_r^{k_r}. \tag{5.2}$$

Equation (5.2) defines the *multinomial distribution*, which is a multivariate distribution, and for which probability-generating functions can be defined in a manner analogous to that presented for univariate distributions. We shall not go into this subject further, but simply call attention to (5.2), which we shall have occasion to use later. Note that (5.2) reduces to the binomial distribution when $r=2$.

Geometric Distribution

In a sequence of Bernoulli trials, the probability that the first success occurs after exactly k failures is $(1-p)^k p$, where p is the probability of success at any particular trial. Let N be the number of failures preceding the first success. (N is often called the waiting time to the first success.) Then N has the *geometric distribution*

$$P\{N=k\} = (1-p)^k p \qquad (k=0,1,\ldots). \tag{5.3}$$

Notice that N has an infinite number of possible realizations, in contrast with the Bernoulli, binomial, and multinomial distributions, which are defined on finite sample spaces. For any $0<p<1$ we have $\sum_{k=0}^{\infty}(1-p)^k p = 1$, so that the distribution (5.3) is proper and there is no need to assign a positive probability to the realization $N=\infty$.

The random variable N has probability-generating function

$$p \sum_{k=0}^{\infty} q^k z^k = \frac{p}{1-qz},$$

mean q/p, and variance q/p^2, where $q=1-p$.

From (5.3) the probability that the waiting time to the first success is at least k is

$$P\{N \geq k\} = \sum_{j=k}^{\infty} q^j p = \frac{pq^k}{1-q} = q^k, \tag{5.4}$$

which is just what we should expect; q^k is the probability that each of the first k trials results in failure.

The formula (5.3) gives the probability that exactly k failures precede the first success in a series of independent trials, where each trial has the same probability p of success. Since the probability of success at each trial is independent of the number of preceding trials, it should be clear that the conditional probability that N will equal $j+k$, given that N is at least j ($j=0,1,2\ldots$), is exactly the same as the unconditional probability (5.3) that N equals k. The definition of conditional probability yields

$$P\{N=j+k|N \geqslant j\} = \frac{P\{N=j+k, N \geqslant j\}}{P\{N \geqslant j\}}$$
$$= \frac{P\{N=j+k\}}{P\{N \geqslant j\}} = \frac{q^{j+k}p}{q^j} = q^k p, \qquad (5.5)$$

or equivalently,

$$P\{N \geqslant j+k|N \geqslant j\} = q^k. \qquad (5.6)$$

Comparison of (5.6) and (5.4) shows that if N has a geometric distribution, then

$$P\{N \geqslant j+k|N \geqslant j\} = P\{N \geqslant k\}. \qquad (5.7)$$

This property of independence of past history, called the lack-of-memory or *Markov property*, plays an important role in queueing theory.

The geometric distribution may be used to describe the duration of service of a customer. For the service time might be viewed as the waiting time for the first success in a sequence of Bernoulli trials, one trial performed per unit of time. (Failure means service continues; success means service is completed.) This crude model will prove to be surprisingly useful.

Negative-Binomial (Pascal) Distribution

The number N of failures preceding the first success in a sequence of Bernoulli trials has the geometric distribution (5.3). Let S_n be the number of failures preceding the nth success. ($S_n + n$ is the number of trials up to and including the nth success.) Then $S_n = N_1 + \cdots + N_n$, where N_j ($j = 1, 2, \ldots, n$) is the number of failures occurring between the $(j-1)$th and jth successes. The $\{N_j\}$ are mutually independent, identically distributed random variables, each with the geometric distribution (5.3) and probability-generating function $p/(1-qz)$. Hence, S_n has probability-generating function $[p/(1-qz)]^n$. The distribution of S_n, the total number of failures preceding the nth success in a sequence of Bernoulli trials, is called the negative-binomial or Pascal distribution. Expansion of the probability-generating function shows that the *negative-binomial distribution* is given by

$$P\{S_n = k\} = \binom{n+k-1}{k} q^k p^n \qquad (k=0,1,\ldots). \qquad (5.8)$$

When $n = 1$, Equation (5.8) reduces to (5.3).

Equation (5.8) is perhaps more easily derived through direct probabilistic reasoning. The nth success will be preceded by exactly k failures if there are exactly k failures and $n-1$ successes in any order among the first $n+k-1$ trials, and the $(n+k)$th trial results in success. Any such sequence has probability $q^k p^{n-1} p$. There are $\binom{n+k-1}{k}$ ways of choosing the k failures out of the first $n+k-1$ trials, and (5.8) follows.

Since S_n can be interpreted as a sum of n mutually independent, identically distributed geometric random variables, S_n has mean nq/p and variance nq/p^2.

Exercises

8. Verify the parenthetical statement of part c of Exercise 6.

9. Repeat Exercise 7, with the phrase "binomial distribution (5.1)" replaced everywhere with "negative-binomial distribution (5.8)."

Uniform Distribution

Let U be a continuous random variable that takes values in $(0, t)$, with distribution function $P\{U \leq x\} = F(x)$,

$$F(x) = \begin{cases} 0 & (x < 0), \\ \dfrac{x}{t} & (0 \leq x \leq t), \\ 1 & (x > t), \end{cases} \qquad (5.9)$$

and corresponding density function $f(x) = (d/dx) F(x)$,

$$f(x) = \begin{cases} \dfrac{1}{t} & (0 < x < t), \\ 0 & (x < 0, \ x > t). \end{cases} \qquad (5.10)$$

Then U has the *uniform distribution*, with mean value

$$E(U) = \int_0^t x f(x)\, dx = \frac{t}{2}$$

and variance

$$V(U) = \int_0^t \left(x - \frac{t}{2} \right)^2 f(x)\, dx = \frac{t^2}{12}.$$

[In general, the uniform distribution is defined over an arbitrary interval (t_1, t_2), but for simplicity we have taken $t_1 = 0$. The standard definition, however, uses the interval $(0,1)$. Of course, all these definitions are equivalent, corresponding only to changes in scale and location.]

According to (5.9), U takes on values between zero and t, and the probability that U lies in any subinterval of $(0,t)$ is proportional by $1/t$ to the length of the subinterval. The random variable U is said to be *uniformly distributed* on $(0,t)$. Since there is no tendency to prefer one point over any other when sampling is performed according to (5.9), a point chosen from a uniform distribution is often said to be chosen *at random*.

The probability is x/t that a point chosen at random in $(0,t)$ will lie in a subinterval of length x, and the probability is $1-(x/t)$ that a randomly chosen point will lie outside the designated subinterval. Thus, if n points are independently chosen at random from $(0,t)$, the probability is $(x/t)^k \cdot [1-(x/t)]^{n-k}$ that k specific points will fall in the designated subinterval of length x and $n-k$ specific points outside it. Since there are $\binom{n}{k}$ ways of specifying k points out of n, then k out of n points chosen at random from $(0,t)$ will fall in any prespecified subinterval of length x with probability

$$\binom{n}{k}\left(\frac{x}{t}\right)^k\left(1-\frac{x}{t}\right)^{n-k}.$$

Reference to (5.1) shows that this is the binomial probability with $p=x/t$.

Similarly, suppose that $(0,t)$ is divided into r subintervals of respective lengths x_i, such that $x_1+\cdots+x_r=t$. Choose n points at random from $(0,t)$, and let N_i be the number of points that fall within the ith subinterval (of length x_i). Then the joint distribution of the numbers of points falling in each subinterval is given by the multinomial probabilities (5.2), with $p_i=x_i/t$.

In the special case $x_1=x_2=\cdots=x_r=t/r$, each subinterval has the same probability of containing k points ($k=0,1,2,\ldots,n$) as each other subinterval. Although the numbers of points falling in each subinterval of equal length are in general unequal, there is no tendency for any particular subinterval to be favored over any other. Thus the uniform distribution has the properties one would desire from any distribution so named.

Let us consider more deeply the question of points uniformly distributed over an interval $(0,t)$. We have shown that when the interval $(0,t)$ is divided into subintervals, the probability that a randomly chosen point will lie in a given subinterval is proportional to the subinterval's length; the probability that k of n randomly chosen points will all lie in a subinterval of length x is given by (5.1). Let us now turn our attention from the discrete distribution describing the number of points in a subinterval to the continuous distribution describing the distances between successive points. We shall find the distribution of the distance between any two consecutive points of n points uniformly distributed over $(0,t)$.

Suppose that n points $U_1,\ldots,U_n$ are chosen at random from $(0,t)$. Let $U_{(k)}$ be the kth smallest; $U_{(1)} \leqslant U_{(2)} \leqslant \cdots \leqslant U_{(n)}$. ($U_{(k)}$ is called the kth *order statistic.*) What is the distribution of the distance $U_{(k+1)} - U_{(k)}$ between two arbitrary successive points?

We have argued previously that points distributed uniformly over an interval are spread evenly throughout the interval in the sense that the probability that a particular point will lie in any given subinterval is proportional to the subinterval's length, regardless of the relative location of the subinterval within the interval $(0, t)$. This suggests that the distribution of the distance between any two successive points should be the same for every pair of consecutive points, even when one of the pair is an end point. Thus, if we take $U_{(0)} = 0$ and $U_{(n+1)} = t$, then the probability $P\{U_{(k+1)} - U_{(k)} > x\}$ should be independent of the index k. Hence it is necessary to evaluate $P\{U_{(k+1)} - U_{(k)} > x\}$ for only one value of k. The evaluation for $k = 0$ is particularly easy: $U_{(1)}$ will exceed x if and only if all n points fall in (x, t). The probability of this occurrence for any one point is $1 - (x/t)$, and since the points are selected independently of each other, the required probability is $P\{U_{(1)} > x\} = [1 - (x/t)]^n$. Therefore, the probability that any two consecutive points are separated by more than x when n points are selected at random in $(0, t)$ is

$$P\{U_{(k+1)} - U_{(k)} > x\} = \left(1 - \frac{x}{t}\right)^n$$

$$(U_{(0)} = 0, \quad U_{(n+1)} = t; \quad k = 0, 1, \ldots, n). \tag{5.11}$$

The result (5.11) has been obtained through intuitive reasoning. Its validity rests on the truth of the symmetry argument that the distribution of $U_{(k+1)} - U_{(k)}$ is independent of k, even for the end points $U_{(0)} = 0$ and $U_{(n+1)} = t$.

We shall now verify (5.11) by a more mathematical argument that is of interest in itself. Choose n points at random in $(0, t)$. We wish to calculate the probability that, of these n points, exactly one point lies in the differential element $(\xi, \xi + d\xi)$, exactly $k - 1$ points lie in $(0, \xi)$, and exactly $n - 1 - (k - 1)$ points lie in $(\xi + x, t)$. The simple diagram in Figure 2.1 shows that this particular arrangement implies $U_{(k+1)} - U_{(k)} > x$.

The probability that any particular point lies in $(\xi, \xi + d\xi)$ is $f(\xi)d\xi = (1/t)d\xi$. The probability that any particular point lies in $(0, \xi)$ is ξ/t, so that the probability that any particular $k - 1$ (independently chosen) points all lie in $(0, \xi)$ is therefore $(\xi/t)^{k-1}$. Similarly, for any particular choice of $n - k$ points, the probability that all lie in $(\xi + x, t)$ is $\{[t - (\xi + x)]/t\}^{n-k}$.

Figure 2.1

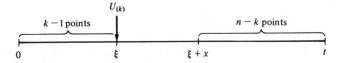

Therefore the probability that, out of n points, a particular point lies in $(\xi, \xi+d\xi)$, $k-1$ particular points lie in $(0,\xi)$, and $n-k$ particular points lie in $(\xi+x, t)$ is given by the product

$$\left(\frac{\xi}{t}\right)^{k-1}\left(\frac{t-(\xi+x)}{t}\right)^{n-k}\frac{1}{t}d\xi.$$

This expression represents the required probability for a particular choice of the points falling in each interval. Since there are n points, there are n different choices for the point that falls in $(\xi, \xi+d\xi)$. And for each such choice, there are $\binom{n-1}{k-1}$ different choices for the $k-1$ points falling in $(0,\xi)$. Once these k points are specified, then the remaining $n-k$ points falling in $(\xi+x, t)$ are automatically specified. Thus the probability that, out of n points, any one lies in $(\xi, \xi+d\xi)$, any $k-1$ lie in $(0,\xi)$, and any $n-k$ lie in $(\xi+x, t)$ is given by

$$n\binom{n-1}{k-1}\left(\frac{\xi}{t}\right)^{k-1}\left(\frac{t-(\xi+x)}{t}\right)^{n-k}\frac{1}{t}d\xi.$$

Finally, the value ξ can be anywhere from zero to $t-x$, so that

$$P\{U_{(k+1)}-U_{(k)}>x\} = n\binom{n-1}{k-1}\int_0^{t-x}\left(\frac{\xi}{t}\right)^{k-1}\left(\frac{t-(\xi+x)}{t}\right)^{n-k}\frac{1}{t}d\xi$$

$$(k=1,2,\ldots,n; \quad U_{(n+1)}=t). \tag{5.12}$$

We have already shown that $P\{U_{(1)}>x\}=[1-(x/t)]^n$. Evaluation of (5.12) for $k=1$ easily shows that $P\{U_{(2)}-U_{(1)}>x\}=[1-(x/t)]^n$. It would now be surprising if our intuitive derivation of (5.11) were incorrect. Further straightforward (although complicated) calculations would show that the result (5.11) is indeed correct.

Exercise

10. (Feller [1971].) Find the distribution function of the length of the arc covering a fixed point x on a circle of circumference c when the endpoints of the arc are independently uniformly distributed over the circumference of the circle. Show that the mean length of the covering arc is $\frac{2}{3}c$. [Observe that this latter result is the same as would be obtained if the point x were not fixed, but were itself chosen randomly. To the extent that this observation could have been made a priori on grounds of symmetry, our conclusion (that the mean length of the covering arc is $\frac{2}{3}c$) is intuitively obvious (because three randomly chosen points would "on the average" partition the circumference into three equal arcs).]

Finally, we note for future reference that if $U_1, U_2, \ldots, U_n$ are independent and uniformly distributed on $(0, t)$, then the joint distribution function of the order statistics $U_{(1)} \leq U_{(2)} \leq \cdots \leq U_{(n)}$ is given by

$$P\{U_{(1)} \leq x_1, \ldots, U_{(n)} \leq x_n\}$$
$$= \sum_{\{j_1, j_2, \ldots, j_n\}} \frac{n!}{j_1! j_2! \cdots j_n!} \left(\frac{x_1}{t}\right)^{j_1} \left(\frac{x_2 - x_1}{t}\right)^{j_2} \cdots \left(\frac{x_n - x_{n-1}}{t}\right)^{j_n}, \quad (5.13)$$

where $0 \leq x_1 \leq x_2 \leq \cdots \leq x_n \leq t$, and the summation is extended over all those sets of nonnegative integers $\{j_1, j_2, \ldots, j_n\}$ for which

$$j_1 + j_2 + \cdots + j_n = n \quad \text{and} \quad j_1 + j_2 + \cdots + j_k \geq k \quad (k = 1, 2, \ldots, n-1).$$

To prove (5.13), observe that its right-hand side is simply a sum of multinomial probabilities, each being the probability that, out of n independent realizations, exactly j_1 values fall in the interval $(0, x_1)$, and jointly j_2 values fall in (x_1, x_2), and so on. The sum is taken over all values of the indices such that, simultaneously, the number of values in $(0, x_1)$ is at least one (that is, $j_1 \geq 1$), the number of values in $(0, x_2)$ is at least two (that is, $j_1 + j_2 \geq 2$), and so on.

Negative-Exponential Distribution

Let X be a continuous random variable with distribution function $P\{X \leq x\} = F(x)$,

$$F(x) = \begin{cases} 1 - e^{-\mu x} & (x \geq 0), \\ 0 & (x < 0), \end{cases} \quad (5.14)$$

and with corresponding density function $f(x) = (d/dx) F(x)$. Thus

$$f(x) = \mu e^{-\mu x} \quad (x \geq 0). \quad (5.15)$$

Then X has the *negative-exponential distribution* with mean $E(X) = \mu^{-1}$ and variance $V(X) = \mu^{-2}$. (For convenience, the adjective "negative" is often omitted from the name.)

In the discussion of the geometric distribution, we showed [see Equation (5.4)] that in a sequence of Bernoulli trials, each trial having probability of success p, the first k trials will all result in failures with probability $(1-p)^k$. That is, $(1-p)^k$ is the probability that the waiting time (number of trials) preceding the first success is at least k.

The mean number of successes in k independent Bernoulli trials is kp. Hence, if k trials are conducted in a time interval of length x, then the success rate—that is, the mean number of successes per unit time, say μ—is

$$\mu = \frac{kp}{x}.$$

Now let $p \to 0$ (and therefore $k \to \infty$) in such a way that the rate μ remains constant. That is, we envision an experiment with an infinite number of trials performed in a finite length of time x, each trial taking vanishingly small time and each trial with vanishingly small probability of success, but with the mean number of successes being equal to the constant $\mu x > 0$. We have

$$\lim_{\substack{p \to 0 \\ k \to \infty \\ pk = \mu x}} (1-p)^k = \lim_{k \to \infty} \left(1 - \frac{\mu x}{k}\right)^k = e^{-\mu x}. \qquad (5.16)$$

This equation shows that the geometric distribution approaches the negative-exponential in the limit as the trials are taken infinitesimally close together, each with vanishingly small probability of success, but with the average number of successes per unit time remaining constant. Thus, if X is the random variable representing the (continuous) time to the first success, rather than the (discrete) number of trials preceding the first success, then $P\{X > x\} = e^{-\mu x}$, where μ^{-1} is the mean time to the first success.

We mentioned previously that the number of trials preceding the first success in a sequence of Bernoulli trials provides a crude but suggestive model for the description of service times. This assertion is further supported by the fact that in the limit the geometric distribution approaches the negative-exponential distribution, and observation has shown the negative-exponential distribution to provide a good statistical description of some service-time distributions. For example, it has long been known that telephone-call durations are well described by the negative-exponential distribution. As we shall see, it is indeed fortunate that this is so; the special properties of the exponential distribution greatly simplify what might otherwise be intractable mathematics. Because of these simplifying mathematical properties, and because data support its use in some important applications such as telephony, it is common in the construction of queueing models to assume exponential service times. We shall now investigate some of these important mathematical properties.

A continuous nonnegative random variable X is said to have the *Markov property* if for every $t > 0$ and every $x > 0$,

$$P\{X > t + x | X > t\} = P\{X > x\}. \qquad (5.17)$$

A random variable with the Markov property (5.17) is often said, for obvious reasons, to have no memory.

Recall from Equation (5.6) that a random variable with a geometric distribution has this characteristic property of lack of memory. As pointed out in the discussion of (5.6), the existence of the Markov property is self-evident in that case. It is easy to show that the Markov property of the (discrete) geometric distribution carries over to the (continuous) exponential distribution. Using the definition of conditional probability, we have

$$P\{X>t+x|X>t\} = \frac{e^{-\mu(t+x)}}{e^{-\mu t}} = e^{-\mu x}. \tag{5.18}$$

This equation implies, for example, that a call with duration distributed according to (5.14) has the same distribution of remaining holding time after it has been in progress for any length of time $t>0$ as it had initially at $t=0$. (This does not mean, of course, that an *individual* call does not "age," but only that the statistical properties of the population of *all* calls are such that one can draw no inference about the future duration of an arbitrary call from this population based on only its past duration.)

Although at first glance one might find the property (5.18) to have little intuitive appeal, it should immediately be stocked in one's inventory of intuition, since the simplifications to which it leads are enormous.

Consider, for example, a two-server system with one waiting position, and suppose that at some time t there are three customers in the system. If the waiting customer waits as long as required for a server to become idle, what is the probability that the waiting customer will be the last of the three to complete service?

We assume that service times are mutually independent random variables, each with the negative-exponential distribution (5.14). The waiting customer will complete service last if and only if his service time exceeds the remaining service time of the customer still in service when the first customer leaves. But because of the Markov property, this remaining time is independent of the previously elapsed time. Therefore, the last two customers have the same distribution (exponential) of time left in the system; that is, there is no bias in favor of either of the customers, so that the required probability is $\frac{1}{2}$.

The problem was solved without knowledge of the instants at which the various customers seized the servers. Even the mean service time was not used in the calculation. It should be clear that without the assumption of negative-exponential service times this problem would have been much more difficult.

We now turn to the relationship between the exponential distribution and the birth-and-death process. Let a random variable X represent the duration of some process, and assume that X has the exponential distribution function given by (5.14). Suppose that at some time t the process is uncompleted. (For example, if X is the duration of a telephone call, then we suppose the call to be in progress at time t.) Then the Markov property (5.18) implies that the process will complete in $(t,t+h)$ with probability

$$1 - e^{-\mu h} = 1 - \left(1 - \mu h + \frac{(\mu h)^2}{2!} - + \cdots\right) = \mu h + o(h) \qquad (h \to 0).$$

Conversely, suppose that if a process whose duration is given by a continuous random variable X is still in progress at time t, then the probability of completion in $(t, t+h)$ is $\mu h + o(h)$ as $h \to 0$ for all $t \geq 0$. That is, assume that

$$P\{X > t+h | X > t\} = 1 - \mu h + o(h) \qquad (t \geq 0, \quad h \to 0). \qquad (5.19)$$

Then, using the definition of conditional probability in the form

$$P\{X > t+h | X > t\} = \frac{P\{X > t+h\}}{P\{X > t\}}$$

in (5.19) rearranging, and letting $h \to 0$, we obtain the differential equation

$$\frac{d}{dt} P\{X > t\} = -\mu P\{X > t\},$$

which has the unique solution

$$P\{X > t\} = e^{-\mu t}$$

satisfying $P\{X > 0\} = 1$. Thus X is exponentially distributed.

Finally, note that the Markov property $P\{X > x+y | X > x\} = P\{X > y\}$ and the definition of conditional probability,

$$P\{X > x+y | X > x\} = \frac{P\{X > x+y\}}{P\{X > x\}},$$

together give the functional equation

$$P\{X > x+y\} = P\{X > x\} P\{X > y\}.$$

It can be shown that the Markov property, as expressed by this functional equation, uniquely characterizes the negative-exponential distribution. More precisely, it can be shown (see, for example, p. 459 of Feller [1968]) that if $u(x)$ is defined for $x > 0$ and bounded in some interval, then the only nontrivial solution of the functional equation

$$u(x+y) = u(x) u(y) \qquad (5.20)$$

is $u(x) = e^{-\mu x}$ for some constant μ. It follows from this theorem (which is stronger than necessary for our conclusion) that if a continuous random variable has the Markov property, then it must be exponentially distributed.

Suppose now that $X_1, X_2, \ldots, X_n$ are independent, identical, exponential random variables, and consider the corresponding order statistics $X_{(1)} \leq$

$X_{(2)} \leq \cdots \leq X_{(n)}$. Observe that $X_{(1)} = \min(X_1, X_2, \ldots, X_n)$ will exceed x if and only if all X_i ($i = 1, 2, \ldots, n$) exceed x. Hence $P\{X_{(1)} > x\} = (e^{-\mu x})^n = e^{-n\mu x}$. We have proved the important fact:

$$P\{\min(X_1, X_2, \ldots, X_n) > x\} = e^{-n\mu x}; \tag{5.21}$$

that is, the distribution of the minimum of n independently chosen values from a negative-exponential distribution is again a negative-exponential distribution, with mean $1/n$ times the mean of the original distribution.

For the sake of example, let us identify the random variables $X_1, \ldots, X_n$ with the durations of n independent simultaneously running time intervals, so that $X_{(j)}$ is the duration of the jth shortest interval. We have just shown that the distribution of the duration of the shortest of the n time intervals is exponential with mean $(n\mu)^{-1}$; that is, $P\{X_{(1)} > x\} = e^{-n\mu x}$. At the point of completion of the shortest interval, the distribution of remaining duration for each of the $n-1$ uncompleted intervals is, by the Markov property, unchanged. Hence, $P\{X_{(2)} - X_{(1)} > y\} = e^{-(n-1)\mu y}$. The general rule should now be obvious.

Suppose that we are given a number of time intervals whose durations $X_1, \ldots, X_n$ are identically distributed, mutually independent, exponential random variables with common mean μ^{-1}, and all durations are longer than x. We seek the probability that exactly j of these intervals are longer than $x + t$. Because of the Markov property, the value of x is irrelevant. Therefore, let us take time x as the origin: $x = 0$.

Now let $N(t)$ be the number of these intervals still uncompleted at time t, and let $P\{N(t) = j\} = P_j(t)$. Since the ith (unordered) interval has duration X_i with the exponential distribution function (5.14), the probability that any particular one of these intervals is still uncompleted after an elapsed time t is

$$P\{X_i > t\} = e^{-\mu t} \quad (i = 1, 2, \ldots, n).$$

Since the $\{X_i\}$ are also mutually independent, it follows that the distribution $\{P_j(t)\}$ is the binomial distribution,

$$P_j(t) = \binom{n}{j}(e^{-\mu t})^j(1 - e^{-\mu t})^{n-j} \quad (j = 0, 1, \ldots, n). \tag{5.22}$$

This problem can also be viewed in the framework of the pure death process. If the system is in state E_j at time t, then the same reasoning that led to (5.22) shows that the conditional probability of transition $E_j \to E_j$ in $(t, t+h)$ is given by

$$P\{N(t+h) = j | N(t) = j\} = e^{-j\mu h} = 1 - j\mu h + o(h) \quad (h \to 0).$$

Similarly, the conditional probability of transition $E_j \to E_{j-1}$ in $(t, t+h)$ is given by

$$P\{N(t+h)=j-1|N(t)=j\} = je^{-(j-1)\mu h}(1-e^{-\mu h})$$
$$= j\mu h + o(h) \quad (h \to 0).$$

Since we must have

$$\sum_{i=0}^{j} P\{N(t+h)=i|N(t)=j\} = 1,$$

it follows that

$$P\{N(t+h) \leq j-2|N(t)=j\} = o(h) \quad (h \to 0).$$

Thus we have shown that $N(t)$ obeys the postulates of the pure death process with death coefficients $\mu_j = j\mu$ $(j=0,1,\ldots,n)$ and initial condition $N(0) = n$. This is precisely the case we considered in Section 2.2 as an example of the pure death process. The corresponding equations are given by (2.9) and (2.10), and the solution by (2.11). The solution (2.11) is identical, of course, with that just obtained by more direct reasoning, the binomial distribution (5.22). [This should not be surprising, since we implicitly assumed (5.22) when we specified the transition probabilities.]

We have thus also shown that if $N(t)$ obeys the postulates of the pure death process with $\mu_j = j\mu$, then the aggregate death process is the same as if it were composed of j mutually independent processes whose durations are identically distributed negative-exponential random variables each with mean μ^{-1}. (But it does not necessarily follow that the aggregate process is in fact such a superposition of independent exponential processes.)

We can calculate from (5.22) the distribution function of the length of time required for all n simultaneously running time intervals to complete. The time required for all the intervals to complete is the time required for $N(t)$ to reach zero; that is, $\{X_{(n)} \leq t\} \Leftrightarrow \{N(t)=0\}$. Hence $P\{X_{(n)} \leq t\} = P_0(t) = (1-e^{-\mu t})^n$. Since $X_{(n)} = \max(X_1, X_2, \ldots, X_n)$, we have proved

$$P\{\max(X_1, X_2, \ldots, X_n) \leq x\} = (1-e^{-\mu x})^n;$$

this result should be compared with (5.21). [Note that $\max(X_1, X_2, \ldots, X_n)$ is *not* exponentially distributed.]

To calculate $E(X_{(n)})$ without performing any integration, let Y_j be the elapsed time between the jth and the $(j-1)$th completions,

$$Y_j = X_{(j)} - X_{(j-1)} \quad (j=1,2,\ldots,n; \quad X_{(0)}=0).$$

Then $X_{(n)} = Y_1 + Y_2 + \cdots + Y_n$, and since Y_j is exponentially distributed with mean $[(n-j+1)\mu]^{-1}$, it follows that

$$E(X_{(n)}) = \frac{1}{n\mu} + \frac{1}{(n-1)\mu} + \cdots + \frac{1}{\mu}.$$

[Thus we have deduced that

$$n\mu \int_0^\infty t(1-e^{-\mu t})^{n-1} e^{-\mu t} \, dt = \frac{1}{\mu}\left(1 + \frac{1}{2} + \cdots + \frac{1}{n}\right).]$$

For example, consider a queueing system with n customers in service and no customers waiting at time zero, and assume that no new customers are allowed into the system. If the service times are independently exponentially distributed according to (5.14), then we let $N(t)$ be the number of servers busy at time t, and $X_{(n)}$ be the length of time until all servers become idle.

In another common application, reliability theory, a device depends on n independent components, each assumed to have a length of life distributed according to (5.14). The device continues to operate as long as at least one of the n components is functioning. Then we can let $N(t)$ be the number of components still functioning at time t, and $X_{(n)}$ be the time at which the device fails.

Finally, suppose that X_1 and X_2 are independent nonnegative random variables with distribution functions $F_1(t)$ and $F_2(t)$. Then the probability that X_2 exceeds X_1 is

$$P\{X_2 > X_1\} = \int_0^\infty [1 - F_2(t)] \, dF_1(t).*$$

If we assume further that X_1 and X_2 are exponential with means μ_1^{-1} and μ_2^{-1}, then the integral on the right-hand side becomes

$$\int_0^\infty e^{-\mu_2 t} \mu_1 e^{-\mu_1 t} \, dt = \frac{\mu_1}{\mu_1 + \mu_2}.$$

We conclude that

$$P\{X_1 < X_2\} = \frac{\mu_1}{\mu_1 + \mu_2}; \tag{5.23}$$

---

*We are using here the notation of the Riemann-Stieltjes integral, which will be discussed in Chapter 5. For the present, the reader should simply interpret the Riemann-Stieltjes integral $\int g(t) \, dh(t)$ as the ordinary Riemann integral with $dh(t) = h'(t) \, dt$; that is, $\int g(t) \, dh(t) = \int g(t) h'(t) \, dt$.

Some Important Probability Distributions

that is, the probability that one exponential variable "wins a race" with another equals the ratio of the winner's rate to the sum of the two rates.

Exercises

11. Let $X_1, \ldots, X_n$ ($n \geq 2$) be mutually independent, identically distributed, exponential random variables. Define the range $R = \max(X_1, X_2, \ldots, X_n) - \min(X_1, X_2, \ldots, X_n)$. Show, with no calculation, that the distribution function $P\{R \leq x\}$ of the range is given by $P\{R \leq x\} = (1 - e^{-\mu x})^{n-1}$.

12. Let $X_1, \ldots, X_n$ be a sequence of mutually independent random variables, and let X_j be exponentially distributed with mean $(j\mu)^{-1}$. Let $S_n = X_1 + \cdots + X_n$. Show that S_n has distribution function $P\{S_n \leq t\} = (1 - e^{-\mu t})^n$. [Hint: No calculation is necessary; compare with the maximum of n exponential variables.]

13. In reliability theory the *failure rate* function $r(t)$ is defined such that $r(t)dt$ can be interpreted as the probability that a component that is still operational at time t will fail in $(t, t + dt)$. Show that if $f(t)$ is the density function of component lifetimes, with corresponding distribution function $F(t)$, then

$$r(t)dt = \frac{f(t)dt}{1 - F(t)}.$$

Conclude that if component lifetimes are exponentially distributed, then the failure rate is constant.

14. Let $X_1, X_2, \ldots, X_n$ be independent exponential random variables with means $\mu_1^{-1}, \mu_2^{-1}, \ldots, \mu_n^{-1}$. Show that

$$P\{X_i = \min(X_1, X_2, \ldots, X_n)\} = \frac{\mu_i}{\mu_1 + \mu_2 + \cdots + \mu_n} \quad (i = 1, 2, \ldots, n).$$

15. Let X_1 and X_2 be independent exponential random variables. Show that

$$P\{\min(X_1, X_2) > t \mid X_1 < X_2\} = P\{\min(X_1, X_2) > t\}$$

or, equivalently,

$$P\{X_i > t \mid X_i = \min(X_1, X_2)\} = P\{\min(X_1, X_2) > t\} \quad (i = 1, 2)$$

[even when $E(X_1) \neq E(X_2)$].

16. At $t = 0$ a customer (the *test customer*) places a request for service and finds all s servers busy and j other customers waiting for service. All customers wait as long as necessary for service, waiting customers are served in order of arrival, and no new requests for service are permitted after $t = 0$. Service times are

assumed to be mutually independent, identically distributed, exponential random variables, each with mean duration μ^{-1}.

a. Let X_1 be the elapsed time from $t=0$ until the customer at the head of the queue enters service, and let X_i ($i=2,3,\ldots,j+1$) be the length of time that the ith customer spends at the head of the queue. (X_{j+1} is the test customer's time spent at the head of the queue.) Show that $X_1, X_2, \ldots, X_{j+1}$ are mutually independent exponential random variables, each with mean $(s\mu)^{-1}$.

b. Find the expected length of time the test customer spends waiting for service in the queue.

c. Find the expected length of time from the arrival of the test customer at $t=0$ until the system becomes completely empty (all customers complete service).

d. Let X be the order of completion of service of the test customer; that is, $X=m$ if the test customer is the mth customer to complete service after $t=0$. Find $P\{X=m\}$ ($m=1,2,\ldots,s+j+1$).

e. Find the probability that the test customer completes service before the customer immediately ahead of him in the queue.

Poisson Distribution

Thus far we have discussed the exponential distribution mainly with respect to its use as a description of service times. But just as it may be useful as a description of durations of service, likewise it may be useful as a description of durations of elapsed time between customer arrival epochs. More generally, we consider points on a line (or customers arriving in time), and we make the assumption that the distances between successive points are independently, identically, exponentially distributed. That is, we label these points $T_1, T_2, \ldots$, and assume that the random variable $X_j = T_j - T_{j-1}$ ($j=1,2,\ldots$; $T_0=0$) has the exponential distribution

$$P\{X_j > t\} = e^{-\lambda t} \quad (j=1,2,\ldots). \tag{5.24}$$

We ask the question: If the distances (lengths of time) $X_j = T_j - T_{j-1}$ between successive points are independently distributed according to (5.24), what is the distribution of the number of points occurring in a fixed interval of length t?

Let $N(t)$ be the number of points occurring in $(0,t)$, with $P\{N(t)=j\} = P_j(t)$. [More formally, we say that $\{N(t), t \geq 0\}$ is a *counting process*.] Let the lengths of time between successive points be independent and have the exponential distribution (5.24). Then the probability that at least one point will occur in $(t, t+h)$ is $1 - e^{-\lambda h} = \lambda h + o(h)$ as $h \to 0$. And the probability that exactly one point will occur in $(t, t+h)$ is

$$\int_0^h e^{-\lambda(h-x)} \lambda e^{-\lambda x} dx = \lambda h e^{-\lambda h} = \lambda h + o(h) \quad (h \to 0).$$

Some Important Probability Distributions

Thus the probability of occurrence of two or more points in $(t, t+h)$ is $1 - e^{-\lambda h} - \lambda h e^{-\lambda h} = o(h)$ as $h \to 0$.

Hence, $N(t)$ satisfies the postulates of the pure birth process with $\lambda_j = \lambda$ ($j = 0, 1, 2, \ldots$). This case has already been discussed in Section 2.2. The problem formulation is given by (2.5) and (2.6), and the solution by (2.7):

$$P_j(t) = \frac{(\lambda t)^j}{j!} e^{-\lambda t} \quad (j = 0, 1, \ldots). \tag{5.25}$$

The probabilities given by (5.25) constitute the ubiquitous *Poisson distribution*. $N(t)$ has probability-generating function $g(z)$ given by (4.3),

$$g(z) = e^{-\lambda t(1-z)}, \tag{5.26}$$

with mean and variance given by (4.7) and (4.10):

$$E(N(t)) = V(N(t)) = \lambda t. \tag{5.27}$$

Because of the Markov property of the exponential distribution, it is not necessary that counting start at any special point $T_0 = 0$; the formula (5.25) gives the distribution of the number of points occurring in any interval or set of disjoint subintervals of total length t.

Conversely, when the distribution of the number of points occurring in any interval of length t (regardless of the number of points occurring in any preceding interval) is given by the Poisson distribution (5.25), then the distances between successive points are mutually independent, identically distributed, exponential variables, with common mean λ^{-1}. Setting $j = 0$ in (5.25) shows that the distance to the first point is exponentially distributed. Suppose that the first point is located at $T_1 = \tau$. Since (5.25) is assumed to hold for *any* interval of length t, then the probability of j points in $(\tau, \tau + t)$ is again given by (5.25). Thus the distance between the first point and the second point is independent of and has the same distribution as the distance to the first point. Repeated use of this reasoning proves the assertion.

Thus Equations (5.24) and (5.25) are equivalent: In a pure birth process with parameter λ (called a *Poisson process* with rate λ), the number of points occurring in a fixed interval of length t has the Poisson distribution (5.25), or equivalently, the lengths of the intervals separating successive points are independent and have identical, negative-exponential distributions (5.24).

Exercise

17. Suppose customers arrive at instants $T_1, T_2, \ldots$, where the interarrival times are independent and have common distribution function $P\{T_j - T_{j-1} \leq x\} = G(x)$ ($j = 1, 2, \ldots$; $T_0 = 0$). (This is called *recurrent* or *renewal input*.) Let $P_j(t)$

be the probability of j arrivals in an interval of length t. Use the fact that

$$P_j(t) = \int_0^t P_{j-1}(t-\xi) \, dG(\xi) \qquad (j=1,2,\ldots)$$

with

$$P_0(t) = 1 - G(t)$$

to prove again: When customer interarrival times are mutually independent random variables, each with the negative-exponential distribution with mean interarrival time λ^{-1}, then the number of customers arriving in a fixed interval of length t has the Poisson distribution with mean λt.

The Poisson process is often used as a description of the input process in queueing systems. That is, (5.24) is assumed to describe the customer interarrival-time distribution, and (5.25) to describe the distribution of the number of customers arriving in a fixed interval of length t. An input process obeying (5.24) and (5.25) is called, naturally enough, *Poisson input*. Because of its mathematical properties, the assumption of Poisson input in the analysis of queueing systems often leads to the easiest mathematics. Specifically, we have, as a consequence of the Markov property, the following important facts:

With Poisson input at rate λ, the distribution of the length of time from an arbitrary instant until the next arrival epoch is the same as the distribution of the length of time between successive arrival epochs, that is, exponential with mean λ^{-1}. Similarly, looking backward from an arbitrary instant, the distribution of the length of time separating an arbitrary instant and the preceding arrival epoch is also exponentially distributed with mean λ^{-1} (assuming, of course, that the process is not measured from a finite origin).

We now use the Markov property to illustrate a point that may be confusing to the beginner. Suppose that, at an arbitrary instant, one begins to observe a Poisson process, and asks for the mean length of time that will elapse until the next arrival epoch. According to the Markov property, the distribution of the length of time preceding the next arrival is the same as the interarrival-time distribution. Therefore, the observer will wait a mean length of time λ^{-1} for the occurrence of the next arrival. But the observer's waiting time is, in general, less than the time between the two bracketing arrival epochs (see Figure 2.2). Hence the mean time separating the

Figure 2.2

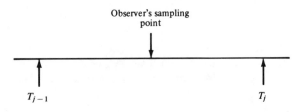

bracketing arrival epochs has mean value larger than λ^{-1}, in apparent contradiction to the assumption that λ^{-1} is the mean duration of an arbitrary interarrival interval.

Of course, there *is* no contradiction. The answer lies in the realization that the interval during which the observer samples is not an *arbitrary* interval. A little reflection should convince the reader that an observer sampling at arbitrary instants is more likely to sample during a long interarrival interval than a short one. Thus the act of observing introduces a bias; the mean length of the sampling interval exceeds that of the arbitrary interarrival interval. This phenomenon, the biasedness of the sampling interval, is true in general, not just for Poisson input. We have mentioned it here because the use of the Markov property makes the existence of the phenomenon clear. We shall discuss the general problem in some detail in Chapter 5.

The geometric distribution describes the waiting time preceding the first success in a sequence of Bernoulli trials; the binomial distribution describes the number of successes in a fixed number of Bernoulli trials. Similarly, the exponential distribution describes a length of time preceding an event, while the Poisson distribution describes the number of such events occurring during a fixed time interval. We have shown [see Equation (5.16)] that the geometric distribution function approaches the negative exponential in the limit as the trials are taken infinitely close together, each with vanishingly small probability of success, but with the average number of successes per unit time remaining constant. We should not be surprised, then, that a similar limiting process produces the Poisson distribution from the binomial:

$$\lim_{\substack{p \to 0 \\ n \to \infty \\ np = \lambda x}} \binom{n}{j} p^j (1-p)^{n-j} = \frac{(\lambda x)^j}{j!} e^{-\lambda x} \quad (j=0,1,\ldots). \quad (5.28)$$

Arrivals occurring according to a Poisson process are often said to occur *at random*. This is because the probability of arrival of a customer in a small interval of length h is proportional to the length h, and is independent of the amount of elapsed time from the arrival epoch of the last customer. That is, when customers are arriving according to a Poisson process, a customer is as likely to arrive at one instant as any other, regardless of the instants at which the other customers arrive. This is similar to the property noted previously for points chosen independently from a uniform distribution on $(0,t)$. However, in that case, the length t of the interval containing the points is known. One might conjecture, therefore, that when it is known that a given number n of arrivals generated by a Poisson process have occurred in an interval $(0,t)$, these arrival epochs are (conditionally) uniformly distributed throughout the interval. This is indeed true.

More precisely, we shall now prove that if the number $N(t)$ of Poisson arrivals in $(0,t)$ is n, then the joint distribution of the arrival epochs $T_1, T_2, \ldots, T_n$ is the same as the joint distribution of the coordinates arranged in increasing order of n independent points, each of which is uniformly distributed over the interval $(0,t)$.

Let $0 \leqslant x_1 \leqslant x_2 \leqslant \cdots \leqslant x_n \leqslant t$. Then, using the definition of conditional probability,

$$P\{T_1 \leqslant x_1, T_2 \leqslant x_2, \ldots, T_n \leqslant x_n | N(t) = n\}$$
$$= \frac{P\{T_1 \leqslant x_1, T_2 \leqslant x_2, \ldots, T_n \leqslant x_n, N(t) = n\}}{P\{N(t) = n\}},$$

we can write, with $x_0 = 0$,

$$P\{T_1 \leqslant x_1, T_2 \leqslant x_2, \ldots, T_n \leqslant x_n | N(t) = n\}$$

$$= \frac{\sum_{\{j_1, j_2, \ldots, j_n\}} \prod_{i=1}^{n} \left\{ \frac{[\lambda(x_i - x_{i-1})]^{j_i}}{j_i!} e^{-\lambda(x_i - x_{i-1})} \right\} e^{-\lambda(t - x_n)}}{\frac{(\lambda t)^n}{n!} e^{-\lambda t}}$$

$$= \sum_{\{j_1, j_2, \ldots, j_n\}} \frac{n!}{j_1! j_2! \cdots j_n!} \left(\frac{x_1}{t}\right)^{j_1} \left(\frac{x_2 - x_1}{t}\right)^{j_2} \cdots \left(\frac{x_n - x_{n-1}}{t}\right)^{j_n}, \quad (5.29)$$

where the summation is extended over all those sets of nonnegative integers $\{j_1, j_2, \ldots, j_n\}$ for which $j_1 + j_2 + \cdots + j_n = n$ and $j_1 + j_2 + \cdots + j_k \geqslant k$ ($k = 1, 2, \ldots, n-1$). Comparison with Equation (5.13) shows that our assertion is proved.

As an application of this result, let us return to the observer who samples a Poisson process at an arbitrary time, say t, and finds that the mean time from t until the next arrival epoch is the mean interarrival time. We consider the problem in more depth: Let $T_0 = 0$, $T_1, T_2, \ldots$ be the arrival epochs of the Poisson process, and define the *forward recurrence time* R_t as the time duration from t until the next arrival epoch; define the *backward recurrence time* or *age* A_t as the time duration between t and the last arrival epoch prior to t; and define the *covering interval* $I_t = A_t + R_t$. I_t is thus the duration of the interarrival interval that contains (covers) the sampling point t. We know from the Markov property that R_t is exponential, that is,

$$P\{R_t > x\} = e^{-\lambda x}; \quad (5.30)$$

and further, A_t and R_t are independent.

Some Important Probability Distributions

Now consider A_t. It follows from the theorem of total probability that for $t > x$,

$$P\{A_t > x\} = \sum_{n=0}^{\infty} P\{t - T_n > x | N(t) = n\} P\{N(t) = n\}.$$

By assumption

$$P\{N(t) = n\} = \frac{(\lambda t)^n}{n!} e^{-\lambda t};$$

and also, since $T_0 = 0$ and $t > x$, we have $P\{t - T_0 > x | N(t) = 0\} = 1$. Therefore,

$$P\{A_t > x\} = e^{-\lambda t} + \sum_{n=1}^{\infty} P\{t - T_n > x | N(t) = n\} \frac{(\lambda t)^n}{n!} e^{-\lambda t}. \quad (5.31)$$

From (5.29) we know that, conditional on the event $N(t) = n$, the joint distribution of $T_1, T_2, \ldots, T_n$ is exactly the same as that which we would obtain if n values were chosen independently from a uniform distribution over $(0, t)$. Therefore, we can apply (5.11). In particular, we have $P\{t - T_n > x | N(t) = n\} = P\{U_{(n+1)} - U_{(n)} > x\}$ where $U_{(n+1)} = t$; thus

$$P\{t - T_n > x | N(t) = n\} = \left(1 - \frac{x}{t}\right)^n. \quad (5.32)$$

Insertion of (5.32) into (5.31) easily yields $P\{A_t > x\} = e^{-\lambda x}$ (when $x < t$). Hence, we conclude that

$$P\{A_t > x\} = \begin{cases} e^{-\lambda x} & (0 \leq x < t), \\ 0 & (x \geq t), \end{cases} \quad (5.33)$$

and therefore,

$$\lim_{t \to \infty} P\{A_t > x\} = e^{-\lambda x}. \quad (5.34)$$

Equation (5.33) could have been obtained with no calculation, simply by observing that $A_t > x$ if and only if no arrival epochs occur in the interval $(t - x, t)$. Equation (5.34) has already been anticipated (at the start of this discussion, in the second paragraph following Exercise 17) on intuitive grounds.

Thus we have shown that

$$\lim_{t \to \infty} P\{A_t > x\} = P\{R_t > x\} \ (= e^{-\lambda x}) \quad (5.35)$$

for the Poisson process. We shall later show that

$$\lim_{t \to \infty} P\{A_t > x\} = \lim_{t \to \infty} P\{R_t > x\} \qquad (5.36)$$

for more general arrival processes (that do not possess the Markov property), a result that one might anticipate on grounds of symmetry.

Let us now consider the covering interval $I_t = A_t + R_t$. We leave it to Exercise 18 for the reader to show that

$$P\{I_t \leq y\} = 1 - e^{-\lambda y} - \lambda \min(y, t) e^{-\lambda y}, \qquad (5.37)$$

and therefore

$$\lim_{t \to \infty} P\{I_t \leq y\} = 1 - e^{-\lambda y} - \lambda y e^{-\lambda y}. \qquad (5.38)$$

Observe that according to (5.37) and (5.38), the interarrival interval I_t that covers the sampling point does not have the same distribution as does an arbitrary interarrival interval. This is in agreement with the observation of the "paradox" that motivated the above analysis.

Exercises

18. Prove Equation (5.37).

19. Let $F(x,y)$ be the limiting joint distribution function of the lengths of the forward recurrence time and the covering interval, that is, $F(x,y) = \lim_{t \to \infty} P\{R_t \leq x, I_t \leq y\}$. We shall show later (without invoking the Markov property) that for a Poisson process

$$F(x,y) = 1 - e^{-\lambda x} - \lambda x e^{-\lambda y} \qquad (0 \leq x \leq y).$$

Using this result,
a. Prove Equation (5.38).
b. Prove that $\lim_{t \to \infty} P\{R_t > x\} = e^{-\lambda x}$. [This result is weaker than (5.30), which we proved using the Markov property.]
c. Prove that $\lim_{t \to \infty} P\{R_t > x, A_t > y\} = e^{-\lambda x} e^{-\lambda y}$.
d. Prove Equation (5.34) and thereby show that

$$\lim_{t \to \infty} P\{R_t > x, A_t > y\} = \lim_{t \to \infty} P\{R_t > x\} P\{A_t > y\}.$$

(This proves independence in the limit; in the text we used the Markov property to prove the stronger result that A_t and R_t are independent for any t.)
e. Prove that, in the limit as $t \to \infty$, the sampling point t is uniformly distributed throughout the covering interval; that is, $\lim_{t \to \infty} P\{R_t \leq x | I_t = y\} = x/y$.

Some Important Probability Distributions

Let us now consider a property of the Poisson process that turns out to be one of the most important in the context of queueing theory. Consider a "system" with states that occur according to some distribution $\{P_j(t)\}$; that is, $P_j(t)$ is the probability that, at time t, the system is in state E_j. Stated differently, $P_j(t)$ is the probability that an *outside observer* who observes the system at time t will find the system in state E_j. Now consider the state distribution as seen by the arriving customers. That is, if $T_1, T_2, \ldots$ are the instants at which customers arrive, let $\Pi_j(t)$ be the probability that the system is in state E_j just prior to time t, where t is now an arrival epoch. $\{\Pi_j(t)\}$ is thus the distribution that represents the viewpoint of the *arriving customer* at time t. In general, these two distributions are different: $\Pi_j(t) \neq P_j(t)$. However, if the arrival epochs $T_1, T_2, \ldots$ follow a Poisson process, then $\Pi_j(t) = P_j(t)$ for all $t \geq 0$ and all $j = 0, 1, \ldots$. In other words, when the input process is Poisson, the *arriving customer's distribution* $\{\Pi_j(t)\}$ and the *outside observer's distribution* $\{P_j(t)\}$ are equal:

$$\Pi_j(t) = P_j(t) \qquad (t \geq 0; \; j = 0, 1, \ldots). \tag{5.39}$$

For the reader who does not think that the equality (5.39) is remarkable, we give the following counterexample. Suppose that a single server is reserved for the exclusive use of a single customer, who uses the server intermittently. Let E_0 represent the state "server idle" and E_1 denote the state "server busy." Then if at time t the customer makes a request for service, he will necessarily find the server idle; hence, $\Pi_0(t) = 1$. However, if an outside observer looks at the server at time t, he may find that the server is busy serving the customer; hence $P_0(t) < 1$. Thus, $\Pi_0(t) \neq P_0(t)$ in this example.

To prove (5.39) we need the following result: If $T_1, T_2, \ldots$ are the arrival epochs in a Poisson process and $N(t)$ is the number of arrivals in $(0, t)$, then

$$P\{T_1 \leq x_1, \ldots, T_n \leq x_n | T_{n+1} = t\} = P\{T_1 \leq x_1, \ldots, T_n \leq x_n | N(t) = n\}. \tag{5.40}$$

Before proving (5.40), we show that (5.40) implies (5.39): Equation (5.40) says that the joint distribution function of the arrival epochs $T_1, T_2, \ldots, T_n$ preceding the point t is the same whether t is an arrival epoch ($T_{n+1} = t$) or an arbitrary point. Now the state of any system at any epoch t is completely determined by the sequence of arrival times prior to t and the realizations of other random variables, such as service times, that do not depend on the arrival process. Therefore, when (5.40) is true, the system looks the same to arrivals as it does to an outside observer; that is, for any system with Poisson arrivals, Equation (5.39) is true. Note that (5.39) holds without regard to whether all, some, or none of the customers in the Poisson arrival stream join the system (thereby causing a change of state).

We shall give further discussion and an alternate proof of (5.39) in Chapter 3. To complete the present proof, it remains only to prove (5.40). An expression for the right-hand side of (5.40) is given in (5.29), namely,

$$P\{T_1 \leq x_1, \ldots, T_n \leq x_n | N(t) = n\}$$

$$= \frac{\sum_{\{j_1, j_2, \ldots, j_n\}} \prod_{i=1}^{n} \left\{ \frac{[\lambda(x_i - x_{i-1})]^{j_i}}{j_i!} e^{-\lambda(x_i - x_{i-1})} \right\} e^{-\lambda(t - x_n)}}{\frac{(\lambda t)^n}{n!} e^{-\lambda t}}, \quad (5.41)$$

where $x_0 = 0$ and the summation is extended over all those sets of nonnegative integers $\{j_1, \ldots, j_n\}$ for which $j_1 + \cdots + j_n = n$ and $j_1 + \cdots + j_k \geq k$ ($k = 1, 2, \ldots, n-1$).

Now consider the probability on the left-hand side of (5.40). If we agree that

$$P\{T_1 \leq x_1, \ldots, T_n \leq x_n | T_{n+1} = t\}$$
$$= \lim_{h \to 0} P\{T_1 \leq x_1, \ldots, T_n \leq x_n | t < T_{n+1} \leq t + h\},$$

then it follows from the definition of conditional probability that

$$P\{T_1 \leq x_1, \ldots, T_n \leq x_n | T_{n+1} = t\}$$
$$= \lim_{h \to 0} \frac{P\{T_1 \leq x_1, \ldots, T_n \leq x_n, t < T_{n+1} \leq t + h\}}{P\{t < T_{n+1} \leq t + h\}}. \quad (5.42)$$

Since the arrival epochs $T_1, T_2, \ldots$ follow a Poisson process, the denominator in (5.42) can be written

$$P\{t < T_{n+1} \leq t + h\} = \frac{(\lambda t)^n}{n!} e^{-\lambda t} [\lambda h + o(h)] \quad (h \to 0); \quad (5.43)$$

and the numerator in (5.42) can be written

$$P\{T_1 \leq x_1, \ldots, T_n \leq x_n, t < T_{n+1} \leq t + h\}$$
$$= \sum_{\{j_1, j_2, \ldots, j_n\}} \prod_{i=1}^{n} \left\{ \frac{[\lambda(x_i - x_{i-1})]^{j_i}}{j_i!} e^{-\lambda(x_i - x_{i-1})} \right\}$$
$$\times e^{-\lambda(t - x_n)} [\lambda h + o(h)] \quad (h \to 0). \quad (5.44)$$

Substitution of (5.43) and (5.44) into (5.42) yields the right-hand side of

(5.41); Equation (5.40) has been established, and consequently, so has (5.39).

Exercise

20. A bus shuttles back and forth between two stations. At each station the bus unloads and all passengers waiting at the station board the bus, which is assumed to be of sufficient capacity so that all passengers can board. Let P_j be the probability that j passengers board the bus at a station; that is, P_j ($j=0,1,2,\ldots$) is the distribution of the number of passengers on the bus from the viewpoint of an outside observer (the bus driver, say). Let Π_j be the probability that an arbitrary passenger shares the bus with j other passengers; that is, Π_j ($j=0,1,2,\ldots$) is the distribution of the number of (other) passengers on the bus from the viewpoint of one of the passengers. Show that $P_j = \Pi_j$ (that is, the bus driver's viewpoint and the passenger's viewpoint are equivalent) if and only if $P_j = (a^j/j!)e^{-a}$ ($j=0,1,2,\ldots$). [The "if" part of the statement provides a discrete analogue of (5.39).]

Finally, we consider the questions of superposition and decomposition of a Poisson process. We have already shown [see Equation (4.19)] that if $N_1(t),\ldots,N_n(t)$ are mutually independent random variables, and $N_i(t)$ has the Poisson distribution with mean $\lambda_i t$ ($i=1,2,\ldots,n$), then the sum $N_1(t) + \cdots + N_n(t)$ also has the Poisson distribution, with mean $\lambda_1 t + \cdots + \lambda_n t$. This implies that the superposition of n streams of Poisson arrivals is again a Poisson stream with mean equal to the sum of the means of the component streams.

The question naturally arises: Can a Poisson process conversely be viewed as the superposition of any number of mutually independent component Poisson processes? That is, can a Poisson stream be decomposed into any number of independent Poisson streams? The answer is yes, as we now show.

Suppose for example, that telephone calls arrive according to a Poisson process with parameter λ at a device that distributes the calls to n machines for processing. If each arrival is directed to machine j randomly with probability λ_j/λ, where $\lambda = \lambda_1 + \cdots + \lambda_n$, then the probability that a call will arrive at machine j in $(t, t+h)$ is $(\lambda_j/\lambda)[\lambda h + o(h)] = \lambda_j h + o(h)$ as $h \to 0$; and likewise the probability of two or more arrivals at machine j is $o(h)$ as $h \to 0$. Hence machine j sees arrivals occurring according to a Poisson process with parameter λ_j. And because of the Markov property, the arrival processes at the different machines are mutually independent.

Let us consider more formally the question for the case of decomposition into two streams. Let $\{N(t), t \geq 0\}$ be the original Poisson process with rate λ; and suppose that whenever this process generates an arrival, with probability λ_i/λ ($i=1,2$) that arrival is counted as having been generated by the process $\{N_i(t), t \geq 0\}$, independently of the assignments of all other arrivals. (Thus, a sequence of assignments is a realization of a sequence of

Bernoulli trials.) We now show that

$$P\{N_1(t)=j, N_2(t)=k\} = \frac{(\lambda_1 t)^j}{j!}e^{-\lambda_1 t}\frac{(\lambda_2 t)^k}{k!}e^{-\lambda_2 t}, \qquad (5.45)$$

from which it follows that $\{N_1(t), t \geqslant 0\}$ and $\{N_2(t), t \geqslant 0\}$ are independent Poisson processes. (The extension to include an arbitrary number of processes is straightforward.)

Now we can write

$$P\{N_1(t)=j, N_2(t)=k\}$$
$$= P\{N_1(t)=j, N_2(t)=k|N(t)=j+k\}P\{N(t)=j+k\}, \quad (5.46)$$

and, by assumption,

$$P\{N(t)=j+k\} = \frac{(\lambda t)^{j+k}}{(j+k)!}e^{-\lambda t}. \qquad (5.47)$$

Since the assignments of arrivals from $\{N(t), t \geqslant 0\}$ to be counted as belonging to $\{N_i(t), t \geqslant 0\}$ constitute a sequence of Bernoulli trials, it follows that the first probability on the right-hand side of (5.46) is the binomial:

$$P\{N_1(t)=j, N_2(t)=k|N(t)=j+k\} = \binom{j+k}{j}\left(\frac{\lambda_1}{\lambda}\right)^j\left(\frac{\lambda_2}{\lambda}\right)^k. \quad (5.48)$$

Insertion of (5.47) and (5.48) into (5.46) yields (5.45).

We conclude: A Poisson process can be decomposed into any number of mutually independent component Poisson processes. A process with this property is called *infinitely divisible* (see Feller [1968]).

The fact that the processes $\{N_1(t), t \geqslant 0\}$ and $\{N_2(t), t \geqslant 0\}$ are independent of each other may seem surprising. Suppose, for example, that $\lambda = 100$ customers per hour, and $\lambda_1 = \lambda_2$. One might argue that if during an hour 99 customers have been observed to have been assigned to stream 1, then one would expect only a small number of customers to have been assigned to stream 2 (because there will be about 100 customers generated on the average, and 99 of them have been assigned to stream 1). On the other hand, one might argue that the number assigned to stream 2 should be relatively large (because a large number have been assigned to stream 1, and each stream receives an equal number of customers on the average). Since one can argue with equal force for either of two opposite conclusions, neither should be believed without further analysis. In this case, "further analysis" shows that the correct conclusion lies precisely midway

Some Important Probability Distributions

between these two extremes: knowledge of the number of customers assigned to stream 1 tells us *nothing* about the number assigned to stream 2.

The process of decomposition can be viewed as a *compound Poisson process*: Let $X_1, X_2, \ldots$ be a sequence of independent, identically distributed random variables, and let $S_N = X_1 + \cdots + X_N$, where N is a random variable, independent of $X_1, X_2, \ldots$. If N has a Poisson distribution, then S_N ($S_0 = 0$) is said to be a compound Poisson process.

According to Exercise 4, if $X_1, X_2, \ldots$ are discrete random variables with probability-generating function $f(z)$, then, in light of (5.26), S_N has probability-generating function $g(f(z)) = e^{-\lambda t [1 - f(z)]}$. In particular, if $X_1, X_2, \ldots$ are Bernoulli random variables with $P\{X_i = 1\} = p$, then S_N has probability-generating function $e^{-\lambda p t (1 - z)}$; that is, S_N also has a Poisson distribution:

$$P\{S_N = j\} = \frac{(\lambda p t)^j}{j!} e^{-\lambda p t}. \tag{5.49}$$

(To see the connection with decomposition of a Poisson process, observe that if we take $p = \lambda_i / \lambda$, then we can interpret N as the number of customers generated by the original stream during an interval of length t, and S_N as the number of customers assigned to stream i during that interval. However, further argument is needed to show that the streams are independent.)

To derive (5.49) directly, observe that

$$P\{S_N = j\} = \sum_{n=j}^{\infty} P\{S_N = j | N = n\} P\{N = n\}, \tag{5.50}$$

where, by assumption,

$$P\{N = n\} = \frac{(\lambda t)^n}{n!} e^{-\lambda t}. \tag{5.51}$$

From the assumption that $X_1, X_2, \ldots$ are Bernoulli variables it follows that, for fixed n, the sum $X_1 + \cdots + X_n$ has the binomial distribution:

$$P\{S_N = j | N = n\} = \binom{n}{j} p^j (1-p)^{n-j} \quad (n \geq j). \tag{5.52}$$

Insertion of (5.51) and (5.52) into (5.50) yields

$$\sum_{n=j}^{\infty} \binom{n}{j} p^j (1-p)^{n-j} \frac{(\lambda t)^n}{n!} e^{-\lambda t} = \frac{(\lambda p t)^j}{j!} e^{-\lambda p t},$$

a result that the reader can easily verify.

We have shown that the Poisson distribution possesses many elegant properties. The simplifications resulting from the exploitation of these properties permit the solution of many problems that might otherwise be intractable.

Exercises

21. Suppose customers arrive according to a Poisson process, with rate λ. Let M be the number of customers that arrive during an interval of length X, where X is a nonnegative random variable, independent of the arrival process, with distribution function $H(t) = P\{X \leqslant t\}$, and mean τ and variance σ^2.

   a. Show that

   $$P\{M=j\} = \int_0^\infty \frac{(\lambda t)^j}{j!} e^{-\lambda t} \, dH(t) \qquad (j=0,1,\ldots),$$

   $$E(M) = \lambda \tau,$$

   $$E(M^2) = \lambda \tau + \lambda^2(\sigma^2 + \tau^2),$$

   $$V(M) = \lambda \tau + \lambda^2 \sigma^2.$$

   b. Show that

   $$E(X|M=j) = \frac{j+1}{\lambda} \frac{P\{M=j+1\}}{P\{M=j\}} \qquad (j=0,1,\ldots).$$

   c. Show that $E(X|M=j) = E(X)$ if and only if

   $$P\{M=j\} = \frac{(\lambda \tau)^j}{j!} e^{-\lambda \tau} \qquad (j=0,1,\ldots).$$

   Note: It can be shown, using the theory of Riemann-Stieltjes integration, which will be discussed in Chapter 5, that

   $$\int_0^\infty \frac{(\lambda t)^j}{j!} e^{-\lambda t} \, dH(t) = \frac{(\lambda \tau)^j}{j!} e^{-\lambda \tau} \qquad (j=0,1,\ldots)$$

   if and only if $H(t) = 0$ when $t < \tau$ and $H(t) = 1$ when $t \geqslant \tau$; that is, when the random variable X always takes the (constant) value τ. Thus, we have proved that if arrivals follow a Poisson process, then $E(X|M=j) = E(X)$ if and only if X is constant.

   d. Show that, if the service time is not a constant, then for all t such that $0 < H(t) < 1$,

   $$P\{X \leqslant t | M=j+1\} < P\{X \leqslant t | M=j\} \qquad (j=0,1,\ldots);$$

   and furthermore, this property of *stochastic dominance* implies that

   $$E(X|M=j+1) > E(X|M=j) \qquad (j=0,1,\ldots).$$

e. Suppose $H(t) = 1 - e^{-\mu t}$. Show, with no calculation, that M has the geometric distribution,

$$P\{M=j\} = \left(\frac{\lambda}{\lambda+\mu}\right)^j \left(\frac{\mu}{\lambda+\mu}\right) \quad (j=0,1,\ldots).$$

f. Conclude that if X is exponential with mean μ^{-1}, then

$$E(X|M=j) = \frac{j+1}{\lambda+\mu} \quad (j=0,1,\ldots).$$

Show how this result could have been obtained with no calculation.

22. Customers request service from a group of s servers according to a Poisson process with mean interarrival time λ^{-1}. Service times are mutually independent, exponentially distributed, random variables with common mean μ^{-1}. At time $t=0$ an observer samples the system and finds all s servers occupied and no customers waiting.
    a. Find the probability that the next arriving customer is blocked.
    b. Let N be the number of customers that arrive prior to the first completion of a customer in service. Find $P\{N=j\}$.
    c. Find the probability that the next arriving customer finds at least two idle servers.

23. *Continuation of Exercise 5 (Section 2.4).* Conclude that procedures (a) and (b) are equivalent if and only if N_ν has a Poisson distribution.

    [*Hint:* Make the substitution $x = x + y$, $y = 0$ to obtain Equation (5.20) of the text.]

24. Consider the single-server queue with an unlimited number of waiting positions. Assume that the input process is Poisson with rate λ. Let the service-time distribution function $H(\xi)$ be arbitrary, with mean τ and variance σ^2, and let

$$\eta(s) = \int_0^\infty e^{-s\xi} dH(\xi).$$

[$\eta(s)$ is called the *Laplace-Stieltjes transform* of $H(\xi)$.] Let N^* be the number of customers remaining in the system just after a service completion epoch, with $\Pi_j^* = P\{N^* = j\}$ and probability-generating function $g(z) = \sum_{j=0}^\infty \Pi_j^* z^j$. It can (and in Chapter 5 will) be shown that the distribution $\{\Pi_j^*\}$ satisfies the equations

$$\Pi_j^* = p_j \Pi_0^* + \sum_{i=1}^{j+1} p_{j-i+1} \Pi_i^* \quad (j=0,1,\ldots), \tag{1}$$

where

$$p_j = \int_0^\infty \frac{(\lambda\xi)^j}{j!} e^{-\lambda\xi} dH(\xi) \quad (j=0,1,\ldots) \tag{2}$$

is the distribution of the number of arrivals during a service time.

a. Show that

$$g(z) = \frac{(z-1)h(z)}{z - h(z)} \Pi_0^*, \qquad (3)$$

where

$$h(z) = \sum_{j=0}^{\infty} p_j z^j.$$

b. Show that

$$g(z) = \frac{(z-1)\eta(\lambda - \lambda z)}{z - \eta(\lambda - \lambda z)} \Pi_0^*. \qquad (4)$$

c. Show that the normalization equation,

$$g(1) = 1, \qquad (5)$$

implies

$$\Pi_0^* = 1 - \rho, \qquad (6)$$

where $\rho = \lambda \tau < 1$, so that Equation (4) becomes

$$g(z) = \frac{(z-1)\eta(\lambda - \lambda z)}{z - \eta(\lambda - \lambda z)} (1 - \rho) \qquad (\rho < 1). \qquad (7)$$

d. Show that

$$E(N^*) = \rho + \frac{\rho^2 [1 + (\sigma^2/\tau^2)]}{2(1-\rho)}. \qquad (8)$$

e. Show that when service times are exponential, then

$$\Pi_j^* = (1-\rho)\rho^j \qquad (j = 0, 1, \ldots), \qquad (9)$$

which is a geometric distribution.

This model, sometimes called the $M/G/1$ queue, will be discussed in detail in Chapter 5.

Erlangian (Gamma) Distribution

Again we consider points on a line (or customers arriving in time) and assume that the distances between successive points are independently, identically, exponentially distributed. That is, we label these points $T_1, T_2, \ldots$ and assume that the random variable $X_j = T_j - T_{j-1}$ ($j = 1, 2, \ldots$; $T_0 = 0$) has the exponential distribution, $P\{X_j > t\} = e^{-\lambda t}$ ($j = 1, 2, \ldots$) [and therefore the number of points occurring in $(0, t)$ has the Poisson distribution with mean λt]. We wish to find the distribution of the distance

Some Important Probability Distributions 65

spanning n consecutive points, that is, the distribution of the sum $S_n = X_1 + \cdots + X_n$ of n mutually independent, identically distributed exponential random variables.

Because S_n is the sum of n interarrival times when the arrivals occur according to the process $\{N(t), t \geq 0\}$, it follows that the events $\{S_n \leq x\}$ and $\{N(x) \geq n\}$ are equivalent; therefore,

$$P\{S_n \leq x\} = P\{N(x) \geq n\} = \sum_{j=n}^{\infty} \frac{(\lambda x)^j}{j!} e^{-\lambda x},$$

where the second equality follows from the assumption that $\{N(t), t \geq 0\}$ is a Poisson process. Hence, if we define $F(x) = P\{S_n \leq x\}$, then we have

$$F(x) = 1 - \sum_{j=0}^{n-1} \frac{(\lambda x)^j}{j!} e^{-\lambda x} \qquad (x \geq 0). \tag{5.53}$$

The distribution defined by (5.53) is called the *Erlangian distribution* of order n ($n = 1, 2, \ldots$). [Observe that the right-hand side of (5.38) is an Erlangian distribution function of order 2.] The component variables $X_1, X_2, \ldots$ are sometimes considered to be *phases* or *stages*, and consequently $F(x)$ is often called the n-phase or n-stage Erlangian distribution function.

The Erlangian density function $f(x) = (d/dx)F(x)$ can also be written down directly as was the distribution function (5.53). Interpret $f(x)dx$ as the probability that the nth point lies in the differential element about the point x. The probability that exactly $n-1$ points lie in $(0, x)$ is $[(\lambda x)^{n-1}/(n-1)!]e^{-\lambda x}$; the (conditional) probability that a new point occurs at x is then λdx. Hence the probability that $n-1$ points lie in $(0, x)$ and that the nth point occurs in $(x, x+dx)$ is $f(x)dx = [(\lambda x)^{n-1}/(n-1)!]e^{-\lambda x}\lambda dx$; the density function is

$$f(x) = \frac{(\lambda x)^{n-1}}{(n-1)!} e^{-\lambda x} \lambda \qquad (\lambda > 0, \quad x \geq 0). \tag{5.54}$$

[Note that this is essentially the same argument that was made in support of Equation (5.43).] A graph of the density function (5.54) is given in Figure 2.3. Of course, (5.54) follows directly from (5.53) by differentiation.

From the interpretation of S_n as a sum of n mutually independent, identical, exponentially distributed phases, each with mean λ^{-1} (and variance λ^{-2}), we immediately obtain the mean $E(S_n) = n\lambda^{-1}$ and variance $V(S_n) = n\lambda^{-2}$. As might be expected, the Erlangian distribution is easier to handle than are distributions in general, but not so easy as the exponential.

As the number of phases increases and the mean phase length decreases in such a way that $E(S_n)$ remains constant $[E(S_n) = c]$, the variance $V(S_n)$

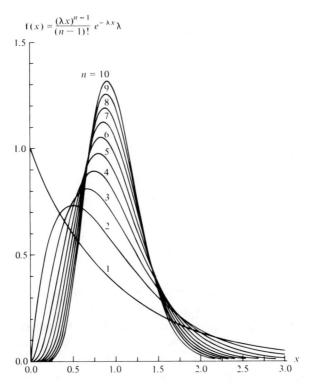

Figure 2.3. Graphs of the Erlangian density function of order n, for $n = 1, 2, \ldots, 10$. The densities have been normalized to have mean unity by taking $\lambda = n$. Adapted from L. Kosten, *Stochastic Theory of Service Systems*, New York: Pergamon Press, 1973.

goes to zero:

$$\lim_{\substack{\lambda \to \infty \\ n \to \infty \\ n\lambda^{-1} = c}} V(S_n) = \lim_{\substack{\lambda \to \infty \\ n \to \infty \\ n\lambda^{-1} = c}} n\lambda^{-1}\lambda^{-1} = 0.$$

Thus, in the limit, the random variable S_n becomes a constant, $S_\infty = c$. Hence, the Erlangian distribution provides a model for a range of input processes (or service times) characterized by complete randomness when $n = 1$ and no randomness when $n = \infty$.

Observe that the Erlangian distribution is the continuous analogue of the negative-binomial distribution, just as the exponential is the continuous analogue of the geometric.

The n-phase Erlangian distribution, where n is a positive integer, is actually a special case of the *gamma distribution*, defined by the density function

$$f(x) = \frac{(\lambda x)^{p-1}}{\Gamma(p)} \lambda e^{-\lambda x} \qquad (p>0, \ \lambda>0, \ x \geqslant 0).$$

The gamma distribution can be viewed as the generalization of the p-phase Erlangian, where the number p of phases is no longer required to be an integer. As expected, we have $E(S_p) = p\lambda^{-1}$ and $V(S_p) = p\lambda^{-2}$. Thus the gamma distribution provides a potentially useful generalization of the Erlangian distribution. In particular, the gamma distribution extends the "sum" S_p to include the case of a fraction of a phase, since p can be less than unity. Hence the gamma distribution provides a model for a range of input processes (or service times), characterized by an arbitrarily large variance when $p \to 0$ and $\lambda \to 0$ but $p\lambda^{-1} = c$, and zero variance when $p \to \infty$ and $\lambda \to \infty$, again holding $p\lambda^{-1} = c$. The term *hyperexponential* is sometimes used to describe this distribution, but this term is most descriptive if applied only when $0 < p < 1$. (The term *hyperexponential* has also been used by Morse [1958] to describe a distribution function that is a weighted sum of exponential distribution functions.)

The n-phase Erlangian distribution arises quite naturally in the consideration of input processes. Suppose, for example, that a device distributes incoming requests for service to two groups of servers on an alternating basis. Then the input to each group of servers is Erlangian with $n = 2$ phases. (But these input processes are not independent.)

Similarly, the Erglangian distribution (5.53) can be used to describe service times that are assumed to be composed of independent phases, each with the same exponential distribution. These phases may correspond to reality in the sense that each phase represents an identifiable component of the service, or they may be hypothesized in order to fit a theoretical distribution to data. In the former case, the number n of phases is equal to the number of components of the service time, whereas in the latter case, the number n of phases is determined from the data.

For another example of the use of the Erlangian distribution in queueing theory, consider again the premise of Exercise 16: At $t = 0$ an arbitrary customer (called the *test customer*) arrives (places a request for service) and finds all s servers busy and $j \geqslant 0$ other customers waiting for service. Customers are served in order of arrival, and all customers wait as long as necessary for service. Service times are assumed to be exponentially distributed with mean μ^{-1}. We wish to find the distribution of the test customer's waiting time. To this end, let X_1 be the elapsed time from $t = 0$ until the customer at the head of the queue enters service; and let X_i ($i = 2, 3, \ldots, j+1$) be the length of time that the ith customer spends at the head of the queue. (X_{j+1} is the time that the test customer spends at the head of the queue.) Clearly, the test customer's waiting time is the sum

$X_1 + \cdots + X_{j+1}$. Now, the length of time that each customer spends at the head of the queue is the elapsed time from his arrival at the head of the queue until the completion of the shortest of the s service times then in progress. Because the service times are independent, identical, exponential variables with mean μ^{-1}, the duration of time from an arbitrary instant until the completion of the shortest remaining service time is also exponentially distributed, with mean $(s\mu)^{-1}$. Hence, $X_1, X_2, \ldots, X_{j+1}$ are all identical, exponential variables, with mean $(s\mu)^{-1}$; furthermore, they are independent. We conclude that the test customer's waiting time W, given that (1) the number Q of customers that he finds waiting ahead of him in the queue is j and (2) $W > 0$ (because he finds all servers busy), has the $(j+1)$-phase Erlangian distribution:

$$P\{W > t | W > 0, Q = j\} = P\{X_1 + \cdots + X_{j+1} > t\}$$
$$= \sum_{i=0}^{j} \frac{(s\mu t)^i}{i!} e^{-s\mu t}. \tag{5.55}$$

Also of importance in applications is the case where the number of exponential variables in the sum is itself a random variable; that is, $S_N = X_1 + \cdots + X_N$, where $X_1, X_2, \ldots$ are independent, identical, exponential variables, and N is a nonnegative discrete random variable, independent of $X_1, X_2, \ldots$. (S_N differs from the compound random variables we have considered previously in that $X_1, X_2, \ldots$ are no longer required to be discrete. In a later chapter we shall develop appropriate transform techniques, analogous to the use of generating functions as in Exercise 4, for studying this kind of compound random variable. For the present we use only direct calculation.) In particular, suppose that $N - 1$ has the geometric distribution,

$$P\{N = 1 + k\} = (1 - p)p^k \quad (k = 0, 1, \ldots). \tag{5.56}$$

Then, applying the theorem of total probability,

$$P\{S_N > t\} = \sum_{j=1}^{\infty} P\{S_N > t | N = j\} P\{N = j\},$$

we can write

$$P\{S_N > t\} = \sum_{j=1}^{\infty} \sum_{i=0}^{j-1} \frac{(\lambda t)^i}{i!} e^{-\lambda t} (1-p) p^{j-1}$$

or, equivalently,

$$P\{S_N > t\} = \sum_{j=0}^{\infty} \sum_{i=0}^{j} \frac{(\lambda t)^i}{i!} e^{-\lambda t} (1-p) p^j. \tag{5.57}$$

Some Important Probability Distributions

If we interchange the order of summation in (5.57) we have

$$P\{S_N > t\} = \sum_{i=0}^{\infty} \sum_{j=i}^{\infty} \frac{(\lambda t)^i}{i!} e^{-\lambda t}(1-p)p^j,$$

and this easily reduces to

$$P\{S_N > t\} = e^{-(1-p)\lambda t}. \tag{5.58}$$

Thus we conclude that if $S_N = X_1 + \cdots + X_N$, where $X_1, X_2, \ldots$ are independent, identical, exponential variables with mean λ^{-1}, and N is a random variable with the geometric distribution (5.56), independent of $X_1, X_2, \ldots$, then S_N has the exponential distribution with mean $[(1-p)\lambda]^{-1}$.

For example, consider a queueing system with *high-priority* customers and *low-priority* customers: High-priority customers will preempt the server from low-priority customers, so that no low-priority customer is ever served when a high-priority customer is present. We wish to find the total length of time a low-priority customer spends in service under two different priority disciplines, *preemptive resume* and *preemptive repeat*. Under the discipline of preemptive resume, whenever a preempted customer reenters service he simply continues his service where he left off. Under the discipline of preemptive repeat, a preempted customer draws a new value of service time from the service-time distribution each time he reenters service.

Clearly, when the priority discipline is preemptive resume, the total service time of a low-priority customer is not affected by the number of times he is preempted, and thus is unaffected by the existence of the high-priority customers.

Now consider the case when the priority discipline is preemptive repeat. We assume that service times of low-priority customers are exponential with mean μ^{-1}, and also that the high-priority customers arrive according to a Poisson process with rate α. We shall show that the total time spent in service by a low-priority customer is (again) exponentially distributed with mean μ^{-1}, independent of the stream of high-priority customers. Thus, in this case, the total time spent in service by a low-priority customer has the same distribution under both priority disciplines, and is not affected by the existence of the high-priority stream. We now show that this surprising result can be obtained by application of the argument leading to (5.58).

Let Y_j be the elapsed time between the start of service of a low-priority customer after his $(j-1)$th preemption and the next arrival of a high-priority customer, and let Z_j be the duration of the new service time drawn by the low-priority customer when he begins service after his $(j-1)$th preemption. If we let $X_j = \min(Y_j, Z_j)$, then the total service time of the low-priority customer is $S_N = X_1 + \cdots + X_N$, where the random variable $N-1$ is the total number of times he is preempted from the server by the arrival of a high-priority customer.

It follows from Equation (5.23) that $P\{Y_j<Z_j\}=\alpha/(\alpha+\mu)$; hence the distribution of N is given by (5.56), with $p=\alpha/(\alpha+\mu)$. Now, observe that $P\{X_j>t\}=P\{\min(Y_j,Z_j)>t|Y_j<Z_j\}$ for $j=1,2,\ldots N-1$; and $P\{X_N>t\}=P\{\min(Y_N,Z_N)>t|Y_N>Z_N\}$. It follows from Exercise 15 that when Y_j and Z_j are independent, exponential random variables, the conditions $Y_j<Z_j$ ($j=1,2,\ldots,N-1$) and $Y_N>Z_N$ are irrelevant; thus $X_1,X_2,\ldots$ are independent identical exponential variables with mean $E(X_j)=(\alpha+\mu)^{-1}$. Exercise 15 further implies that N is independent of the random variables $X_1,X_2,\ldots$ (because no inference can be drawn about the value of X_j from knowledge of whether $j<N$ or $j=N$).

Therefore, Equation (5.58) applies, with $p=\alpha/(\alpha+\mu)$ and $\lambda=\alpha+\mu$; hence, $P\{S_N>t\}=e^{-\mu t}$, and our assertion is proved.

This result could have been anticipated on intuitive grounds: Since the service times are exponential, they have the Markov property, which implies that it makes no difference whether a low-priority customer who reenters service continues with his old service time (preemptive resume) or selects a new one (preemptive repeat). Note that this intuitive argument also shows that the assumption of a Poisson arrival process for the high-priority customers is superfluous; regardless of the probabilistic structure of the process of preemption, the distribution function of the total service time for the low-priority customers remains the same.

Exercises

25. An operations research consultant is retained to help a highway engineering department study the relationship between the flow of vehicles on an expressway and the delays suffered by other vehicles that want to merge onto the expressway from an entrance ramp. The consultant proposes the following model. Let U_1 be the time from the instant the merging vehicle moves to the head of the entrance ramp until the first car in the expressway stream passes the entrance point, and let U_k ($k=2,3,\ldots$) be the times between the $(k-1)$th and kth vehicles in the expressway stream. The merging vehicle will start to enter the expressway immediately if its required acceleration time V_1, say, is less than U_1; in that case his merging time is, by definition, V_1. If $V_1>U_1$ the merging vehicle waits until the first expressway vehicle passes, and then begins to enter the expressway if its *new* required acceleration time V_2 is less than U_2; in this case the merging time is U_1+V_2. Thus, the merging time of the vehicle on the ramp is $U_1+\cdots+U_{n-1}+V_n$ if $V_k>U_k$ for all $k=1,2,\ldots,n-1$ and $V_n<U_n$.

Further, asserts the consultant, we can assume that $U_1,U_2,\ldots$ are independent, identical, exponential variables with mean α^{-1}, say, and $V_1,V_2,\ldots$ are independent, identical, exponential variables with mean β^{-1}, say. Of course, the consultant admits, the accuracy of numerical values for the merging times that will be derived from this model will depend on the accuracy of the underlying assumptions; nevertheless, the model should give a good qualitative picture of the functional dependence of merging times on the parameters α and β.

Remarks

A critic of the model argues that such qualitative information is not helpful, since a qualitative description is intuitively obvious to even the most casual observer: Merging times increase as α increases relative to β^{-1}. The model will not give useful qualitative information because the assumption that the vehicles on the expressway can be described by a Poisson process is unrealistic. Clearly, asserts the critic, the Poisson assumption is valid only when the expressway traffic is light; otherwise, a distribution that implies less variability in the interarrival times $U_1, U_2, \ldots$ is required. Therefore, concludes the critic, the model will be useful only when a more accurate description of the expressway traffic is provided.

Evaluate the arguments of the consultant and the critic.

26. In the model of Exercise 25, let X be the merging time of a vehicle; that is, X is the time from the instant the vehicle reaches the head of the ramp until it is merged into the mainstream of expressway traffic. Then, as a consequence of the assumptions of the model of Exercise 25, X has the same distribution as does the acceleration time V. Suppose now that each vehicle has a characteristic acceleration time V; that is, the merging time is $U_1 + \cdots + U_{n-1} + V$ if $V > U_k$ for all $k = 1, 2, \ldots, n-1$ and $V < U_n$.

   a. Show that

   $$E(X|V=c) = c + (e^{\alpha c} - 1) E(U | U \leqslant c).$$

   b. Assume that V is exponentially distributed with mean β^{-1}, and show that $E(X) = (\beta - \alpha)^{-1}$ when $\beta > \alpha$ and $E(X) = \infty$ when $\beta \leqslant \alpha$.

   c. Show, for the assumption of part b, that the variance $V(X)$ of the merging times is

   $$V(X) = \frac{\beta + 2\alpha}{\beta - 2\alpha} \left(\frac{1}{\beta - \alpha} \right)^2$$

   when $\beta > 2\alpha$, and $V(X) = \infty$ when $\beta \leqslant 2\alpha$.

   Feller [1971] uses this model of "gaps" as an illustration in his exposition of renewal theory. For further references and a discussion in the context of vehicular traffic theory, see Ashton [1966] and Haight [1963]. It is interesting to observe that if one were asked to make an a priori choice between the hypotheses of this model and those of Exercise 25, one could make a good argument in favor of either; nevertheless, the conclusions that follow from these two sets of apparently similar hypotheses are radically different.

2.6. Remarks

We have reviewed some topics from the theory of probability and stochastic processes that are particularly important in queueing theory. There are many excellent books in which these and related topics are covered. We mention a few for reference.

Feller [1968] is perhaps the most widely referenced book on probability. It contains valuable material on discrete probability distributions, generating functions, and birth-and-death processes. The first chapter of a second volume, Feller [1971], treats the uniform and negative-exponential distributions. These two books comprise a most thorough and lucid treatment of probability theory and its applications.

Fisz [1963] provides a firm background in probability, covering many topics of interest to applied workers. A recent textbook in probability that is especially recommended is Neuts [1973], which includes a good discussion of probability-generating functions and the method of collective marks.

Some recent textbooks that treat the theory of stochastic processes with an eye toward applications in queueing theory and related areas are Bhat [1972], Çinlar [1975], Clarke and Disney [1970], Cox and Miller [1965], Karlin [1968], and Ross [1970]. Another book by this last author, Ross [1972], gives a good elementary treatment of probability and stochastic processes.

Takács [1960a] provides short summaries of some important topics in stochastic processes, followed by problems with complete solutions.

Finally, Cohen [1969] and Syski [1960] are two important reference books in queueing theory that contain substantial preliminary material on probability and stochastic processes.

There are many other books on queueing theory that do not include substantial preliminary material on probability and stochastic processes. These books will be referenced in the present text when appropriate and will be discussed briefly in an annotated bibliography in the final chapter.

[3]
Birth-and-Death Queueing Models

3.1. Introduction

In Chapter 1 we discussed the characteristics of a queueing model, and in Chapter 2 we discussed the birth-and-death process. In this chapter we shall develop a correspondence between certain queueing models and certain birth-and-death processes, and then specialize the results of our previous birth-and-death analysis to provide an analysis of these queueing models.

Recall from Chapter 1 that a queueing model can be defined (roughly speaking) in terms of three characteristics: the input process, the service mechanism, and the queue discipline. The *input process* describes the sequence of requests for service. For example, a common assumption for the input process, and one that we shall make in this chapter, is that of *Poisson* (or *random*) *input*, where the customers are assumed to arrive according to a Poisson process. Another input process we shall study in this chapter is called *quasirandom input*, where each idle source generates requests independently and with the same exponentially distributed inter-request time.

The *service mechanism* includes such characteristics as the number of servers and the lengths of time that customers hold the servers. In this chapter we shall study primarily models with an arbitrary number of parallel servers and with independent, identically distributed *exponential service times*. We shall also consider some variants, such as one in which the servers are selected in a prespecified order.

The *queue discipline* specifies the disposition of blocked customers (customers who find all servers busy). In this chapter we shall consider two different queue disciplines. When blocked customers do not wait, but return immediately to their prerequest state, the queue discipline is said to be *blocked customers cleared* (BCC). And when blocked customers wait as long as necessary for service, the queue discipline is said to be *blocked customers delayed* (BCD). In this latter case, it is sometimes necessary to specify the order in which waiting customers are selected from the queue for service. We shall concern ourselves primarily with service in order of arrival, although we will also outline some results for service in random order.

Our strategy is to exploit the fact that certain queueing processes can be modeled as birth-and-death processes. Thus, if we let $\{N(t), t \geq 0\}$ be the number of customers present at time t, then the statistical-equilibrium state probabilities, $P_j = \lim_{t \to \infty} P\{N(t) = j\}$, are given by Equation (3.10) of Chapter 2, which we repeat for convenience:

$$P_j = \begin{cases} \dfrac{1}{1 + \sum_{k=1}^{\infty} (\lambda_0 \lambda_1 \cdots \lambda_{k-1} / \mu_1 \mu_2 \cdots \mu_k)} & (j = 0), \\[1em] \dfrac{\lambda_0 \lambda_1 \cdots \lambda_{j-1}}{\mu_1 \mu_2 \cdots \mu_j} P_0 & (j = 1, 2, \ldots). \end{cases} \quad (1.1)$$

As we showed in Chapter 2, this equation is equivalent to the "rate up = rate down" equations

$$\lambda_j P_j = \mu_{j+1} P_{j+1} \quad (j = 0, 1, 2, \ldots), \quad (1.2)$$

augmented by the normalization equation

$$\sum_{j=0}^{\infty} P_j = 1. \quad (1.3)$$

The birth rates λ_j $(j = 0, 1, 2, \ldots)$ and death rates μ_j $(j = 1, 2, \ldots)$ must be chosen to reflect the assumptions about the input process, the service mechanism, and the queue discipline.

For example, consider an infinite-server system, and suppose that arrivals occur according to a Poisson process, with rate λ, and service times are exponentially distributed, with mean μ^{-1}, independently of the arrival process and each other. (That is, we have Poisson input and exponential service times.) Suppose we observe the system at some time t, and suppose $N(t) = j$. Interarrival times are exponential with mean λ^{-1}, and service times are exponential with mean μ^{-1}; thus the time from t until the next event, whether an arrival or a service completion, is also exponentially distributed, with mean $(\lambda + j\mu)^{-1}$. (This follows because the time to the next event is the minimum of the exponential interarrival time and the j

Introduction

exponential service times.) Thus, the probability that an event will occur in $(t, t+h)$ is $(\lambda + j\mu)h + o(h)$ as $h \to 0$. When an event does occur, the probability that it will be caused by an arrival rather than a service completion is $\lambda/(\lambda + j\mu)$. [According to Exercise 15 of Section 2.5, the length of time required for the event to occur and the type of the event (whether arrival or service completion) are independent.] Therefore, the probability of occurrence of the transition $E_j \to E_{j+1}$ in $(t, t+h)$ is

$$[(\lambda + j\mu)h + o(h)] \frac{\lambda}{\lambda + j\mu} + o(h) = \lambda h + o(h) \qquad (h \to 0),$$

where the second $o(h)$ on the left-hand side of this equation is the probability of multiple arrivals and service completions that could effect this transition. Thus,

$$P\{N(t+h) = j+1 | N(t) = j\} = \lambda h + o(h) \qquad (h \to 0). \tag{1.4}$$

By a similar argument,

$$P\{N(t+h) = j-1 | N(t) = j\} = j\mu h + o(h) \qquad (h \to 0). \tag{1.5}$$

The probability that no event occurs in $(t, t+h)$ is $e^{-(\lambda + j\mu)h}$; and because the probability that more than one event will occur in $(t, t+h)$ is $o(h)$ as $h \to 0$, it follows that

$$\begin{aligned} P\{N(t+h) = j | N(t) = j\} &= e^{-(\lambda + j\mu)h} + o(h) \\ &= 1 - (\lambda + j\mu)h + o(h) \qquad (h \to 0). \end{aligned} \tag{1.6}$$

Again, because the probability of more than one event occurring in $(t, t+h)$ is $o(h)$ as $h \to 0$, it follows that

$$P\{N(t+h) = k | N(t) = j\} = o(h) \qquad (h \to 0, \ |k-j| \geq 2). \tag{1.7}$$

Thus, we have shown that in the infinite-server system having Poisson arrivals with rate λ and exponential service times with mean μ^{-1}, the number $N(t)$ of customers present at time t obeys the postulates of the birth-and-death process with birth rates $\lambda_j = \lambda$ $(j = 0, 1, 2, \dots)$ and death rates $\mu_j = j\mu$ $(j = 1, 2, \dots)$. [This model has been considered previously—see Equation (4.20) of Chapter 2.] Thus, Equation (1.1) applies, and we have

$$P_j = \frac{(\lambda/\mu)^j}{j!} e^{-\lambda/\mu} \qquad (j = 0, 1, 2, \dots). \tag{1.8}$$

We return our attention to the general s-server case. One of the quantities of interest that can be calculated from the equilibrium probabilities

(1.1) is the *carried load a'*, which is defined as the mean number of busy servers at an arbitrary instant in equilibrium:

$$a' = \sum_{j=1}^{s-1} jP_j + s \sum_{j=s}^{\infty} P_j. \qquad (1.9)$$

The first sum on the right-hand side of (1.9) reflects the assumption that if the number of customers in the system is less than the number of servers, then all these customers will be in service; the second sum reflects the fact that all s servers will be busy if there are at least s customers in the system.

In view of the definition of carried load as the mean number of busy servers, it can be shown (using Little's theorem, which is discussed and proved in Section 5.2) that the carried load equals the product of the mean service completion rate and the mean service time.

The *server occupancy* ρ is defined as the load carried per server in equilibrium:

$$\rho = \frac{a'}{s}. \qquad (1.10)$$

The server occupancy ρ measures the degree of utilization of a group of servers, and is therefore sometimes called the *utilization factor*. Clearly, we must always have $\rho \leq 1$.

Another important quantity is the *offered load a*, which is defined as the product of the arrival rate and the mean service time. Thus, if λ is the arrival rate and τ is the mean service time, then

$$a = \lambda \tau. \qquad (1.11)$$

The offered load is a dimensionless quantity whose numerical values, which provide a measure of the demand placed on the system, are expressed in units called *erlangs* (abbreviated *erl*).

In the case of Poisson input, the offered load a is equal to the mean number of arrivals per service time. That is, let X be an arbitrary service time, with distribution function $P\{X \leq t\} = H(t)$ and mean value $\tau = \int_0^\infty t\, dH(t)$; and let M be the number of arrivals during the service time X. If we let $p_j = P\{M = j\}$, then

$$p_j = \int_0^\infty \frac{(\lambda t)^j}{j!} e^{-\lambda t} dH(t) \qquad (j = 0, 1, 2, \ldots). \qquad (1.12)$$

Now $E(M) = \sum_{j=1}^\infty j p_j = \int_0^\infty \lambda t\, dH(t)$, where the last equality follows after using (1.12) and interchanging the order of summation and integration. Comparison with (1.11) shows that $E(M) = a$ for Poisson input, as asserted.

For example, consider again the infinite-server model with Poisson arrivals, with rate λ, and exponential service times, with mean μ^{-1}. To calculate the carried load, substitute (1.8) into (1.9). Note that in this case ($s=\infty$), the definition (1.9) reduces to the mean value of the (Poisson) distribution (1.8); hence, $a'=\lambda/\mu$. To calculate the offered load, use the fact that when $H(t)=1-e^{-\mu t}$, the mean service time τ is given by $\tau=\mu^{-1}$; then (1.11) yields $a=\lambda/\mu$. Thus we see that in this example $a'=a$. This equality of carried and offered loads reflects the fact that in this model, all arrivals are served (because $s=\infty$), and therefore the mean number of service completions per unit time equals the mean number of arrivals per unit time.

Before proceeding to the application of Equation (1.1) to queues with particular choices of the input process, service mechanism, and queue discipline, we direct our attention to the relationship between the probabilities given by Equation (1.1), which describe the system at an arbitrary instant during equilibrium, and the probabilities that describe the system at the instants at which the customers arrive.

3.2. Relationship between the Outside Observer's Distribution and the Arriving Customer's Distribution

In Section 2.5 we showed [see Equation (5.39)] that for systems with Poisson input, the arriving customer's distribution $\{\Pi_j(t)\}$ and the outside observer's distribution $\{P_j(t)\}$ are equal:

$$\Pi_j(t) = P_j(t) \qquad (t \geq 0; \; j = 0, 1, \ldots). \tag{2.1}$$

That is, when customers arrive in a Poisson stream, the probability that there are j customers present in the system at any time t is the same whether or not t is an arrival point. We also showed by counterexample that (2.1) is not true in general.

We now give an alternative argument in support of (2.1) for systems with Poisson input. Let $C(t, t+h)$ be the event {a customer arrives in the interval $(t, t+h)$}, and define

$$\Pi_j(t) = \lim_{h \to 0} P\{N(t) = j \mid C(t, t+h)\}, \tag{2.2}$$

where, as before, $N(t)$ is the number of customers present at time t, and we use the notation $P_j(t) = P\{N(t) = j\}$. [Note that the occurrence of the event $\{C(t, t+h)\}$ does not imply that the arriving customer necessarily enters the system; it includes the case where the arrival departs immediately, without causing a state transition.]

Consider the probability $P\{C(t, t+h) \mid N(t) = j\}$. By assumption, the customers arrive according to a Poisson process; hence, the probability that an arrival occurs in any interval of length h is $\lambda h + o(h)$ as $h \to 0$, regardless of the value of t, j, or anything else. Thus, for Poisson input,

$$P\{C(t,t+h)|N(t)=j\} = P\{C(t,t+h)\};$$

that is, the events $\{C(t,t+h)\}$ and $\{N(t)=j\}$ $(j=0,1,...)$ are independent. It follows from the independence of these events that

$$P\{N(t)=j|C(t,t+h)\} = P\{N(t)=j\}.$$

Thus, for Poisson input, Equation (2.2) reduces to (2.1), as was to be shown.

For the purpose of making calculations for systems with non-Poisson input, it is often useful to express (2.2) in the form commonly referred to as Bayes' rule:

$$\Pi_j(t) = \lim_{h \to 0} \frac{P\{C(t,t+h)|N(t)=j\}P_j(t)}{\sum_{k=0}^{\infty} P\{C(t,t+h)|N(t)=k\}P_k(t)}. \quad (2.3)$$

Of interest in most applications is the equilibrium arriving customer's distribution $\{\Pi_j\}$, which we define as

$$\Pi_j = \lim_{t \to \infty} \Pi_j(t) \quad (j=0,1,...). \quad (2.4)$$

In particular, Equation (2.1) yields (for systems with Poisson input)

$$\Pi_j = P_j \quad (j=0,1,...). \quad (2.5)$$

An interpretation of (2.5) [and (2.1)] is: Since Poisson arrivals occur randomly in time, the proportion Π_j of arrivals that occur when the system is in state E_j equals the proportion P_j of time that the system spends in state E_j.

Note that we have nowhere assumed that the process $\{N(t), t \geq 0\}$ is necessarily a birth-and-death process. If it is, and if $P\{C(t, t+h)|N(t)=j\} = \lambda_j h + o(h)$ as $h \to 0$, then Equations (2.3) and (2.4) yield

$$\Pi_j = \frac{\lambda_j P_j}{\sum_{k=0}^{\infty} \lambda_k P_k} \quad (j=0,1,...), \quad (2.6)$$

where P_j $(j=0,1,...)$ is given by (1.1).

Exercises

1. **a.** A single-server queueing system with an unlimited number of waiting positions is modeled as a birth-and-death process with $\lambda_j = (j+1)^{-1}\lambda$ for

The Erlang Loss System

$j \geq 0$ and $\mu_j = \mu$ for $j \geq 1$, where births correspond to arrivals and deaths to service completions. Show that

$$P_j = \frac{(\lambda\tau)^j}{j!} e^{-\lambda\tau} \quad (j=0,1,\ldots)$$

and

$$\Pi_j = (1 - e^{-\lambda\tau})^{-1} P_{j+1} \quad (j=0,1,\ldots),$$

where $\tau = \mu^{-1}$. Calculate the carried load a'. Convince yourself that the arrival rate equals $\sum_{j=0}^{\infty} \lambda_j P_j$, and calculate the offered load a. Verify that $a = a'$.

b. A single server with exponential service times with mean μ^{-1} serves customers who arrive in a Poisson stream at rate λ. An arriving customer who finds j other customers in the system will, with probability $j/(j+1)$, depart (*renege*) immediately; all customers who do not depart immediately wait as long as necessary for service. Find $\{P_j\}$ and $\{\Pi_j\}$. Find the carried load a' and the offered load a, and verify that the probability that an arbitrary arrival does not receive service equals the ratio $(a - a')/a$.

c. Customers arrive in a Poisson stream at a single server with an infinite number of waiting positions. Observation indicates that the server works faster as the queue size increases. If this system is described by a birth-and-death model with $\lambda_j = \lambda$ and $\mu_j = j\mu$ ($j=0,1,\ldots$), find $\{P_j\}$ and $\{\Pi_j\}$. Calculate the carried load a'. By considering the queue discipline in which (i) each arrival enters service immediately, preempting the customer in service (if any) and moving him back to the head of the queue, and (ii) the customers in the queue are served in reverse order of arrival, convince yourself that the mean service time equals $\sum_{j=0}^{\infty} \Pi_j \mu_{j+1}^{-1}$. Calculate the offered load a, and verify that $a = a'$.

For a list of references and discussion of the models of this exercise from a different point of view, see Conolly and Chan [1977], and also Brumelle [1978].

2. Customers arrive at a two-chair shoeshine stand at rate 10 per hour. The average length of a shoeshine is 6 minutes. There is only one attendant, so that one chair is used as a waiting position. Customers who find both chairs occupied go away.
   a. Assuming Poisson input and exponential service times, write and solve the statistical-equilibrium probability state equations.
   b. Find the mean number of customers served per hour.
   c. Repeat the analysis to calculate the mean number of customers served per hour when there are two attendants at the stand (and no waiting positions).

3.3. Poisson Input, s Servers, Blocked Customers Cleared: The Erlang Loss System

In this model, we assume that customer arrivals follow a Poisson process with rate λ; that service times are exponentially distributed with mean service time μ^{-1}, independent of each other and the arrival process; and that customers who find all s servers busy leave the system and have no effect upon it, that is, *blocked customers cleared* (BCC). (However, the assumption of exponential service times will be seen to be unnecessary.) An application of the BCC assumption arises in telephone traffic engineering, where calls that find all trunks busy are given a busy signal. Other applications should be apparent.

Since customers arrive at random with rate λ, but effect state changes only when $j < s$ (because blocked customers are cleared), we write

$$\lambda_j = \begin{cases} \lambda & \text{when } j = 0, 1, \ldots, s-1, \\ 0 & \text{when } j = s. \end{cases} \tag{3.1}$$

Because service times are exponential, when there are j customers in service the rate μ_j at which service completions occur is

$$\mu_j = j\mu \quad (j = 1, 2, \ldots, s). \tag{3.2}$$

The rates (3.1) and (3.2), in conjunction with (1.1), give for the statistical-equilibrium probability of j busy servers

$$P_j = \frac{\dfrac{(\lambda/\mu)^j}{j!}}{\sum_{k=0}^{s} \dfrac{(\lambda/\mu)^k}{k!}} \quad (j = 0, 1, \ldots, s) \tag{3.3}$$

and $P_j = 0$ for $j > s$. The distribution (3.3) is called the *truncated Poisson distribution*. Note that the rates λ and μ appear in (3.3) only through the ratio λ/μ, which, according to Equation (1.11), is the offered load a. We call a system with Poisson arrivals and blocked customers cleared [whose equilibrium-state probabilities are given by (3.3) for any service-time distribution] an *Erlang loss system*. (A. K. Erlang, 1878–1929, was a Danish mathematician who laid the foundations of modern teletraffic and queueing theory—see Brockmeyer et al. [1948].)

The probability that all s servers are busy is given by (3.3) with $j = s$. Formula (3.3) with $j = s$ is called the *Erlang loss formula* or *Erlang B formula* in the United States and is denoted by

$$B(s, a) = \frac{a^s/s!}{\sum_{k=0}^{s} a^k/k!}. \tag{3.4}$$

Likewise, the truncated Poisson distribution (3.3) is also known as the *Erlang loss distribution*. In Europe, the right-hand side of (3.4) is called *Erlang's first formula* and is denoted by $E_{1,s}(a)$. The formula (3.4) was first published by Erlang in 1917. As already suggested, the Erlang loss formula has found extensive application in the field of telephone traffic engineer-

The Erlang Loss System

ing. Curves of (3.4) for fixed values of s are plotted against increasing values of a in Figures A.1 and A.2 of the appendix. Equation (3.4) is also tabulated in Dietrich et al. [1966]. Further mathematical and numerical properties of the Erlang loss formula are discussed in Jagerman [1974].

We emphasize that (3.4) gives both the proportion of time that all s servers are busy and the proportion of arriving customers who find all s servers busy (and thus are lost). Note also that (3.4), which we have derived from first principles, is the same as (1.6) of Chapter 1, which was derived through a heuristic argument. We have now answered some, but not all, of the questions raised in Chapter 1 about the validity of this formula.

For the Erlang loss system, the carried load a', calculated directly from the definition (1.9) and the probabilities (3.3), is easily shown (see Exercise 3) to be

$$a' = a[1 - B(s,a)]. \qquad (3.5)$$

This result can be interpreted to say that the carried load a' equals that portion of the offered load a that is not lost. (In particular, when $s = \infty$, then $a' = a$; that is, the offered load is the load that would be carried if there were no blocking.) Similarly, we can interpret $aB(s,a)$ as the *lost load*. Then it follows (in this case) that the proportion of arrivals that is lost equals the ratio of the lost load to the offered load.

Exercise

3. Derive (3.5) from the definition (1.9) and the probabilities (3.3).

Investigation of (3.4) and (3.5) leads to the following important conclusion: As the number of servers is increased and the offered load is increased in such a way that the probability of blocking remains constant, the server occupancy [defined by Equation (1.10) as the carried load per server] increases. In other words, large server groups are more efficient than small ones.

Unfortunately, in practice hardware limitations sometimes preclude exploitation of this fact. Moreover, high-occupancy server groups are more vulnerable to service degradation during periods of overload than are smaller server groups with the same blocking probability but lower occupancy.

For a server group of fixed size, occupancy increases with increasing load, thereby increasing server-group efficiency; unfortunately, the probability of blocking also increases with increasing load. Hence, efficient use of equipment must be balanced against the provision of acceptable service for the customers.

Exercises

4. Consider an Erlang loss system with 10 servers. Measurements show that about 1% of the arriving customers are lost (denied service). It is estimated that the rate of requests for service will double over the next year. How many servers must be added so that the system will provide the same grade of service (that is, the same probability that an arrival will find an idle server) to the customers in the face of the increased demand?

5. An entrepreneur offers services that can be modeled as an s-server Erlang loss system. Suppose the arrival rate is 4 customers per hour; the average service time is 1 hour; the entrepreneur earns $2.50 for each customer served; and the entrepreneur's operating cost c is $1.00 per server per hour (whether the server is busy or idle). What is the optimal number of servers, and what is the hourly profit earned when the optimal number of servers is provided? What is the maximum value of c beyond which it is unprofitable for the entrepreneur to remain in business?

6. Show that

$$B(s,a) = \frac{aB(s-1,a)}{s+aB(s-1,a)}.$$

How might this recurrence be computationally useful?

7. Consider an Erlang loss system with *retrials*. Any customer who finds all s servers busy will return and place another request for service at some later time, and will persist in this manner until he receives service. Let $a = \lambda\tau$, where τ is the average service time and λ is the rate at which the original requests for service occur. (That is, λ is the arrival rate with retrials not counted.) Since all customers are served, the carried load equals the original offered load $a = \lambda\tau$. Therefore, it can be argued, the probability of blocking (the probability that a customer requesting service will find all servers busy and be forced to retry) is given by $B(s,\hat{a})$, where $\hat{a}$ satisfies the equation

$$\hat{a}[1 - B(s,\hat{a})] = a.$$

Discuss the strengths and weaknesses of this approximate analysis of the Erlang loss system with retrials. (See Riordan [1962].)

8. Consider an equilibrium s-server Erlang loss system with exponential service times and offered load a erl. A statistician observes the system at a random instant and waits until the next customer arrives. Show that the probability p that this customer is blocked is

$$p = \frac{a}{a+s} B(s,a).$$

Explain why $p \neq B(s,a)$.

The Erlang Loss System

9. *The Erlang loss system as a semi-Markov process.* A (stationary) *Markov chain* is a stochastic process $\{X_n, n = 0, 1, 2, \ldots\}$ whose values are countable, and for which the transition probabilities satisfy

$$P\{X_{n+1}=j|X_n=i, X_{n-1}=i_{n-1},\ldots, X_1=i_1, X_0=i_0\}$$
$$= P\{X_{n+1}=j|X_n=i\} = p_{ij}; \quad (1)$$

that is, for each n, the conditional probability that the process will have value j (be in state E_j) after the $(n+1)$th step depends only on the state of the process after the nth step, and is independent of the past history of the process prior to the nth step. If we define

$$P_j^* = \lim_{n \to \infty} P\{X_n = j | X_0 = i\}, \quad (2)$$

then it is known that, under certain conditions, the limiting distribution P_j^* $(j = 0, 1, 2, \ldots)$ exists and is the unique solution of the equations

$$P_j^* = \sum_{i=0}^{\infty} P_i^* p_{ij} \quad (j = 0, 1, 2, \ldots) \quad (3)$$

and

$$\sum_{j=0}^{\infty} P_j^* = 1. \quad (4)$$

A *semi-Markov process* is a Markov chain in which the times between transitions are random variables. It can be shown that (a) if the Markov chain $\{X_n, n = 0, 1, 2, \ldots\}$ takes only the values $0, 1, \ldots, k$, and (b) if, for each n, m_i is the mean length of continuous time that $X_n = i$ before making a transition, then the proportion P_j of time that the process spends in state E_j is the weighted average of the probabilities $\{P_j^*\}$:

$$P_j = \frac{m_j P_j^*}{\sum_{i=0}^{k} m_i P_i^*} \quad (j = 0, 1, \ldots, k). \quad (5)$$

(These important topics are discussed in most books on probability and stochastic processes, such as those referenced in Section 2.6.)

Now consider the equilibrium, s-server Erlang loss system with exponential service times. Then the distribution $\{P_j\}$ above corresponds to the equilibrium birth-and-death probabilities. Assume that the conditions required for the above results are satisfied, and
a. find p_{ij} $(i, j = 0, 1, \ldots, s)$;
b. find m_i $(i = 0, 1, \ldots, s)$;
c. show that Equation (5) agrees with (3.3).

An important theorem is that all of the results derived above are true for any service-time distribution function with finite mean, even though the Markov property of the exponential distribution function was used explicitly in the above derivation. [Specifically, in using the birth-and-death formulation we have assumed that if at any instant t there are j customers in service, each with mean service time μ^{-1}, then the probability that a customer will complete service in $(t, t+h)$ is $j\mu h + o(h)$ as $h \to 0$, independent of the value of t and the lengths of time that the customers have been in service.]

The remarkable fact that the truncated Poisson distribution (3.3) is valid for an arbitrary service-time distribution function was conjectured by Erlang himself in 1917, and proved by Sevast'yanov [1957]. Subsequently, many investigators studied this problem, and a variety of proofs have been offered. (See pp. 271–278 of Syski [1960]; see also Takács [1969], Cohen [1976], Oakes [1976], and Brumelle [1978].) It is worthwhile at this point to give proofs for the special cases $s=1$ and $s=\infty$. A proof for the case of arbitrary s will be outlined in Chapter 4. Before proceeding with the proofs, however, let us consider the following example, which illustrates the utility of the theorem.

First, consider a server group that serves two independent streams of Poisson traffic (for example, a telephone trunk group serving eastbound and westbound traffic) on a BCC basis. That is, suppose that arrivals from stream i occur according to a Poisson process with rate λ_i, and assume first that the customers in each stream have a common (but arbitrary) service-time distribution function with mean τ. Then the two streams can be viewed as composing a single Poisson stream with rate $\lambda_1 + \lambda_2$, so that the outside observer's distribution $\{P_j\}$ is given by (3.3) with $\lambda = \lambda_1 + \lambda_2$. Since each stream is Poisson, each stream sees the (same) outside observer's distribution, and hence each stream sees the same arriving customer's distribution; thus the probability of blocking is the same for each stream.

Suppose now that the (arbitrary) service-time distribution functions of the two Poisson streams are not identical, with stream i having mean service time τ_i. Arrivals still occur according to a Poisson process with rate $\lambda_1 + \lambda_2$. It follows from our theorem that, regardless of the form of the composite service-time distribution function, the state probabilities are still given by (3.3), where the composite mean service time τ remains to be determined. Again, since each stream is Poisson, each stream sees the same state distribution. Thus, the proportion of served customers who are from stream i is the same as the proportion of arriving customers who are from stream i, namely λ_i/λ. Therefore, the (composite) mean service time τ for all customers who are served is

$$\tau = \frac{\lambda_1}{\lambda}\tau_1 + \frac{\lambda_2}{\lambda}\tau_2.$$

Multiplying through by λ, we obtain

$$a = a_1 + a_2,$$

where $a = \lambda \tau$ and $a_i = \lambda_i \tau_i$.

We conclude that when two streams of Poisson traffic of magnitudes a_1 and a_2 erlangs, respectively, are offered to a single group of s serves on a BCC basis, each stream suffers a blocking probability given by the Erlang loss formula (3.4) with $a = a_1 + a_2$. The generalization of this result to include an arbitrary number of independent Poisson streams is clearly true.

Exercise

10. Two independent Poisson streams of traffic, called low-priority and high-priority traffic, are handled by a primary group of 10 servers. Low-priority customers who are blocked on (overflow from) the primary group are cleared from the system, whereas high-priority customers who overflow the primary group are routed to a backup of overflow servers. According to measurements, the high-priority customers require an average of 12 minutes for service and arrive at rate 20 per hour, of which about 2 per hour overflow.
  a. If the arrival rate of low-priority customers doubles, by what factor will the overflow rate of the high-priority customers increase?
  b. If the overflow group handles the overflow traffic on a BCC basis, is it legitimate to use the Erlang loss formula to decide how many servers to provide in the overflow group?

We now give a heuristic argument for the case $s = 1$ of the theorem: For a system with Poisson input and blocked customers cleared, the distribution $\{P_j\}$ of the number of customers present at an arbitrary point in equilibrium is the Erlang loss distribution (3.3).

We suppose that customers arrive according to a Poisson process with rate λ. An arrival who finds the server idle will hold it for a time interval whose duration is a random variable with mean τ. An arrival who finds the server occupied will leave the system immediately. Hence the server alternates between busy and idle states, each busy period of mean length τ and each idle period of mean length λ^{-1}. This is illustrated in Figure 3.1.

Consider a single cycle composed of an idle period and an adjacent busy period. The cycle has mean length $\lambda^{-1} + \tau$, and thus the ratio of the mean busy period to the mean cycle length is

$$\frac{\tau}{\lambda^{-1} + \tau} = \frac{a}{1 + a}, \tag{3.6}$$

where the right-hand side follows from (1.11). This is the same as P_1, given

Figure 3-1.

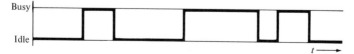

by (3.3) when $s=1$. Since the ratio of mean values $\tau/(\lambda^{-1}+\tau)$ is indeed the same as the proportion of time that the server is busy, we conclude that $P_1 = a/(1+a)$ is the probability that the server is occupied, for any service-time distribution function with mean τ, as asserted. Since we require that $P_0 + P_1 = 1$, it follows that $P_0 = 1/(1+a)$, again in agreement with (3.3). Hence, the theorem is proved for $s=1$.

Let us consider this single-server model again, this time from the point of view of arriving customers. We have argued that the proportion of time (or probability) the server is busy is $a/(1+a)$, that is, $P_0 = 1/(1+a)$ and $P_1 = a/(1+a)$, and we know that for systems with Poisson input $\{\Pi_j\} = \{P_j\}$. We now calculate Π_0 and Π_1 directly by application of Equation (1.4) of Chapter 2:

$$\Pi_1 = \frac{E(N)}{1+E(N)}, \qquad (3.7)$$

where N is the number of customers who arrive during an arbitrary busy period. When blocked customers are cleared, the busy period is the same as the service time; hence $E(N) = a$. Thus, we conclude from comparison of (3.6) and (3.7) that, as promised, $\Pi_1 = P_1$ and, since $\Pi_0 + \Pi_1 = 1 = P_0 + P_1$, therefore $\Pi_0 = P_0$.

We now turn to the case $s = \infty$. We shall calculate the (transient) distribution $\{P_j(t)\}$ of the number of customers present at time t, and show that for any service-time distribution function $H(x)$ with finite mean $\mu^{-1} = \int_0^\infty x \, dH(x)$,

$$\lim_{t \to \infty} P_j(t) = \frac{(\lambda/\mu)^j}{j!} e^{-\lambda/\mu} \qquad (j = 0, 1, 2, \ldots); \qquad (3.8)$$

that is, $\lim_{t \to \infty} P_j(t) = P_j$, where P_j is given by (3.3) with $s = \infty$.

To begin, consider an arbitrary customer (the test customer) who is assumed to have arrived at some time t_0 during the interval $(0, t)$. We shall calculate the probability $p(t)$ that the test customer is still present at time t.

Let X be the service time of a customer, with distribution function $H(x)$. The test customer will be present at time t if either (a) $X > t$ or (b) $X < t$ and $t_0 + X > t$. Event (a) has probability $1 - H(t)$. To calculate the probability of event (b), note that if $X = x < t$, then event (b) occurs if and only if t_0 lies in the interval $(t - x, t)$. (See Figure 3.2.) Since by assumption the arrivals occur according to a Poisson process, it follows [see the argument surrounding Equation (5.29) of Chapter 2] that the location of the arrival epoch t_0 is uniformly distributed in $(0, t)$. Therefore the probability that t_0 lies in the interval $(t - x, t)$ is x/t. Hence, event (b) has probability $\int_0^t (x/t) \, dH(x)$, and $p(t)$ is given by

$$p(t) = 1 - H(t) + \int_0^t \frac{x}{t} \, dH(x). \qquad (3.9)$$

The Erlang Loss System

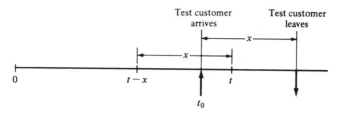

Figure 3-2.

Equation (3.9) gives the probability $p(t)$ that an arbitrary customer, who is assumed to have arrived in $(0,t)$, is still in service at time t. Thus, if n customers arrive in $(0,t)$ according to a Poisson process, the probability that exactly j of them will still be present at time t is the binomial

$$\binom{n}{j} p^j(t)[1-p(t)]^{n-j};$$

it now follows from the theorem of total probability that

$$P_j(t) = \sum_{n=j}^{\infty} \binom{n}{j} p^j(t)[1-p(t)]^{n-j} \frac{(\lambda t)^n}{n!} e^{-\lambda t}. \qquad (3.10)$$

Reference to Equation (5.52) of Chapter 2 shows that

$$P_j(t) = \frac{[\lambda t p(t)]^j}{j!} e^{-\lambda t p(t)} \qquad (j=0,1,\dots). \qquad (3.11)$$

Hence, for arbitrary service-time distribution function, the number of customers in service at time $t>0$ has the Poisson distribution with time-dependent mean $\lambda t p(t)$. [The reader should compare (3.11) with Equation (4.26) of Chapter 2.] For another derivation of (3.11), see Exercise 44.

We now consider the statistical-equilibrium distribution (3.8). We leave it to Exercise 11 for the reader to show that for any distribution function $H(x)$ $(0 \leq x < \infty)$ with finite mean $\mu^{-1} = \int_0^\infty x \, dH(x)$,

$$\lim_{t \to \infty} t[1 - H(t)] = 0. \qquad (3.12)$$

It follows from (3.9) and (3.12) that $\lim_{t \to \infty} tp(t) = \mu^{-1}$. Hence, taking limits through (3.11) yields (3.8), as asserted.

Exercises

11. Prove Equation (3.12).

12. *Blocked customers held.* The following model is widely used in teletraffic engineering: Customers arrive according to a Poisson process at a group of s

servers. An arriving customer is willing to spend an amount of time T (called the *sojourn time*) in the system, where T is a random variable, after which he will depart regardless of whether or not at the expiration of his sojourn time he is in service or is still waiting in the queue. In other words, a blocked customer will wait for service as long as time T; if he receives service before the expiration of T, he then holds the server for the remainder of the time T. There is no limit to the number of customers who may be waiting for service.

a. Show that the distribution $\{P_j(t)\}$ of the number of customers present at time t is given by (3.11), where $H(x)$ is the distribution function of the sojourn times.

b. Show that if the sojourn times are exponentially distributed with mean μ^{-1}, then so are the service times.

c. Let q be the equilibrium probability that an arriving customer does not receive service. Show that, if the sojourn times are exponential and the customers are served from the queue in order of arrival, then

$$q = P(s,a) - \frac{s}{a} P(s+1,a),$$

where $a = \lambda/\mu$ is the offered load and $P(s,a) = \sum_{j=s}^{\infty} P_j$. Show that $q = 1 - a'/a$, where a' is the carried load, defined by (1.9).

An interesting variation of the Erlang loss model is obtained with the additional assumption of *ordered hunt*. That is, we assume that the servers are numbered $1, 2, \ldots, s$, and each arriving customer takes the lowest-numbered idle server. As before, we assume that customers arrive according to a Poisson process, blocked customers are cleared, and the system is in statistical equilibrium. Again, it is not necessary to assume that service times are exponentially distributed. For this model we calculate the load carried by each server.

Real systems that might be described by such a model include a parking lot in which the parking places are deemed less desirable as they get farther from the exit, or a telecommunications system in which the trunks can be financed on either a flat-rate or measured-rate (usage) basis (see Exercise 13).

Let N_k denote the number of servers among the first k ($k = 1, 2, \ldots, s$) ordered servers that are busy at an arbitrary instant in equilibrium. Because blocked customers are cleared, the values realized by the random variable N_k do not depend on the disposition of any customer who overflows the first k ordered servers. Thus, for each value of the index k the distribution of N_k is the Erlang loss distribution (3.3); that is,

$$P\{N_k = j\} = \frac{\dfrac{a^j}{j!}}{\sum_{i=0}^{k} \dfrac{a^i}{i!}} \qquad (j = 0, 1, \ldots, k; \quad k = 1, 2, \ldots, s), \qquad (3.13)$$

The Erlang Loss System   89

where a is the offered load. Also, it follows from (3.5) that the mean number $E(N_k)$ of servers busy among the first k ordered servers is

$$E(N_k) = a[1 - B(k,a)] \qquad (k=1,2,\ldots,s). \tag{3.14}$$

To find the load carried by the jth ordered server in the k-server group, let us first define the random variables X_j ($j=1,2,\ldots,s$), with $X_j = 0$ when the jth ordered server is idle and $X_j = 1$ when the jth ordered server is busy. Then $E(X_j)$ is the mean number of busy servers in the single-server group that consists of only the jth ordered server; therefore, by definition $E(X_j)$ is equal to the load carried by the jth ordered server. If we define $\tilde{p}_j = P\{X_j = 1\}$, then $E(X_j) = \tilde{p}_j$; that is, the load carried by a single server equals the probability that the server is busy at an arbitrary instant (in equilibrium).

Now, clearly,

$$N_j = X_1 + \cdots + X_j, \tag{3.15}$$

from which it follows that

$$X_j = N_j - N_{j-1}. \tag{3.16}$$

Taking expected values in (3.16) yields

$$E(X_j) = E(N_j) - E(N_{j-1}), \tag{3.17}$$

which, together with Equation (3.14), implies that the load carried by the jth ordered server is given by

$$\tilde{p}_j = a[B(j-1,a) - B(j,a)]. \tag{3.18}$$

Equation (3.18) can be interpreted to say that the load carried by the jth ordered server is the difference between the load that overflows the $(j-1)$th ordered server and that which overflows the jth ordered server. It is intuitively obvious that $\tilde{p}_1 > \tilde{p}_2 > \tilde{p}_3 > \cdots$; for a proof see Messerli [1972].

Exercises

13. Suppose that a company with a private telephone network can choose from two classes of telephone trunks, of which the first is paid for by a flat monthly rental that is equivalent to $14 per hour, and the second is charged for according to usage at the rate of $.50 per minute. (These numbers are for illustrative purposes only, and were not chosen to reflect any existing rate structure.) If it is required that the total number of trunks should be sufficient to handle the offered load, estimated at 2 erlangs, with a loss no higher than 2%, what is the most economical division between measured-rate and flat-rate trunks, and what is the associated cost?

14. Prove that in an Erlang loss system with ordered hunt the equilibrium probability that an arbitrary customer will be served by the jth ordered server equals the ratio of the load carried by that server to the total offered load.

15. Prove that the variance v of the Erlang loss distribution (3.3) is given by

$$v = a'(1 - \tilde{p}_s),$$

where a' is the total load carried by an s-server Erlang loss system and $\tilde{p}_s$ is the load carried by the last ordered server in the group.

3.4. Poisson Input, s Servers with Exponential Service Times, Blocked Customers Delayed: The Erlang Delay System

Consider now the case in which customers who find all s servers busy join a queue and wait as long as necessary for service, that is, *blocked customers delayed* (BCD). No server can be idle if a customer is waiting. The number of waiting positions in the queue is assumed to be infinite. Applications of this model should be self-evident. (We note for future reference that in the terminology of queueing theory, the s-server BCD queue with Poisson input and exponential service times is called the $M/M/s$ queue. This terminology will be discussed in Chapter 5.) Since customers arrive according to a Poisson process with rate λ, and every arrival effects a change of system state, we have

$$\lambda_j = \lambda \qquad (j = 0, 1, \ldots). \tag{4.1}$$

The assumption of exponential service times implies that if at any time all the j customers in the system are in service, the rate at which service completions occur is $j\mu$; if all s servers are busy, only those customers that are in service are eligible to leave, so that the service completion rate is $s\mu$. Hence

$$\mu_j = \begin{cases} j\mu & (j = 1, 2, \ldots, s), \\ s\mu & (j = s+1, s+2, \ldots). \end{cases} \tag{4.2}$$

Using (4.1) and (4.2) in Equation (1.1), we can write

$$P_j = \frac{a^j}{j!} P_0 \qquad (j = 1, 2, \ldots, s-1) \tag{4.3}$$

and

$$P_j = \frac{a^j}{s! \, s^{j-s}} P_0 \qquad (j = s, s+1, \ldots), \tag{4.4}$$

The Erlang Delay System

where $a = \lambda/\mu$ and where P_0 is given by

$$P_0 = \left(\sum_{k=0}^{s-1} \frac{a^k}{k!} + \sum_{k=s}^{\infty} \frac{a^k}{s^{k-s}} \right)^{-1}. \tag{4.5}$$

If $a < s$, the infinite geometric sum on the right-hand side converges, and

$$P_0 = \left(\sum_{k=0}^{s-1} \frac{a^k}{k!} + \frac{a^s}{s!(1-a/s)} \right)^{-1} \quad (0 \leq a < s). \tag{4.6}$$

If $a \geq s$, the infinite geometric sum diverges to infinity. Then $P_0 = 0$ and hence $P_j = 0$ for all finite j. For $a \geq s$, therefore, the queue length is infinite (greater than any prespecified finite value) with probability 1. In this case, we say that no statistical-equilibrium distribution exists.

Let $C(s,a)$ denote the probability that all servers are occupied; $C(s,a) = \sum_{j=s}^{\infty} P_j$. It follows from (4.4) that

$$\sum_{j=s}^{\infty} P_j = \frac{a^s}{s!(1-a/s)} P_0; \tag{4.7}$$

hence

$$C(s,a) = \frac{\dfrac{a^s}{s!(1-a/s)}}{\sum_{k=0}^{s-1} \dfrac{a^k}{k!} + \dfrac{a^s}{s!(1-a/s)}} \quad (0 \leq a < s). \tag{4.8}$$

Formula (4.8) is called the *Erlang delay formula* (because blocked customers are delayed until service commences) or *Erlang C formula* in the United States; in Europe it is called *Erlang's second formula* and is denoted by $E_{2,s}(a)$. Like the Erlang loss formula (3.4), the Erlang delay formula (4.8) was first published by Erlang in 1917 (see Brockmeyer et al. [1948]). Curves for (4.8) are given in Figures A.3 and A.4 of the appendix. See also Dietrich et al. [1966] and Descloux [1962].

Since we have Poisson input, the arriving customer's distribution $\{\Pi_j\}$ is identical to the calculated outside observer's distribution $\{P_j\}$. Thus, (4.8) gives both the proportion of time that all servers are busy and the proportion of customers who find all servers busy.

Unlike the Erlang loss probabilities, the Erlang delay probabilities are *not* valid for an arbitrary service-time distribution.

Again unlike the Erlang loss probabilities, the Erlang delay probabilities constitute a proper distribution only when the offered load is less than the number of servers. This restriction follows mathematically from the convergence criterion for the infinite geometric sum and does not apply if the number of queue positions is finite (in which case customers finding all queue positions busy are cleared from the system).

From the definition (1.9) of carried load we can easily calculate that

$$a' = a; \tag{4.9}$$

that is, the carried and offered loads are equal. This is intuitively clear, because in the Erlang delay model all blocked customers wait as long as necessary for service, and therefore all arriving customers are served. Indeed, (4.9) provides an intuitive partial explanation for the restriction $a < s$ (or, equivalently, $\rho < 1$), for $a > s$ would require the carried load (the mean number of busy servers) to exceed the number of servers, which is clearly impossible.

Exercises

16. **a.** Show that for every integer $s > a$,

$$C(s,a) = \frac{sB(s,a)}{s - a[1 - B(s,a)]},$$

and use this result to show that

$$C(s,a) > B(s,a) \quad \text{when} \quad a > 0.$$

b. Show that for every integer $s > a$,

$$C(s,a) = \frac{1}{1 + (s-a)[aB(s-1,a)]^{-1}},$$

where $B(0,a) = 1$.

c. Show that for every integer $s > a + 1$,

$$C(s,a) = \frac{1}{1 + \left(\frac{s-a}{a}\right)\frac{s - 1 - aC(s-1,a)}{(s-1-a)C(s-1,a)}}.$$

How might these formulas be computationally useful?

17. Review and reconsider Exercises 4 and 5 of Chapter 1. In the loss-delay system, what proportion of customers are denied service, what proportion wait for service, and what proportion are served immediately?

Let us briefly consider the single-server case without the requirements of Poisson input and exponential service times. We have argued that the carried load equals the offered load by virtue of the assumption that all arrivals get served (although some may have to wait), and this is true for arbitrary interarrival-time and service-time distribution functions. Also, by the definition (1.9), in the single-server case the carried load equals the proportion of time the server is busy, $a' = \sum_{j=1}^{\infty} P_j$ (where, in general, we may not know how to calculate the $\{P_j\}$). We conclude that the proportion of time that a single-server BCD system is busy is always equal to the offered load; that is, $1 - P_0 = a = \lambda\tau$. [As a check, note that $C(1,a) = a$.]

A quantity of some interest for the single-server queue is the *busy period*, defined as the length of time from the instant the (previously idle) server is seized until it next becomes idle and there is no one waiting in the queue. The calculation of this quantity is difficult in general, and we shall investigate it in some detail in Chapter 5. But with the simple tools now at hand, we can easily calculate the mean busy period for systems with Poisson input (and arbitrary distribution of service times). Denote by b the mean length of the busy period. Then the ratio of the mean busy period to the total cycle time (contiguous idle period and busy period) is $b/(\lambda^{-1} + b)$, where λ^{-1} is the mean interarrival time. But this ratio, which is independent of the form of the service-time distribution function, is simply the proportion of time that the server is busy; therefore,

$$1 - P_0 = \frac{b}{\lambda^{-1} + b}. \qquad (4.10)$$

Since we are assuming that all arrivals ultimately get served, it follows that

$$1 - P_0 = a = \lambda\tau. \qquad (4.11)$$

We conclude from (4.10) and (4.11) that the mean busy period b is given by

$$b = \frac{\tau}{1 - a} \qquad (0 \leq a < 1). \qquad (4.12)$$

Observe that b increases rapidly as a approaches unity. Thus, for example, when the server occupancy is 90%, the mean busy period is 10 service times; when the server occupancy is 99%. the mean busy period is 100 service times.

Exercise

18. Is the analysis leading to (4.12) valid for arrival processes other than the Poisson? Why?

We now turn to consideration of the Erlang delay system with ordered hunt: The servers are numbered $1, 2, \ldots, s$, and each arriving customer takes the lowest-numbered idle server. Blocked customers wait until served, and no customers can be waiting in the queue if there is an idle server. As usual, we assume Poisson input, exponential service times, and statistical equilibrium.

Recall that we previously considered the Erlang loss system with ordered hunt, and we obtained Equation (3.18) for the probability $\tilde{p}_j$ that the jth ordered server is busy at an arbitrary instant in equilibrium. (Equivalently, $\tilde{p}_j$ is the load carried by the jth ordered server.) The derivation of (3.18) depended critically on the assumption of blocked customers cleared. Even though we are no longer making this assumption, it turns out, surprisingly, that we can nevertheless solve the corresponding problem using only elementary considerations. (However, in contrast with the Erlang loss model, we must now require that service times be exponentially distributed.)

To calculate the probability p_j that the jth ordered server is busy (under the assumptions of the Erlang delay model) we partition time into two mutually exclusive and exhaustive sets of intervals, namely, those time intervals during which there is at least one customer waiting and those time intervals during which there are no customers waiting.

We let N be the total number of customers in the system at an arbitrary time (at which the system is assumed to be in statistical equilibrium) and as before, define $X_j = 0$ when the jth ordered server is idle and $X_j = 1$ when it is busy. Then, from the theorem of total probability we can write

$$p_j = P\{X_j = 1 | N \leq s\} P\{N \leq s\} + P\{X_j = 1 | N > s\} P\{N > s\}$$

$$(j = 1, 2, \ldots, s). \quad (4.13)$$

If no server can be idle when customers are waiting, then clearly $P\{X_j = 1 | N > s\} = 1$. We will now argue that, remarkably,

$$P\{X_j = 1 | N \leq s\} = \tilde{p}_j \quad (4.14)$$

where $\tilde{p}_j$, given by (3.18), is the load carried by the jth ordered server in an Erlang loss system.

To see the truth of (4.14), observe that because of the Markov property the behavior of the Erlang delay system during the time intervals when $N \leq s$ is unaffected by the behavior of the system when $N > s$; therefore, the load carried by the jth ordered server during the time intervals when no customers are waiting is the same as that for the corresponding Erlang loss system. This clever intuitive argument was first advanced by Vaulot [1925]. We shall give a generalization and a more rigorous proof in Chapter 4.

Now, it follows easily from (4.4) that $\sum_{j=s+1}^{\infty} P_j = (a/s)C(s, a)$; hence

$$P\{N > s\} = \frac{a}{s} C(s, a). \quad (4.15)$$

Therefore, we conclude from Equations (4.13)–(4.15) and (3.18) that the load p_j carried by the jth ordered server in an s-server Erlang delay system

The Erlang Delay System

with ordered hunt is given by

$$p_j = a[B(j-1,a) - B(j,a)][1 - \rho C(s,a)] + \rho C(s,a), \qquad (4.16)$$

where $\rho = a/s$ is the average load carried per server (the server occupancy).

Exercises

19. Consider again the premise of Exercise 13. Now, however, assume that (a) sufficient waiting positions are provided so that no calls will be lost and (b) the total number of trunks in the group is 4. What is the most economical division between measured-rate and flat-rate trunks, and what is the associated hourly cost?

20. Prove that for an s-server Erlang delay system, the statistical-equilibrium probability that j customers are in service, given that $j \leqslant s$, is the same as the corresponding unconditional probability for the Erlang loss system. Generalize this theorem so that it applies to general birth-and-death queueing models.

21. Reconsider Exercise 14 with "Erlang loss system" replaced by "Erlang delay system."

We now turn to the analysis of waiting times in the equilibrium Erlang delay system in which the queue discipline is service in order of arrival. In particular, we will calculate, for each $t \geqslant 0$, the probability $P\{W > t\}$ that an arbitrary customer will wait in excess of t before entering service. To this end, let us write

$$P\{W > t\} = P\{W > 0\} P\{W > t | W > 0\}. \qquad (4.17)$$

We shall call a queue discipline *nonbiased*[*] if it is such that whenever a customer is selected from the queue for service, his service time is a random variable with the same distribution function as an arbitrary service time. For example, nonbiased queue disciplines include service in order of arrival (first come, first served), service in random order, and service in reverse order of arrival (last come, first served). An example of a queue discipline that is not nonbiased is one sometimes called shortest-processing-time-first, in which the customer selected for service when a server becomes available is that customer whose service time is least among all those waiting (where, of course, it is assumed that these values can be ascertained).

---

[*]In the first edition, we used the word *nonscheduled* instead of *nonbiased*; the latter seems to be a more descriptive term.

If two queueing models are the same except for their queue disciplines, and if their queue disciplines are nonbiased, then they will have the same distribution of state probabilities. Hence, all other things being equal, the probability $P\{W>0\}$ is the same regardless of the (nonbiased) queue discipline. In particular, for the Erlang delay system in equilibrium we have, for any nonbiased queue discipline, from (4.7),

$$P\{W>0\} = C(s,a) = \frac{a^s}{s!(1-a/s)} P_0. \qquad (4.18)$$

It follows that to find $P\{W>t\}$ for the Erlang delay system with order-of-arrival service, it is sufficient to calculate the conditional probability $P\{W>t|W>0\}$.

Let Q be the number of customers waiting for service at the instant of arrival of an arbitrary customer (the test customer). Then, it follows from the theorem of total probability that (for any queueing system)

$$P\{W>t|W>0\} = \sum_{j=0}^{\infty} P\{W>t|W>0, Q=j\} P\{Q=j|W>0\}. \qquad (4.19)$$

Recall from Equation (5.55) of Chapter 2 that for exponential service times,

$$P\{W>t|W>0, Q=j\} = \sum_{i=0}^{j} \frac{(s\mu t)^i}{i!} e^{-s\mu t}. \qquad (4.20)$$

Now, from the definition of conditional probability we can write

$$P\{Q=j|W>0\} = \frac{P\{Q=j, W>0\}}{P\{W>0\}}. \qquad (4.21)$$

Note that although Q and W are defined from the viewpoint of the arriving customer, we can (because the input process is Poisson) adopt the viewpoint of the outside observer. Thus,

$$P\{Q=j, W>0\} = P_{s+j} = \frac{a^s}{s!}\left(\frac{a}{s}\right)^j P_0, \qquad (4.22)$$

where the second equality follows from (4.4). Substitution of (4.22) and (4.18) into (4.21) yields the geometric distribution

$$P\{Q=j|W>0\} = (1-\rho)\rho^j \qquad (j=0,1,\dots), \qquad (4.23)$$

where $\rho = a/s$ is the server occupancy. Now substitute (4.20) and (4.23) into (4.19) and compare with Equations (5.57) and (5.58) of Chapter 2. We

The Erlang Delay System

conclude that

$$P\{W>t|W>0\} = e^{-(1-\rho)s\mu t}; \quad (4.24)$$

that is, for the equilibrium Erlang delay system with service in order of arrival, the conditional (given that the customer is not served immediately) waiting-time distribution function is the negative-exponential.

To summarize, we have argued that (1) the waiting time for an arbitrary customer who finds all servers busy is the sum $X_1 + \cdots + X_{j+1}$ of the times that he and each of the j customers ahead of him spend at the head of the queue, that these times are independent exponential random variables with mean $(s\mu)^{-1}$, and hence their sum has the $(j+1)$-phase Erlangian distribution (4.20). This is the analysis leading to Equation (5.55) of Chapter 2. The next step in our argument is that (2) the number of customers that the (assumed blocked) test customer finds ahead of him in the queue has the geometric distribution (4.23). Thus, the test customer's waiting time is a sum of independent, identical, exponential random variables, where the number of variables in the sum is geometric; therefore, as shown by Equation (5.58) of Chapter 2, the test customer's waiting time is exponentially distributed.

Equation (4.24) is particularly suited to graphical presentation; for each value of ρ, Equation (4.24) gives a straight line when $P\{W>t|W>0\}$ is plotted against $s\mu t$ on semilog paper, as in Figure A.5 of Appendix A. The unconditional waiting-time probability $P\{W>t\}$ is easily obtained from Equations (4.17), (4.18), and (4.24):

$$P\{W>t\} = C(s,a)e^{-(1-\rho)s\mu t}. \quad (4.25)$$

Numerically, we can easily calculate $P\{W>t\}$ by evaluating the factor $e^{-(1-\rho)s\mu t}$ from its graph in Appendix A, evaluating $C(s,a)$ from its graph in Appendix A or from one of the tabulations of the Erlang delay formula, and forming the required product.

Since the conditional waiting time for blocked customers is exponentially distributed according to (4.24), it follows that the conditional mean waiting time $E(W|W>0)$ (the mean wait suffered by those customers who are blocked) is given by

$$E(W|W>0) = \frac{1}{(1-\rho)s\mu}, \quad (4.26)$$

and thus the overall mean waiting time $E(W)$ is given by

$$E(W) = \frac{C(s,a)}{(1-\rho)s\mu}. \quad (4.27)$$

These formulas are tabulated or plotted in Dietrich et al. [1966] and Descloux [1962].

It is an important fact, which we shall prove later, that the formulas (4.26) and (4.27) are valid for an Erlang delay system with any nonbiased queue discipline, even though the derivation given here assumes service in order of arrival.

Finally, a note on terminology: Some queueing theorists call the waiting time that would be experienced by an outside observer, were he in fact to join the system as a customer, the *virtual waiting time*; and they call the waiting time experienced by an arriving customer the *actual waiting time*. We have shown that the arriving customer's distribution $\{\Pi_j\}$ and the outside observer's distribution $\{P_j\}$ are equal for systems with Poisson arrivals, from which it follows that the actual waiting times and the virtual waiting times have the same distribution for systems with Poisson arrivals. In particular, it follows that Equation (4.25) describes both actual and virtual waiting times in the equilibrium $M/M/s$ queue.

Exercises

22. Repeat Exercise 5 with "Erlang loss system" replaced by "Erlang delay system with order-of-arrival service," and with the additional requirement that the entrepreneur must pay $10 for each customer whose wait for service exceeds $\frac{1}{2}$ hour. If the entrepreneur has control over the order of selection from the queue, should he select in order of arrival?

23. Consider a 10-server Erlang delay system that handles an offered load of 6 erlangs. If the arrival rate increases by $\frac{1}{3}$, what is the corresponding increase in

a. the fraction of customers who must wait for service, and

b. the mean waiting time for those customers who must wait for service?

Repeat the exercise for the case where, instead, the mean service time increases by a factor of $\frac{1}{3}$.

24. In an Erlang delay system with service in order of arrival, what proportion of those customers who are blocked wait longer than the average waiting time for blocked customers?

25. Consider a telephone system in which the central office equipment that provides dial tone (the dial-tone machines) can be modeled as an Erlang delay system with service in order of arrival. (That is, the dial-tone machines are the servers, and a subscriber who goes "off hook" to place a call is a customer.) If a subscriber goes off hook and still has not received dial tone after 30 seconds, what should she do?

26. Show that in the Erlang delay system with order-of-arrival service, the customer waiting-time distribution function has variance

$$V(W) = \frac{1 - [1 - C(s,a)]^2}{(s\mu)^2(1-\rho)^2}.$$

27. Let W be the waiting time and T the sojourn time (waiting time plus service time) of an arbitrary customer. Show that in the single-server Erlang delay system with service in order of arrival,

$$P\{T>t\} = P\{W>t|W>0\}.$$

28. **a.** Consider an Erlang delay system, and denote by L the mean queue length, W the mean waiting time for service, and λ the arrival rate. Verify the equation $L = \lambda W$. (This well-known equation, often called *Little's theorem*, is true under very general conditions and is often written using the present symbols. See Section 5.2.)

    b. Redefine L to be the mean number of customers in the system (waiting and in service) and W to be the mean time spent in the system. (W is the mean sojourn time.) Verify $L = \lambda W$ using these redefinitions.

29. Prove that in an Erlang delay system with order-of-arrival service, the equilibrium conditional probability that a blocked customer will still be waiting in the queue when the next customer arrives is equal to the server occupancy or utilization factor ρ. (This is a special case of a more general result. See Exercise 33 of Chapter 5.)

30. Let N be the number of customers found by an arrival at an s-server Erlang delay system. Using the fact that the events $\{Q=j,\ W>0\}$ and $\{N=s+j\}$ are equivalent, show that Equation (4.19) can be written (for any nonbiased queue discipline)

$$P\{W>t|W>0\} = (1-\rho)\sum_{j=0}^{\infty} \rho^j P\{W>t|N=s+j\},$$

where ρ is the server occupancy.

31. Consider the differential-difference equations

$$\frac{d}{dt} F_j(t) = c F_{j-1}(t) - c F_j(t)$$

$$[t \geq 0;\ j=0,1,\ldots;\ F_{-1}(t)=0], \qquad (1)$$

where c is an arbitrary constant. Define

$$F(x,t) = \sum_{j=0}^{\infty} F_j(t) x^j. \qquad (2)$$

a. Show that $F(x,t)$ satisfies

$$\frac{\partial}{\partial t} F(x,t) = c(x-1) F(x,t), \qquad (3)$$

which has solution

$$F(x,t) = F(x,0) e^{-(1-x)ct}. \qquad (4)$$

b. Let $N(t)$ be the number of events occurring in $(0, t)$ according to a Poisson process with rate λ. Then, according to Equation (2.5) of Chapter 2, we can set $F_j(t) = P\{N(t) = j\} = P_j(t)$ and $c = \lambda$ in Equation (1). Show that Equation (4) yields the Poisson distribution,

$$P_j(t) = \frac{(\lambda t)^j}{j!} e^{-\lambda t} \quad (j = 0, 1, \ldots).$$

c. In an Erlang delay system with s servers, mean service time μ^{-1}, and service in order of arrival, let N be the number of customers waiting or in service when an arbitrary customer (the test customer) arrives, let W be the test customer's waiting time, and define $W_j(t) = P\{W > t | N = s + j\}$. Show that

$$W_j(h + t) = s\mu h W_{j-1}(t) + (1 - s\mu h) W_j(t) + o(h) \quad (h \to 0),$$

and therefore,

$$\frac{d}{dt} W_j(t) = s\mu W_{j-1}(t) - s\mu W_j(t)$$

$$[t \geq 0; \quad j = 0, 1, \ldots; \quad W_{-1}(t) = 0]. \tag{5}$$

d. Define

$$W(x, t) = \sum_{j=0}^{\infty} W_j(t) x^j, \tag{6}$$

and show by comparison with Equation (4) that

$$W(x, t) = W(x, 0) e^{-(1-x)s\mu t}. \tag{7}$$

e. Show that

$$W(x, 0) = \frac{1}{1 - x}. \tag{8}$$

f. Show that the result of Exercise 30 implies

$$P\{W > t | W > 0\} = (1 - \rho) W(\rho, t). \tag{9}$$

g. Deduce from Equations (7)–(9) that

$$P\{W > t | W > 0\} = e^{-(1-\rho)s\mu t}. \tag{10}$$

Compare with Equation (4.24).

h. Show directly from Equations (6)–(8) that

$$W_j(t) = \sum_{i=0}^{j} \frac{(s\mu t)^i}{i!} e^{-s\mu t}. \tag{11}$$

Compare Equation (11) with Equation (4.20).

The Erlang Delay System

32. *Service in random order.* Consider the equilibrium s-server Erlang delay system with service in random order. Let N be the number of customers waiting or in service when an arbitrary customer (the test customer) arrives, let W be the test customer's waiting time, and define $W_j(t) = P\{W > t | N = s+j\}$.

a. Show that

$$W_j(h+t) = \lambda h W_{j+1}(t) + \frac{j}{j+1} s\mu h W_{j-1}(t)$$

$$+ [1 - (\lambda + s\mu)h] W_j(t) + o(h)$$

$$[h \to 0;\ j = 0, 1 \ldots;\ W_{-1}(t) = 0], \tag{1}$$

and therefore,

$$\frac{d}{dt} W_j(t) = \lambda W_{j+1}(t) + \frac{j}{j+1} s\mu W_{j-1}(t) - (\lambda + s\mu) W_j(t)$$

$$[j = 0, 1, \ldots;\ W_{-1}(t) = 0], \tag{2}$$

where

$$W_j(0) = 1 \quad (j = 0, 1, \ldots).$$

b. Define

$$W_j^{(\nu)} = \left(\frac{d^\nu}{dt^\nu} W_j(t)\right)_{t=0} \quad (j = 0, 1, \ldots;\ \nu = 0, 1, \ldots;\ W_j^{(0)} = 1), \tag{3}$$

and assume that $W_j(t)$ has the Maclaurin series representation

$$W_j(t) = \sum_{\nu=0}^{\infty} \frac{t^\nu}{\nu!} W_j^{(\nu)} \quad (j = 0, 1, \ldots). \tag{4}$$

Using the result of Exercise 30, show that

$$P\{W > t | W > 0\} = 1 + (1-\rho) \sum_{\nu=1}^{\infty} \frac{t^\nu}{\nu!} \sum_{j=0}^{\infty} \rho^j W_j^{(\nu)}. \tag{5}$$

c. Show that the right-hand side of (5) can be evaluated to give

$$P\{W > t | W > 0\} = 1 - s\mu t \frac{1-\rho}{\rho} \ln \frac{1}{1-\rho}$$

$$+ \frac{(s\mu t)^2}{2!}(1-\rho)\left[2 - \frac{1-\rho}{\rho} \ln \frac{1}{1-\rho}\right] - + \cdots. \tag{6}$$

For further discussion of this method see Riordan [1953]. Pollaczek [1946] gives a closed-form solution for the probability $P\{W > t | W > 0\}$. Summaries of Pollaczek's and other studies are given in Riordan [1962] and Syski [1960].

33. Let $B(t)$ be the distribution function of the busy period in the single-server Erlang delay system, with mean $b = \int_0^\infty t\, dB(t)$; and let W be the equilibrium waiting time of an arbitrary customer. Convince yourself that when service is in reverse order of arrival, then $P\{W \leq t | W > 0\} = B(t)$. Conclude, therefore, that $E(W|W>0) = b$, where b is given by (4.12). Note that this agrees with (4.26) when $s=1$, thereby providing an illustration of the theorem that the mean waiting time does not depend on the (nonbiased) order of service.

3.5. Quasirandom Input

In all the birth-and-death queueing models considered so far, we have assumed that requests for service occur according to a Poisson process with rate λ. With Poisson input, the probability that a request for service will occur in any interval $(t, t+h)$ is independent of the state of the system at time t. This important property is at the heart of the proof in Section 3.2 of the equality of the outside observer's distribution $\{P_j\}$ and the arriving customer's distribution $\{\Pi_j\}$ for systems with Poisson input.

Consider now a system in which the requests for service are generated by a finite number of sources. In such a system the probability of an arrival in an interval $(t, t+h)$ will not be independent of the system state at time t, since it will, in general, depend directly on the number of sources idle (and therefore available to generate new requests) at time t. Consequently, the outside observer's and the arriving customer's distributions are not equal.

These properties are illustrated by the finite-source system with an equal number of sources and servers. When all the servers are occupied no new requests can occur, since there are no idle sources to generate new requests. Therefore the probability of blocking is zero, whereas the proportion of time all servers are busy can take any value between zero and one.

The particular kind of finite-source input we shall consider is often called quasirandom input (as opposed to Poisson or completely random input). We say that a finite number n of (identical) sources generate *quasirandom input* if the probability that any particular source generates a request for service in any interval $(t, t+h)$ is $\gamma h + o(h)$ as $h \to 0$ if the source is idle at time t, and zero if the source is not idle (waiting or being served) at time t, independently of the states of any other sources.

It follows from this definition that if a particular source is idle at time t, the distribution of time from t until the source next generates a request for service is exponential with mean γ^{-1}; that is, with probability $e^{-\gamma x}$ the source will not originate a request for service in $(t, t+x)$.

If the number of sources idle at time t is n, the probability that exactly j of them generate requests for service in $(t, t+x)$ is

$$\binom{n}{j}(1-e^{-\gamma x})^j (e^{-\gamma x})^{n-j}.$$

Now let $n \to \infty$ and $\gamma \to 0$ in such a way that $n\gamma = \lambda$. Then, it can be shown that

$$\lim_{\substack{n \to \infty \\ \gamma \to 0 \\ n\gamma = \lambda}} \binom{n}{j} (1 - e^{-\gamma x})^j (e^{-\gamma x})^{n-j} = \frac{(\lambda x)^j}{j!} e^{-\lambda x}. \tag{5.1}$$

Thus, assuming that each idle source can generate at most one request in $(t, t+x)$, we have shown that the distribution of the number of requests generated is, in the limit, Poisson. It should be intuitively clear that because $\gamma \to 0$ the restriction of at most one request per source can be removed without affecting the conclusion. Then, more formally, we conclude:

If the number n of independent sources generating quasirandom input increases to infinity and the request rate γ per idle source decreases to zero in such a way that the overall idle-source request rate $n\gamma$ remains constant ($n\gamma = \lambda$), then in the limit the input process is Poisson with rate λ.

In the case of Poisson input the number of sources, or potential customers, is infinite (and is therefore often called *infinite-source input*). However, the sources affect the system only when they make requests for service, that is, when they become customers. Therefore, with Poisson input, there is no need to distinguish between sources and customers. This is not true, however, in the case of finite-source input. With finite-source input, a source assumes the role of a customer when it places a request for service, and it remains a customer as long as it is either waiting for service or in service. As soon as it becomes eligible to generate a new request for service, it again assumes the role of a source. We shall refer to a source as a customer only when we wish to stress that it is making a request for service. Any source can become a customer more than once, so the number of blocked customers is the number of requests (generated by sources) that are blocked. To avoid this confusion, we shall say that a source generates a request rather than that a source becomes a customer. The meanings of the words customer, source, and request should be clear from the context in which they are used.

Let us consider a system with n sources, s servers, exponential service times, and quasirandom input, where each idle source generates requests for service at rate γ. If at any time t the number of idle sources is $n-j$, then the probability that a request for service will occur in $(t, t+h)$ is $(n-j)\gamma h + o(h)$ as $h \to 0$. Thus queueing models with quasirandom input and exponential service times can be studied in the framework of the birth-and-death process, in a manner analogous to that used for the corresponding systems with Poisson input; we simply have a different specification of the birth coefficients $\{\lambda_j\}$.

As with the infinite-source models considered previously, it is often useful to think in terms of carried and offered loads. In Section 3.1 we defined the carried load a', for any input process, by Equation (1.9). We can also retain the definition of offered load a, given by Equation (1.11) as the product of the arrival rate (the mean number of requests per unit time) λ and the mean service time τ. With Poisson input, the arrival rate λ is independent of the system to which it is offered and thus provides a simple characterization of the amount of traffic that is to be handled by any system that may be provided to handle it. The same is not true of quasirandom input, because the instantaneous arrival rate λ_j depends on the state of the system, and the proportion of time the system spends in each state depends on the number of servers and the disposition of blocked customers. Therefore, although the traffic can be characterized by the number of sources n and the idle request γ per source, independently of the system to which it is offered, the mean number of requests per unit time λ (and hence also the offered load a) that a system receives depends directly on the characteristics of the particular system that receives it.

For each of the quasirandom input models that we consider, we shall derive a relationship between the *offered load per idle source* $\hat{a} = \gamma/\mu$ and the total offered load a. Unfortunately these relationships involve state probabilities that depend on the number of servers. For this reason, it is sometimes useful to consider the *intended offered load* a^*, defined as the load that the sources would offer to the system if there were enough servers so that no blocking could occur (that is, $s = n$). Then, as we shall show later, the intended offered load a^* is given, for any n-source system, by

$$a^* = n \frac{\hat{a}}{1+\hat{a}}, \qquad (5.2)$$

where $\hat{a} = \gamma/\mu$ is the offered load per idle source.

Finally, let us note for future reference that for systems in which the queue discipline is blocked customers cleared or blocked customers delayed, the carried load a' can be calculated from the formula

$$a' = \frac{1}{\mu} \sum_{j=0}^{n-1} \lambda_j P_j. \qquad (5.3)$$

To prove (5.3), we sum both sides of the basic "rate up = rate down" equations (1.2), where we take

$$\mu_j = \begin{cases} j\mu & (j=0,1,\ldots,s), \\ s\mu & (j=s+1,s+2,\ldots,n). \end{cases} \qquad (5.4)$$

[The death rates defined by (5.4) describe both the delay model and the loss model. In the latter case, the values for $j > s$ are irrelevant because $\lambda_s = 0$, which implies that $P_j = 0$ for $j > s$.] Using the fact that, for n-source systems with quasirandom input, $\lambda_n = 0$ and $P_j = 0$ for $j > n$, we obtain

$$\sum_{j=0}^{n-1} \lambda_j P_j = \mu \sum_{j=1}^{s-1} j P_j + \mu s \sum_{j=s}^{n} P_j. \tag{5.5}$$

Comparison of (5.5) with (1.9) proves (5.3).

3.6. Equality of the Arriving Customer's n-Source Distribution and the Outside Observer's $(n-1)$-Source Distribution for Birth-and-Death Systems with Quasirandom Input

We have already shown by example that in systems with finite-source input, in contrast to the infinite-source case, the outside observer's distribution $\{P_j\}$ and the arriving customer's distribution $\{\Pi_j\}$ are, in general, unequal. In the case of the birth-and-death queueing models with quasirandom input, we can calculate $\{P_j\}$ directly from the birth-and-death probabilities (1.1). But $\{\Pi_j\}$ is the distribution of direct relevance to one concerned with the quality of service provided by the servers to the customers. Hopefully, the two distributions will be simply related. Happily, they are.

For example, consider the simple system composed of one source and one server. Suppose that the source generates requests at rate γ when idle and rate zero when not idle, and let the mean service time be τ. Let $\gamma\tau = \hat{a}$; $\hat{a}$ is the load offered by the source when idle. The server alternates between busy and idle states, with each busy period of mean length τ. Consider a single cycle composed of an idle period and an adjacent busy period. The cycle has mean length $\gamma^{-1} + \tau$, and thus the ratio of the mean busy period to the mean cycle length is

$$\frac{\tau}{\gamma^{-1} + \tau} = \frac{\hat{a}}{1 + \hat{a}}.$$

As we did with a similar example for Poisson input, we interpret this ratio as the proportion of time (or probability) that the server is busy when the system is in equilibrium. This ratio is not the proportion of requests blocked, which we have already argued is zero when (as in this example) the number of sources does not exceed the number of servers. Let $P_j[n]$ and $\Pi_j[n]$ be the n-source outside observer's and arriving customer's state probabilities, respectively. We then have $P_1[1] = \hat{a}/(1 + \hat{a})$ and $P_0[1] = 1/(1 + \hat{a})$ for the outside observer's distribution, in contrast with the arriving customer's distribution, $\Pi_1[1] = 0$ and $\Pi_0[1] = 1$.

Consider now the same system with the addition of an identical source. Since blocking can occur in this two-source, one-server system, we must consider the disposition of blocked requests. Let us assume that blocked customers are cleared. (That is, any source that finds all servers busy when placing a request returns immediately to its previous state as an idle source.) The server again alternates between busy and idle states. The busy period still has mean length τ. But since there are now two sources, each bidding for service at rate γ when idle, the overall bid rate when both sources are idle is 2γ, and the idle portion of the cycle has mean length $(2\gamma)^{-1}$. Forming the ratio of the mean busy period to the mean cycle length, we obtain

$$\frac{\tau}{(2\gamma)^{-1}+\tau} = \frac{2\hat{a}}{1+2\hat{a}}.$$

Thus the outside observer's distribution is $P_1[2] = 2\hat{a}/(1+2\hat{a})$ and $P_0[2] = 1/(1+2\hat{a})$.

Blocking can occur only when the server is occupied by one of the sources. The number of requests blocked during a cycle is the number generated by the idle source during the service time of the busy source. Since we have assumed blocked customers cleared, the mean number of requests generated by a source during the service time of the other source is $\gamma\tau = \hat{a}$. On the other hand, the total number of requests generated during a cycle is simply the sum of the number generated during the busy period plus the number generated during the idle period. The latter number is 1, namely, the arrival who finds the server idle, ending the idle period and starting the busy period. Forming the ratio of the mean number of requests blocked to the mean number generated per cycle, we obtain $\hat{a}/(1+\hat{a})$. That is, $\Pi_1[2] = \hat{a}/(1+\hat{a})$ and $\Pi_0[2] = 1/(1+\hat{a})$.

Comparison with our calculations for the one-source case shows that $\Pi_0[2] = P_0[1]$ and $\Pi_1[2] = P_1[1]$. In other words, the state distribution seen by a source in the two-source system when placing a request is the same as the state distribution the source would see if it placed no requests but instead acted the part of an outside observer of the one-source system.

It is true in general that for equilibrium n-source birth-and-death systems with quasirandom input, the state distribution seen by a source when placing a request is the same as if that source were not contributing load to the servers, but instead were only observing the corresponding $(n-1)$-source system continuously or at random instants. For many years the $(n-1)$-source outside observer's viewpoint was considered only an approximation to the n-source arriving customer's viewpoint, since (it was argued) no account is taken of the fact that the particular source places a load on the system, interacting with other sources, and thus affecting the state distribution. This argument neglects to mention that the requests generated by any particular source do not follow a Poisson process (why?); apparently these two effects cancel each other exactly.

We now give a formal proof of the general theorem: In any equilibrium n-source birth-and-death queueing system with quasirandom input, the

Quasirandom Input: $\Pi_j[n] = P_j[n-1]$

arriving customer's distribution is the same as the outside observer's distribution for the corresponding $(n-1)$-source system.

We consider a system with n sources, each source originating requests at rate γ when idle and rate 0 otherwise (quasirandom input). Then the request rate when j sources are busy (in service or waiting for service) is

$$\lambda_j = (n-j)\gamma \qquad (j=0,1,\ldots,n). \tag{6.1}$$

To calculate the arriving customer's distribution $\{\Pi_j\}$, substitute (6.1) into (2.6); this yields

$$\Pi_j = \frac{(n-j)P_j}{\sum_k (n-k)P_k}. \tag{6.2}$$

In order to emphasize the dependence on the number n of sources, we write $P_j = P_j[n]$ and $\Pi_j = \Pi_j[n]$. Then (6.2) becomes

$$\Pi_j[n] = \frac{(n-j)P_j[n]}{\sum_{k=0}^{n-1} (n-k)P_k[n]} \qquad (j=0,1,\ldots,n-1). \tag{6.3}$$

Now, according to Equations (1.1) and (6.1), the outside observer's distribution can be written

$$P_j[n] = \frac{n(n-1)\cdots(n-j+1)\gamma^j}{\mu_1 \mu_2 \cdots \mu_j} P_0[n] \qquad (j=1,2,\ldots,n) \tag{6.4}$$

and

$$P_0[n] = \left(1 + \sum_{k=1}^{n} \frac{n(n-1)\cdots(n-k+1)\gamma^k}{\mu_1 \mu_2 \cdots \mu_k}\right)^{-1}. \tag{6.5}$$

Substitution of (6.4) into (6.3) yields

$$\Pi_j[n] = \frac{\dfrac{n(n-1)\cdots(n-j+1)(n-j)\gamma^j}{\mu_1 \mu_2 \cdots \mu_j} P_0[n]}{nP_0[n] + \sum_{k=1}^{n-1} \dfrac{n(n-1)\cdots(n-k+1)(n-k)\gamma^k}{\mu_1 \mu_2 \cdots \mu_k} P_0[n]}$$

$$(j=1,2,\ldots,n-1). \tag{6.6}$$

After cancellation of the factor $nP_0[n]$ in (6.6), we have

$$\Pi_j[n] = \frac{\dfrac{(n-1)\cdots(n-j)\gamma^j}{\mu_1\mu_2\cdots\mu_j}}{1 + \sum_{k=1}^{n-1} \dfrac{(n-1)\cdots(n-k)\gamma^k}{\mu_1\mu_2\cdots\mu_k}} \qquad (j=1,2,\ldots,n-1). \quad (6.7)$$

Comparison of Equation (6.7) with (6.4) and (6.5) shows that $\Pi_j[n] = P_j[n-1]$ for $j=1,2,\ldots,n-1$. Since we must have

$$\sum_{j=0}^{n-1} P_j[n-1] = \sum_{j=0}^{n-1} \Pi_j[n] = 1,$$

we conclude that

$$\Pi_j[n] = P_j[n-1] \qquad (j=0,1,\ldots,n-1). \quad (6.8)$$

The theorem is proved. For further discussion, see Exercises 1 and 5 of Chapter 4 and Exercise 1 of Chapter 5.

Taking limits through (6.8) as $n\to\infty$ and $\gamma\to 0$ while holding $n\gamma=\lambda$, the dependence on n vanishes and, since quasirandom input becomes Poisson input in the limit, we again illustrate the equality of the arriving customer's distribution and the outside observer's distribution for systems with Poisson input.

3.7. Quasirandom Input, s Servers, Blocked Customers Cleared: The Engset Formula

We assume that the number of sources is $n \geqslant s$ and that each source independently generates requests at rate γ when idle and rate zero otherwise. Since blocked customers are assumed cleared, a new request will effect a change of state only when at least one server is idle. Hence, we have birth rates

$$\lambda_j = \begin{cases} (n-j)\gamma & (j=0,1,\ldots,s-1), \\ 0 & (j=s). \end{cases} \quad (7.1)$$

Let us assume for the present that service times are exponential. Then we have death rates

$$\mu_j = j\mu \qquad (j=1,2,\ldots,s), \quad (7.2)$$

where μ^{-1} is the mean service time. As with the Erlang loss model, the assumption of exponential service times can be shown to be superfluous.

The Engset Formula

The rates (7.1) and (7.2), in conjunction with (1.1), give for the statistical-equilibrium probability of j busy servers (where $\hat{a} = \gamma/\mu$)

$$P_j[n] = \frac{\binom{n}{j}\hat{a}^j}{\sum_{k=0}^{s}\binom{n}{k}\hat{a}^k} \qquad (j = 0, 1, \ldots, s) \tag{7.3}$$

and $P_j[n] = 0$ for $j > s$. If the substitution

$$\hat{a} = \frac{p}{1-p} \tag{7.4}$$

is made in (7.3), then, after multiplication of numerator and denominator by $(1-p)^n$, Equation (7.3) takes the form of the *truncated binomial distribution*

$$P_j[n] = \frac{\binom{n}{j}p^j(1-p)^{n-j}}{\sum_{k=0}^{s}\binom{n}{k}p^k(1-p)^{n-k}} \qquad (j = 0, 1, \ldots, s). \tag{7.5}$$

Solving (7.4) for p gives

$$p = \frac{\hat{a}}{1+\hat{a}}, \tag{7.6}$$

which is immediately recognized as $P_1[1]$. In other words, p is the probability that an arbitrary source would be busy if there were no interaction among the sources. If $n = s$, in which case each source assumes busy and idle states unaffected by the states of the other sources, then the denominator on the right-hand side of (7.5) sums to unity, and the distribution of the number of customers in service is the simple binomial. In this case the mean number of busy servers is np; thus, the intended offered load a^*, defined as the load that would be offered (and carried) if $s = n$, is given by (5.2), as asserted.

Expression (7.3)—or (7.5)—gives the proportion of time the system spends in each state E_j. The probability $\Pi_j[n]$ that one of the n sources finds the system in state E_j when placing a request is obtained by replacing n by $n-1$ in (7.3):

$$\Pi_j[n] = \frac{\binom{n-1}{j}\hat{a}^j}{\sum_{k=0}^{s}\binom{n-1}{k}\hat{a}^k} \qquad (j = 0, 1, \ldots, s). \tag{7.7}$$

In particular, the proportion of requests that find all s servers busy is $\Pi_s[n]$; the right-hand side of (7.7) with $j=s$ is often called the *Engset formula*.

Exercise

34. Show that
$$\lim_{\substack{n\to\infty \\ \gamma\to 0 \\ n\gamma=\lambda}} \Pi_j[n] = P_j,$$

where $\Pi_j[n]$ is given by (7.7) and P_j is given by (3.3).

To calculate the carried load a' we use Equation (5.3). Substitution of (7.3) into (5.3) gives, after some calculation,

$$a' = a^*\left[1 - \left(1 - \frac{s}{n}\right)P_s[n]\right], \qquad (7.8)$$

where a^* is the intended offered load, defined by (5.2). In view of the fact that the carried load a' is that portion of the offered load a that is not cleared, we have

$$a' = a(1 - \Pi_s[n]). \qquad (7.9)$$

Equations (7.8) and (7.9) give the following relationship between the offered load per idle source $\hat{a} = \gamma/\mu$ and the total offered load a:

$$\hat{a} = \frac{a(1 - \Pi_s[n])}{n(1 - P_s[n]) - a(1 - \Pi_s[n]) + sP_s[n]}. \qquad (7.10)$$

Thus we see that the offered load generated by a finite number of sources cannot be specified without first calculating its effect on the particular system to which it is offered.

Recall from Section 3.5 that we defined the intended offered load a^* as the load that the sources would offer to the system if there were as many servers as sources (in which case no blocking would occur, and the carried and offered loads would be equal). Clearly, when no blocking occurs the queue discipline is irrelevant; therefore, we can make the calculation for the model at hand (blocked customers cleared) and apply it to all systems that differ from the present one only in the queue discipline. Thus, we have established (5.2) in general.

The point to be made here is simply that the offered load a is a most useful and simple measure of demand for Poisson traffic, but not nearly so useful or simple for quasirandom traffic. When blocked customers are cleared, the offered load a is bounded from below by the intended offered load a^* (why?); thus Equation (5.2) provides a low-side estimate of the offered load when the probability of blocking is small.

The results of this section were calculated here on the assumption of exponential service times. But, as with Poisson input, when blocked customers are cleared the birth-and-death equations $(n-j)\gamma P_j = (j+1)\mu P_{j+1} (j=0,1,\ldots,s-1)$ are valid for any service-time distribution function with finite mean μ^{-1}; and hence (7.7) also remains valid. (Note that in the heuristic discussion of single-server systems with one and two sources, no assumption was made about the form of the service-time distribution function.) Cohen [1957] (see also pp. 278–284 of Syski [1960]) has derived a "generalized Engset formula," which shows that for finite-source BCC systems with nonidentical sources, the state probabilities depend only on the mean service times and mean idle-source interarrival times for each source, and not on the forms of their respective distribution functions.

Exercise

35. Four sources share access to t o servers. Assume that (1) the input generated by the sources can be described as quasirandom, with each source characterized by a mean time between requests when idle of $\gamma^{-1}=27$ minutes and a mean service time $\mu^{-1}=3$ minutes, and (2) blocked customers are cleared. If an additional identical source is given access to the servers, by what percentage will the probability of blocking increase? If a statistician were observing the system and tabulating the total number of requests for service per hour generated by the sources, what values would he obtain (when $n=4$ and $n=5$)?

3.8. Quasirandom Input, s Servers with Exponential Service Times, Blocked Customers Delayed

We now turn to the finite-source analogue of the Erlang delay model: We assume that n sources generate quasirandom input to a group of s exponential servers, and that blocked customers wait as long as necessary for service. This model has found wide application. For example, Scherr [1967] has used it to predict the performance of a computer system composed of a central processing unit (server) and a number of teletypewriter terminals (sources), using either batch processing or time sharing. Teletraffic engineers have used it to describe systems in which a small number of lines (sources) in a central office have access to a common outgoing trunk group (servers). Another important example is provided by the so-called *machine interference* or *repairman* models. Here the sources are machines that can break down, thereby requiring the attention of a repairman (server). References, numerical examples, and related models are given in Cox and Smith [1961], Feller [1968], and Page [1972].

In this model we have quasirandom input, where every request for service effects a change of state; hence

$$\lambda_j = (n-j)\gamma \qquad (j=0,1,\ldots,n), \tag{8.1}$$

where n is the number of sources and γ is the request rate for an idle source. The rate μ_j at which busy sources become idle is

$$\mu_j = \begin{cases} j\mu & (j=1,2,\ldots,s), \\ s\mu & (j=s+1,s+2,\ldots,n), \end{cases} \tag{8.2}$$

where μ^{-1} is the mean service time.

With $\hat{a} = \gamma/\mu$ the rates (8.1) and (8.2), in conjunction with Equation (1.1), give for the statistical-equilibrium probability of j customers present

$$P_j[n] = \begin{cases} \binom{n}{j} \hat{a}^j P_0[n] & (j=1,2,\ldots,s-1), \\ \dfrac{n!}{(n-j)!\, s!\, s^{j-s}} \hat{a}^j P_0[n] & (j=s,s+1,\ldots,n), \end{cases} \tag{8.3}$$

where $P_0[n]$ is given by

$$P_0[n] = \left[\sum_{k=0}^{s-1} \binom{n}{k} \hat{a}^k + \sum_{k=s}^{n} \frac{n!}{(n-k)!\, s!\, s^{k-s}} \hat{a}^k \right]^{-1}. \tag{8.4}$$

Note that, in contrast with the infinite-source Erlang delay model, the number of terms in the second summation on the right-hand side of (8.4) is finite. Therefore the equilibrium-state distribution (8.3) and (8.4) is nondegenerate for all values of the parameters γ and μ. That is, since the number of sources n is finite, the number of customers waiting for service can never exceed the finite value $n-s$, so that ever-increasing queue lengths cannot occur. In effect, finite-source systems are self-regulating, since the request rate gets smaller as the system gets busier, and the stream of requests shuts off completely when there are no idle sources available to generate new requests.

The probability $P\{W>0\}$ that a customer must wait for service is the probability that a source finds all servers busy when placing a request. Therefore

$$P\{W>0\} = \sum_{j=s}^{n-1} \Pi_j[n] = \sum_{j=s}^{n-1} P_j[n-1]. \tag{8.5}$$

To calculate the waiting-time distribution function, we can write

$$P\{W>t\} = \sum_{j=0}^{n-s-1} P\{W>t \mid N=s+j\} P\{N=s+j\}. \tag{8.6}$$

The Delay System with Quasirandom Input

$P\{N=s+j\}$ is the probability that a source finds all s servers busy and j other sources waiting for service when it places a request for service. Hence

$$P\{N=s+j\} = \Pi_{s+j}[n] = P_{s+j}[n-1]. \tag{8.7}$$

If requests are served in order of arrival, the conditional probability $P\{W>t|N=s+j\}$ that a customer waits beyond t when placing a request, given that he finds $s+j$ other customers present, is exactly the same in a finite-source system as in its infinite-source counterpart; the waiting time of a customer who finds all s servers busy and j other waiting customers ahead of him is the sum of the lengths of time that each of the $j+1$ customers spends at the head of the queue. Since these time intervals are independent exponential variables with common mean $(s\mu)^{-1}$, we have [see Equation (4.20)]

$$P\{W>t|N=s+j\} = e^{-s\mu t} \sum_{i=0}^{j} \frac{(s\mu t)^i}{i!}. \tag{8.8}$$

By the same argument

$$E(W|N=s+j) = \frac{j+1}{s\mu}. \tag{8.9}$$

Substituting (8.7) and (8.8) into (8.6) we obtain (see Exercise 36)

$$P\{W>t\} = c \sum_{j=0}^{n-s-1} \frac{[\phi(t)]^j}{j!} e^{-\phi(t)}, \tag{8.10}$$

where

$$\phi(t) = \frac{s\mu}{\gamma} + s\mu t \tag{8.11}$$

and

$$c = \Pi_0[n] \frac{(n-1)! \hat{a}^s}{s!} \left(\frac{\hat{a}}{s}\right)^{n-s-1} e^{s\mu/\gamma}. \tag{8.12}$$

Observe that the distribution of W is essentially Erlangian of order $n-s$.

The order-of-arrival waiting-time distribution function (8.10) and some related quantities have been tabulated by Descloux [1962], who also gives a concise discussion covering much of the same material presented here. Tabulations are also available in Peck and Hazelwood [1958].

Exercise

36. Verify Equation (8.10).

Using (8.9), we can directly calculate the mean wait for service (for arbitrary order of service) suffered by an arbitrary customer when placing a request:

$$E(W) = \sum_{j=0}^{n-s-1} E(W|N=s+j)P\{N=s+j\}. \tag{8.13}$$

Although (8.13) may be computationally useful, unfortunately it does not yield algebraically neat results, and we shall not pursue it further.

As was true in the BCC case, the offered load a cannot be specified in the BCD case without first calculating its effect on the particular system to which it is offered. With BCD, the right-hand side of Equation (5.3) becomes

$$\frac{1}{\mu} \sum_{j=0}^{n-1} \lambda_j P_j[n] = \hat{a}\left(n - \sum_{j=1}^{n} jP_j[n]\right). \tag{8.14}$$

Equations (5.3) and (8.14) imply

$$\hat{a} = \frac{a'}{n - \sum_{j=1}^{n} jP_j[n]}. \tag{8.15}$$

In BCD systems, the carried load a' and the offered load a are equal:

$$a' = a. \tag{8.16}$$

Equations (8.15) and (8.16) give the following relationship between the offered load per idle source $\hat{a} = \gamma/\mu$ and the total offered load a:

$$\hat{a} = \frac{a}{n - \sum_{j=1}^{n} jP_j[n]}. \tag{8.17}$$

Note that (8.17) could have been anticipated on intuitive grounds: The total offered load a is the product of the offered load per idle source $\hat{a}$ and the average number $n - \sum_{j=1}^{n} jP_j[n]$ of idle sources.

It can be shown (see Exercise 37) that

$$\sum_{j=1}^{n} jP_j[n] = a\left(1 + \frac{E(W)}{\mu^{-1}}\right). \tag{8.18}$$

The Delay System with Quasirandom Input

Thus the offered load $\hat{a}$ per idle source, the total offered load a, the mean wait $E(W)/\mu^{-1}$ for service in units of mean service time, and the number n of sources are related according to the equation

$$\hat{a} = \frac{a}{n - a[1 + \mu E(W)]}. \tag{8.19}$$

The offered load a is bounded from above by the intended offered load $a^* = n[\hat{a}/(1+\hat{a})]$ (why?), which provides an approximation to a when the probability of blocking is small.

Further insight can be gained from Equation (8.19). Let T be the sojourn time (the sum of the waiting time and the service time) of a source. Then (8.19) can be written

$$n = \lambda(\gamma^{-1} + E(T)), \tag{8.20}$$

where $\lambda = a\mu$ is the actual rate at which requests for service are made by the sources.

Equation (8.20) is especially useful when this model is used for computer performance evaluation. In this application, n terminals (sources) request the use of a computer (server) to process *transactions*. The length of time that a terminal takes to generate a request for the computer to process a transaction is called the *think time*; that is, in this application γ^{-1} is the mean think time. The length of time from the instant a terminal generates a transaction (that is, requests service from the computer) until the computer completes the transaction (and instantaneously responds by communicating this fact to the user at the terminal) is called the *response time*; hence $E(T)$ is the mean response time. The rate at which transactions are processed (which equals the rate at which they are generated) is called the *throughput*; hence λ is the throughput, which is a measure of the computer system's processing power. In this application, then, Equation (8.20) relates the number of terminals, mean think time, mean response time, and throughput to each other.

For example, suppose that 30 terminals with a mean think time of 20 seconds are serviced by a single central processing unit (CPU) whose mean time to process a transaction is 1 second. A tedious but straightforward calculation [using (8.13) or its equivalent, with $n = 30$, $\gamma^{-1} = 20$, $\mu^{-1} = 1$, and $s = 1$] gives for the mean response time $E(T) = E(W) + \mu^{-1} = 10.4$ seconds. Substituting this value into (8.20), we calculate the throughput $\lambda = 30/(20 + 10.4) = .987$ transactions per second (which, for $s = 1$ server with mean service time unity, also equals the utilization of the CPU). If a second CPU were added, then $s = 2$ and the mean response time would drop to 1.68 seconds; simultaneously, according to (8.20), the throughput would rise to $30/(20 + 1.68) = 1.38$ transactions per second and the utilization of each of the CPUs would drop to $1.38/2 = 69\%$.

The derivation of Equations (8.19) and (8.20) outlined here is based on the assumption of quasirandom input and exponential service times; the derivation of formulas (8.3) and (8.4) for the probabilities $\{P_j[n]\}$ uses these assumptions explicitly. Nevertheless, Equations (8.18) and (8.19) are true whether or not these assumptions hold and, in fact, can be shown to be true for very general models that describe a finite number of sources circulating through a closed network. As we shall discuss in Chapter 5, and as is hinted at in Exercise 37, Equation (8.18) can be written immediately as a special case of the very general theorem known as Little's theorem. In fact, Equation (8.20) can itself be written immediately from this theorem, without any other analysis; that argument will be given as an example in Section 5.2. (For an application of these results in the context of computer performance evaluation, see Chapters 4.11 and 4.12 of Kleinrock [1976] and Chapter 3.9 of Kobayashi [1978].)

Exercises

37. Verify Equation (8.18) by direct calculation. (This result can also be obtained, immediately, by application of the Little's theorem $L=\lambda W$; see Exercise 28 and Section 5.2.)

38. Reconsider Exercise 35, but instead of assumption (2), assume that blocked customers are delayed. For $n=4$ and $n=5$, find also the server occupancy, the mean waiting time, the proportion of requests that wait more than 45 seconds for service, and the proportion of time that an arbitrary source is idle (neither waiting nor in service).

39. Using Equations (6.3) and (8.17), show that

$$a\Pi_j[n] = (n-j)\hat{a}P_j[n] \qquad (j=0,1,\ldots,n)$$

for systems with quasirandom input, exponential service times, and blocked customers delayed. Does this equation remain valid for other queue disciplines?

40. Consider a single-server queueing system with quasirandom input and blocked customers delayed, and assume that the service completion process is a pure death process with rate $\mu_j = j\mu$ (thus the server speeds up as the queue size increases). Find $\Pi_j[n]$ $(j=0,1,\ldots,n-1)$.

3.9. Summary

In this chapter we have shown that certain queueing models can be treated within the framework of the birth-and-death process. In particular, we have studied four queueing models, with two types of input (Poisson and quasirandom), two different queue disciplines (BCC and BCD), and exponential service times. We have stated that although the assumption of exponential service times was used to justify the formulation as a birth-and-death process, the results derived for BCC are valid for any service-time distribution function with finite mean. We have also derived the waiting-time distribution function for the BCD case in which customers are served in their order of arrival, and we have outlined the derivation for service in random order. Each model was discussed in detail, and ways in which the various models relate to each other were pointed out. (It should be noted that in a common classification scheme, to be discussed later, the s-server Erlang delay model is referred to as $M/M/s$.)

Systems with quasirandom input differ in some basic respects from systems with Poisson input (even though every Poisson-input model can be obtained, by a limiting process, directly from its quasirandom-input counterpart). In particular, in contrast to systems with Poisson input, (1) finite-source systems can admit only a finite number of states, so that questions of convergence do not arise; (2) the magnitude of the offered load depends on the state probabilities of the system to which it is offered; and (3) the arriving customer's distribution and the outside observer's distribution are unequal, but the former agrees with the latter for the corresponding system with one less source.

Finite-source models are discussed in detail by Syski [1960], who covers most of the material given here, including a summary of Cohen's paper on the generalization of the Engset formula. (As previously mentioned, tabulations of finite-source formulas are given in Descloux [1962] and Peck and Hazelwood [1958].)

Poisson input models have been more widely studied. Several references were given in this chapter, and many others are available. Of particular historical interest, Fry [1928, second edition 1965] discusses both Poisson and finite-source models in the context of teletraffic theory.

From a practical viewpoint, the Erlang loss and Erlang delay models are the most important, and situations for which they may be realistic models should be self-evident. Some interesting papers that are notable for their integration of theoretical and practical (numerical) results concerning the use of the Erlang loss model and some of its variants in teletraffic studies are Beneš [1959, 1961], Descloux [1965, 1973], Jagerman [1975], Kuczura and Neal [1972], and Neal and Kuczura [1973]. The finite-source models are more difficult to understand and calculate but, as noted above, have found several important applications.

Some of the material presented in this chapter is not well known outside of the literature of teletraffic theory. In particular, the concepts of offered load and carried load, and the relationship between the arriving customer's distribution and the outside observer's distribution, have not received the attention in the general literature of queueing theory that they seem to warrant.

Exercises

41. *Queue with feedback.* Customers request service on a BCD basis from a group of s servers according to a Poisson process with rate λ. Upon seizing an idle server, a customer holds it for an exponentially distributed length of time with mean μ^{-1}. Upon leaving the server, the customer rejoins the queue (as if he were a new arrival) with probability p, and leaves the system with probability $1-p$. Find the statistical-equilibrium probability that a new arrival finds all servers busy.

42. A single server serves customers of two priority classes on a preemptive basis: High-priority customers will preempt the server from low-priority customers, so that no low-priority customer is ever served when a high-priority customer is present. High-priority customers arrive according to a Poisson process at rate λ_1 and have exponential service times with mean μ_1^{-1}. Low-priority customers arrive according to a Poisson process at rate λ_2, and have exponential service times with mean μ_2^{-1}. Assume preemptive-repeat priority discipline (a new value of the service time is chosen from the exponential distribution each time a preempted customer enters the server). All customers wait as long as necessary for service.
  a. Find the probability that a low-priority customer is preempted from the server n times.
  b. Find the distribution function of the total length of time that a low-priority customer occupies the server.
  c. Find the mean length of time from the instant a low-priority customer first enters service until he leaves the system, assuming that when a customer is preempted no other low-priority customer can enter service until the preempted customer completes his service.
  d. Find the waiting-time distribution function for high-priority customers, assuming service in order of arrival within priority classes.
  e. What criteria must be met to ensure bounded delays for customers of each class?
  f. Repeat parts a–e assuming preemptive-resume priority discipline (preempted customers resume service where they left off).

43. *Priority reservation.* A group of s telephone trunks is offered a_1 erl of eastbound Poisson traffic and a_2 erl of westbound Poisson traffic. Both eastbound and westbound call holding times are exponentially distributed with common mean. To give the eastbound traffic better service than the westbound traffic, any westbound call that finds less than $n+1$ ($n<s$) trunks idle is cleared, whereas eastbound requests are cleared only when all s trunks are busy.
  a. Write the state equations that determine P_j, the statistical-equilibrium probability that j calls are in the system ($j=0,1,2,\ldots,s$). Include all boundary conditions.
  b. Solve the equations for P_j ($j=0,1,2,\ldots,s$).
  c. Write the expressions (in terms of P_j) for the proportions of eastbound and westbound calls lost.

44. In order to minimize its telephone bill, a company decides to investigate leasing a number of flat-rate telephone trunks between two locations. Management has decided that nonadministrative personnel will be given access to the flat-rate trunks only, while administrators will have access to the regular toll network in addition to the flat-rate trunks. Suppose that the situation can be described by the following model.

High-priority calls originate according to a Poisson process at rate λ_1 attempts per minute. These calls are first offered to the group of s flat-rate trunks. If a high-priority call finds all s flat-rate trunks busy, it will immediately be offered to the toll network where, it is assumed, there will always be at least one idle trunk. Low-priority calls originate according to a Poisson process at rate λ_2 attempts per minute. These calls are offered to the flat-rate trunks only; if a low-priority call finds all flat-rate trunks busy, it waits in a queue until a flat-rate trunk becomes available. The holding times for both types of call are exponentially distributed with mean length μ^{-1} minutes.

Show that if the company leases the flat-rate trunks at a cost of c dollars per trunk per minute, the overall cost per minute $c(s)$ is given by

$$c(s) = cs + \lambda_1 \left[r_0 + re^{-\mu}(1 - e^{-\mu})^{-1} \right] B(s),$$

where r_0 is the cost of a toll call for the first minute or any fraction thereof, r is the cost of a toll call for each additional minute or any fraction thereof, and

$$B(s) = \frac{\dfrac{(a_1 + a_2)^s}{s!} \dfrac{1}{1 - a_2/s}}{\displaystyle\sum_{k=0}^{s-1} \dfrac{(a_1 + a_2)^k}{k!} + \dfrac{(a_1 + a_2)^s}{s!} \dfrac{1}{1 - a_2/s}} \qquad (a_2 < s)$$

and $B(s) = 1$ when $a_2 \geq s$, where $a_1 = \lambda_1/\mu$ and $a_2 = \lambda_2/\mu$. Find the mean waiting time for low-priority calls and the occupancy of the flat-rate trunks.

45. *Time-varying Poisson input.*

a. Consider a generalization of the Poisson process in which the parameter λ can vary with time (time-varying Poisson input); that is, assume that as $h \to 0$ the probability of an arrival in any interval $(x, x+h)$ is $\lambda(x)h + o(h)$ and the probability of more than one arrival is $o(h)$. Let $P_j(t)$ be the probability of exactly j arrivals in $(0, t)$, and show that

$$P_j(t) = \frac{[\Lambda(t)]^j}{j!} e^{-\Lambda(t)} \qquad (j = 0, 1, \ldots), \qquad (1)$$

where

$$\Lambda(t) = \int_0^t \lambda(x) dx. \qquad (2)$$

b. Consider the transient analysis of the infinite-server queue with (ordinary) Poisson input; the probability $P_j(t)$ that exactly j customers are present at

any time t is given by Equation (3.11). Show that the number of customers present at time t can be interpreted as the number of arrivals in $(0, t)$ according to a time-varying Poisson process in which the effective arrival rate at every time x $(0 \leq x \leq t)$ is $\lambda(x) = \lambda[1 - H(t-x)]$. Conclude that (1) applies; show that (2) yields $\Lambda(t) = \lambda t p(t)$, where $p(t)$ is given by (3.9); and compare with (3.11).
This exercise was suggested by Børge Tilt.

46. *Transient analysis of the single-server Erlang delay model.*
  a. Show that if we adopt the time scale in which the mean service time is the time unit, then the probability $P_j(t)$ that exactly j customers are present at time t satisfies the equations

$$\frac{d}{dt} P_0(t) = -aP_0(t) + P_1(t) \tag{1}$$

and

$$\frac{d}{dt} P_j(t) = aP_{j-1}(t) - (1+a)P_j(t) + P_{j+1}(t) \qquad (j=1,2,\ldots), \tag{2}$$

where a is the offered load.
  b. In order to solve the equations (1) and (2), consider the auxiliary system of equations

$$\frac{d}{dt} \hat{P}_j(t) = a\hat{P}_{j-1}(t) - (1+a)\hat{P}_j(t) + \hat{P}_{j+1}(t) \qquad (j=0, \pm 1, \pm 2, \ldots) \tag{3}$$

and

$$\hat{P}_0(t) = a\hat{P}_{-1}(t), \tag{4}$$

and show that

$$P_j(t) = \hat{P}_j(t) \qquad (j=0,1,2,\ldots); \tag{5}$$

that is, any solution of (3) and (4) for $j = 0, \pm 1, \pm 2, \ldots$ will also be a solution of (1) and (2) for $j = 0, 1, 2, \ldots$. (The values of $\hat{P}_j(t)$ for $j = -1, -2, \ldots$ are irrelevant.)
  c. Define the generating function

$$\hat{P}(z,t) = \sum_{j=-\infty}^{\infty} \hat{P}_j(t) z^j, \tag{6}$$

and show that $\hat{P}(z,t)$ satisfies the differential equation

$$\frac{d}{dt} \hat{P}(z,t) = [az - (1+a) + z^{-1}]\hat{P}(z,t), \tag{7}$$

the most general solution to which is

$$\hat{P}(z,t) = G(z) \exp[-(1+a)t + (az + z^{-1})t], \tag{8}$$

where $G(z)$ is an arbitrary function of z.

d. Using the fact that the modified Bessel functions $I_k(y)$ are given by the well-known expansion

$$\exp[\tfrac{1}{2}y(x+x^{-1})] = \sum_{k=-\infty}^{\infty} I_k(y)x^k, \qquad (9)$$

show that

$$\hat{P}(z,t) = G(z)e^{-(1+a)t} \sum_{k=-\infty}^{\infty} I_k(2a^{1/2}t)a^{k/2}z^k. \qquad (10)$$

e. Now assume that $G(z)$ can be represented by an expansion of the form

$$G(z) = \sum_{j=-\infty}^{\infty} c_{-j}z^j, \qquad (11)$$

and show, by equating coefficients in (6) and (10), that

$$\hat{P}_j(t) = e^{-(1+a)t} \sum_{k=-\infty}^{\infty} c_k a^{\frac{1}{2}(j+k)} I_{j+k}(2a^{1/2}t) \qquad (12)$$

(where the coefficients $\{c_k\}$ remain to be determined).

f. Show that Equation (9) implies that $I_0(0)=1$ and $I_k(0)=0$ when $k\neq 0$; and therefore, if the number of customers present at $t=0$ is i, then $c_{-i}=1$ and $c_k=0$ when $k\leq 0$ and $k\neq -i$. Thus, Equation (12) can be written

$$P_j(t) = e^{-(1+a)t}\left[a^{\frac{1}{2}(j-i)}I_{j-i}(2a^{1/2}t) + \sum_{k=1}^{\infty} c_k a^{\frac{1}{2}(j+k)} I_{j+k}(2a^{1/2}t)\right]. \qquad (13)$$

g. Show that Equations (13) and (4) imply

$$a^{-\frac{1}{2}i}I_{-i} + \sum_{k=1}^{\infty} d_k I_k = a^{-\frac{1}{2}i+\frac{1}{2}}I_{-(i+1)} + a^{1/2}\sum_{k=1}^{\infty} d_k I_{k-1}, \qquad (14)$$

where we have written $d_k = c_k a^{k/2}$ and $I_k = I_k(2a^{1/2}t)$.

h. Using the fact that $I_k = I_{-k}$ [as is easily verified by replacing x by x^{-1} in both sides of (9)], show that (14) implies

$$d_k = 0 \quad (k=1,2,\ldots,i),$$

$$d_{i+1} = a^{-\frac{1}{2}i-\frac{1}{2}},$$

$$d_{i+1+m} = a^{-\frac{1}{2}i-\frac{1}{2}m-\frac{1}{2}}(1-a) \quad (m=1,2,\ldots).$$

i. Conclude that

$$P_j(t) = a^{\frac{1}{2}(j-i)} e^{-(1+a)t} \left[I_{j-i}(2a^{1/2}t) + a^{-1/2} I_{j+i+1}(2a^{1/2}t) \right.$$

$$\left. + (1-a) \sum_{k=2}^{\infty} a^{-k/2} I_{j+i+k}(2a^{1/2}t) \right]. \qquad (15)$$

j. For what values of the offered load a is Equation (15) valid?

This derivation of (15) is that of Cox and Smith [1961]. For further discussion, generalizations, and references, see also Cohen [1969], Riordan [1962], Syski [1960], and Takács [1962a].

[4]
Multidimensional Birth-and-Death Queueing Models

In this chapter we shall discuss methods of analysis of queueing models for which the natural definition of a state requires more than one variable. Our approach will be through the use of examples that illustrate various techniques and often represent systems important in themselves.

4.1. Introduction

In Section 2.2 we studied the one-dimensional birth-and-death process, the essential characteristic of which is that the system occupies "states" E_0, $E_1,\ldots$ and the rates at which changes of state occur depend only on the instantaneous state of the system and not on the past history of the process (the Markov property). In Section 2.3 we showed that, under appropriate conditions often assumed to hold in practice, the state probabilities $\{P_j\}$ are related to each other according to the equations (3.11) of Chapter 2, which we restate for convenience:

$$(\lambda_j + \mu_j)P_j = \lambda_{j-1}P_{j-1} + \mu_{j+1}P_{j+1} \qquad (j=0,1,\ldots), \tag{1.1}$$

where $\{\lambda_j\}$ and $\{\mu_j\}$ are the rates of transition upward and downward, respectively, and $\lambda_{-1} = \mu_0 = 0$.

We noted that the equations (1.1) have the following simple and important interpretation: For each state, the rate at which the system leaves the state equals the rate at which the system enters the state. In other words, these equations express a law of *conservation of flow*, that is, rate out = rate in. The equations (1.1) are easily solved by recurrence [the normalized solution is given by the equations (3.10) of Chapter 2].

In this chapter we consider systems that have the Markov property, but whose states require more than one variable for definition. For example, consider the infinite-server queue with Poisson input at rate λ and two types of customers, where the service times of customers of type i ($i = 1, 2$) are independent, identically distributed, exponential random variables with mean duration τ_i. If $\tau_1 = \tau_2 = \tau$, then the equilibrium probabilities P_j ($j = 0, 1, \ldots$) that there are j customers in the system satisfy the equations (1.1) with $\lambda_j = \lambda$ and $\mu_j = j\tau^{-1}$. However, if $\tau_1 \neq \tau_2$, one cannot specify the death rate μ_j without knowledge of the composition of the state E_j. But if we define the two-dimensional states $\{E_{j_1, j_2}\}$, where j_i is the number of customers of type i in the system, with corresponding equilibrium distribution $\{P(j_1, j_2)\}$, then clearly the Markov property still holds, and the death rate corresponding to the state E_{j_1, j_2} is

$$\mu(j_1, j_2) = j_1 \tau_1^{-1} + j_2 \tau_2^{-1}.$$

Note that the one-dimensional equations (1.1) are not incorrect in this case; they are simply inapplicable because the death rates cannot be specified.

In what follows, we consider systems that can be described by a multidimensional analogue of the simple birth-and-death equations (1.1). More precisely, we *assume* that systems that have the Markov property when the states are properly defined have equilibrium state distributions that satisfy "rate out = rate in" equations analogous to (1.1). Unlike the equation for the one-dimensional case, however, the analogous set of multidimensional difference equations cannot, in general, be solved by recurrence. We shall devote this chapter to a discussion of queueing models, many of practical importance, that can be formulated as multidimensional birth-and-death processes, and to special techniques for solving the equations of conservation of flow that are assumed to describe these systems in equilibrium.

As an example of a multidimensional birth-and-death queueing model, we consider a finite-source system with nonidentical sources. In particular, imagine a single-server system with two sources. Assume that source i generates requests at constant rate γ_i when idle and rate 0 otherwise, and has exponential service times with mean μ_i^{-1}. Assume that the queue discipline is blocked customers delayed. Let $S_i = 0$ if source i is idle, $S_i = 1$ if source i is being served, and $S_i = 2$ if source i is waiting for service. Finally, let $P\{S_1 = j, S_2 = k\} = P(j, k)$ ($j, k = 0, 1, 2$) be the statistical-equilibrium state distribution. Now, the reader should be convinced that the conservation-of-flow equations, which equate the rate the system leaves

Introduction

each state to the rate at which it enters that state, are

$$(\gamma_1 + \gamma_2)P(0,0) = \mu_1 P(1,0) + \mu_2 P(0,1),$$
$$(\gamma_2 + \mu_1)P(1,0) = \gamma_1 P(0,0) + \mu_2 P(2,1),$$
$$(\gamma_1 + \mu_2)P(0,1) = \gamma_2 P(0,0) + \mu_1 P(1,2),$$
$$\mu_1 P(1,2) = \gamma_2 P(1,0),$$
$$\mu_2 P(2,1) = \gamma_1 P(0,1).$$

Observe that the sum of the terms on the left-hand side is identical to the sum of the terms on the right-hand side; thus any one of these equations is redundant and can be ignored. The four remaining equations, together with the normalization equation

$$P(0,0) + P(1,0) + P(0,1) + P(1,2) + P(2,1) = 1,$$

uniquely determine the unknown probabilities.

Exercise

1. *Finite-source systems with nonidentical sources.* We proved in Section 3.6 that in any equilibrium n-source (one-dimensional) birth-and-death queueing system with quasirandom input (identical sources), the state distribution seen by a source when placing a request is the same as would be seen by an outside observer of the corresponding $(n-1)$-source system. This suggests that for (multidimensional) birth-and-death systems with nonidentical sources, the state distribution seen by a particular source when placing a request is the same as that particular source would see if he placed no requests, but instead acted as an outside observer of the system. Descloux [1967] has observed that this generalization cannot always be made, and has given the necessary and sufficient conditions for its validity. To illustrate this, consider a system composed of a single server and two nonidentical sources, as above, but, unless stated otherwise, make no assumption about the queue discipline.

   a. Show that the probability b_2 that source 2 is blocked (finds the server occupied when placing a request for service) is given by

   $$b_2 = \frac{P(1,0)}{P(0,0) + P(1,0)}.$$

   b. Show that if source 2 makes no requests for service but instead acts the part of an outside observer, then the probability b_2' that source 2 sees the server occupied is given by

   $$b_2' = \frac{\gamma_1/\mu_1}{1 + (\gamma_1/\mu_1)}.$$

c. Assume that blocked customers are cleared. Find the distribution $\{P(j,k)\}$ and show that $b_2 = b_2'$.

d. Assume that blocked customers are delayed. Find the distribution $\{P(j,k)\}$ and show that

$$b_2 = \frac{(\gamma_1/\mu_2)(\mu_2+\gamma_1+\gamma_2)}{\mu_1+(\mu_1/\mu_2)\gamma_1+\gamma_2+(\gamma_1/\mu_2)(\mu_2+\gamma_1+\gamma_2)}.$$

Show that $b_2 = b_2'$ if and only if $\mu_1 = \mu_2$.
This exercise was suggested by P. J. Burke.

4.2. Product Solutions (Mixed Traffic, Queues in Tandem)

Infinite-Server Group with Two Types of Customers

Consider an infinite-server group that receives two distinct types of Poisson traffic (such as ordinary telephone conversations and data) with arrival rates λ_1 and λ_2. Assume that customers of type 1 have exponentially distributed service times with mean μ_1^{-1}, that customers of type 2 have exponentially distributed service times with mean μ_2^{-1}, and that each value of a service time is obtained from its corresponding distribution independently of all others and the arrival process. Let $P(j_1, j_2)$ be the statistical-equilibrium joint probability that at any instant there are j_1 customers of type 1 and j_2 customers of type 2 in service. Then, equating rate out to rate in for each state, the statistical-equilibrium state equations are

$$(\lambda_1+\lambda_2+j_1\mu_1+j_2\mu_2)P(j_1,j_2)$$
$$= \lambda_1 P(j_1-1,j_2) + \lambda_2 P(j_1,j_2-1)$$
$$+ (j_1+1)\mu_1 P(j_1+1,j_2) + (j_2+1)\mu_2 P(j_1,j_2+1)$$
$$[P(-1,j_2)=P(j_1,-1)=0; \quad j_1=0,1,\ldots; \quad j_2=0,1,\ldots]. \quad (2.1)$$

In addition, of course, the probabilities must satisfy the normalization equation

$$\sum_{j_1=0}^{\infty}\sum_{j_2=0}^{\infty} P(j_1,j_2) = 1. \quad (2.2)$$

In this case, we already know the answer. Since the server group is infinite, the two types of customers do not affect one another. Thus the marginal distribution of the number of customers of each type is that which would be obtained by solving the corresponding one-dimensional problem, namely the Poisson distribution:

$$P_1(j) = \sum_{k=0}^{\infty} P(j,k) = \frac{(\lambda_1/\mu_1)^j}{j!} e^{-(\lambda_1/\mu_1)}, \quad (2.3)$$

$$P_2(j) = \sum_{k=0}^{\infty} P(k,j) = \frac{(\lambda_2/\mu_2)^j}{j!} e^{-(\lambda_2/\mu_2)}. \quad (2.4)$$

Product Solutions (Mixed Traffic, Queues in Tandem)

Since the number of customers present of each type is independent of the number present of the other type, therefore $P(j_1,j_2) = P_1(j_1)P_2(j_2)$; that is,

$$P(j_1,j_2) = \frac{(\lambda_1/\mu_1)^{j_1}}{j_1!} \frac{(\lambda_2/\mu_2)^{j_2}}{j_2!} e^{-[(\lambda_1/\mu_1)+(\lambda_2/\mu_2)]}. \qquad (2.5)$$

Let us substitute the solution (2.5) into the equations (2.1). We see immediately that, as far as (2.1) is concerned, the factor

$$\exp\left\{-\left[\frac{\lambda_1}{\mu_1} + \frac{\lambda_2}{\mu_2}\right]\right\}$$

is irrelevant, since it divides out. More importantly, each factor on the left of (2.1) corresponds to a single factor on the right and vice versa. Specifically, it is easy to verify that (2.5) satisfies the equations

$$\begin{aligned}\lambda_1 P(j_1,j_2) &= (j_1+1)\mu_1 P(j_1+1,j_2), \\ \lambda_2 P(j_1,j_2) &= (j_2+1)\mu_2 P(j_1,j_2+1), \\ j_1 \mu_1 P(j_1,j_2) &= \lambda_1 P(j_1-1,j_2), \\ j_2 \mu_2 P(j_1,j_2) &= \lambda_2 P(j_1,j_2-1).\end{aligned} \qquad (2.6)$$

[Observe that the first and third equations of the set (2.6) are identical, as are the second and fourth.] Thus we have shown that, in this case, the conservation-of-flow equation (2.1) can be decomposed into the four separate conservation-of-flow equations (2.6). However, there seems to be no intuitive reason to predict that this decomposition should be possible. (For the reader interested in further theoretical considerations, we mention that this decomposition property is related to *Kolmogorov's criterion for reversibility*—see Keilson [1965, 1979] and Kelly [1979] for discussion and references.) The important point to be made here is that, as we shall demonstrate, the decomposition property holds for a very large variety of models that do not appear similar at first glance; and as a practical matter, this property of decomposition implies that a "simple" solution of the basic conservation-of-flow equations exists.

For example, the first equation of the set (2.6) shows that

$$P(j_1,j_2) = \frac{(\lambda_1/\mu_1)^{j_1}}{j_1!} P(0,j_2);$$

similarly, it follows from the second equation of (2.6) that

$$P(j_1,j_2) = P(j_1,0)\frac{(\lambda_2/\mu_2)^{j_2}}{j_2!}.$$

Thus we see that the decomposition implies (and is implied by) the statement that the solution $P(j_1,j_2)$ is a product of two factors, one depending functionally on only the index j_1, the other on only j_2:

$$P(j_1,j_2) = P_1(j_1)P_2(j_2), \tag{2.7}$$

or equivalently, in this case,

$$P(j_1,j_2) = \frac{(\lambda_1/\mu_1)^{j_1}}{j_1!} \frac{(\lambda_2/\mu_2)^{j_2}}{j_2!} c, \tag{2.8}$$

where the constant c is determined from the normalization equation (2.2).

In practice, a good strategy is to assume a product solution of the form (2.8) (or some other form) and see if it satisfies the original conservation-of-flow equations. If it does, then the solution has been obtained, except for a multiplicative constant required to make the probabilities sum to unity. (That this solution is unique can be argued on physical grounds.) If it doesn't, not much effort has been expended, and a different approach can be undertaken. (The reader may recognize this strategy as the method of *separation of variables*, widely used in solving partial differential equations.) The extension to models with more than two variables should be clear. The question of exactly what form of product solution should be tried is not easily answered, except to say that the examples of the following sections should be helpful.

Finally, with regard to the particular model used as an example here, we remark that although we have assumed that the service times for each type of customer are exponentially distributed, we know from previous considerations that the marginal distributions are of the form (2.3) and (2.4) for any distribution of service times with means μ_1^{-1} and μ_2^{-1}. It is probably true that in any system in which customers of a certain type are cleared when blocked, or when no blocking can occur, the results are independent of the form of the service-time distribution function for that type. (See Fortet and Grandjean [1963].)

Finite-Server Group with Two Types of Customers and Blocked Customers Cleared

As in the previous example, we assume two types of customers, with customers of type i characterized by parameters λ_i and μ_i; but the number of servers is now $s < \infty$, and blocked customers are cleared. The system is again described by (2.1), which is valid now only for $j_1 + j_2 < s$. When $j_1 + j_2 = s$, then the states E_{j_1+1, j_2} and E_{j_1, j_2+1} cannot occur. Therefore, again equating the rate of flow out of state E_{j_1, j_2} to the rate of flow into

E_{j_1,j_2}, we have

$$(j_1\mu_1+j_2\mu_2)P(j_1,j_2)=\lambda_1 P(j_1-1,j_2)+\lambda_2 P(j_1,j_2-1)$$
$$(j_1+j_2=s). \qquad (2.9)$$

Observe that (2.9) can be obtained from (2.1) by deleting the first two terms on the left and the last two terms on the right. It follows that the product solution (2.8) also satisfies (2.9), since each term deleted (remaining) on the left corresponds to a term deleted (remaining) on the right.

Thus, in this case the boundary conditions do not change the form of the solution, which is again given by Equation (2.8). The probability $P(j)$ that there are j customers in service is given by $P(j)=\Sigma_{j_1+j_2=j}P(j_1,j_2)$, which, with the help of the binomial theorem, reduces to

$$P(j)=\frac{1}{j!}\left(\frac{\lambda_1}{\mu_1}+\frac{\lambda_2}{\mu_2}\right)^j c. \qquad (2.10)$$

The normalization equation, $\Sigma_{k=0}^{s}P(k)=1$, implies that

$$c=\left[\sum_{k=0}^{s}\frac{1}{k!}\left(\frac{\lambda_1}{\mu_1}+\frac{\lambda_2}{\mu_2}\right)^k\right]^{-1}.$$

We conclude that

$$P(j)=\frac{\dfrac{a^j}{j!}}{\displaystyle\sum_{k=0}^{s}\frac{a^k}{k!}} \qquad (j=0,1,\ldots,s), \qquad (2.11)$$

where $a=(\lambda_1/\mu_1)+(\lambda_2/\mu_2)$. This result agrees, of course, with that obtained by different reasoning in Section 3.3: When two streams of Poisson traffic of magnitudes a_1 erl and a_2 erl are simultaneously served by a group of s servers on a BCC basis, each stream sees the state distribution, given by the Erlang loss (truncated Poisson) probabilities with the parameter $a=a_1+a_2$.

Exercises

2. **a.** Three cities, A, B, and C, are interconnected by two trunk groups, group 1 consisting of s_1 trunks connecting A and B, and group 2 consisting of s_2 trunks between B and C (Figure 4.1). A call between A and C simultaneously holds a trunk in each group. Blocked calls are cleared. Write and solve the equilibrium state equations for this network.

Figure 4.1.

  b. Consider the network of part a augmented by the addition of group of s trunks directly connecting A and C. In the augmented network, all calls between A and C are carried on the new group if a trunk is available. Calls between A and C that find all s trunks in the new group busy are alternate-routed via groups 1 and 2 if there is an idle trunk in each group, and are cleared otherwise. Whenever a direct trunk between A and C becomes available and there is at least one overflow call in progress, an overflow call will be switched down from the alternate route to the direct route, thus using one less trunk to complete the call. Write and solve the equilibrium state equations for this network.
  c. Consider the same network as in part b, but without the capability of switching alternate-routed calls down to the direct route. Is this system easier or harder to analyze than that of part b? Why?
  d. Suppose now that calls offered to any of the three direct groups can be alternate-routed via the other two, and that any alternate-routed call can be switched back to its direct route. Write and solve the equilibrium state equations for this network.

3. A group of s trunks serves two types of Poisson traffic on a BCC basis. Type 1 is ordinary telephone traffic, with arrival rate λ_1 and exponential holding times with mean μ_1^{-1}. Type 2 is "wideband" data traffic, with arrival rate λ_2 and exponential holding times with mean μ_2^{-1}. Each wideband call requires the simultaneous use of k trunks for transmission. If a wideband call finds less than k idle trunks, it is cleared from the system. Find the proportion of ordinary calls and the proportion of wideband calls that are lost.

4. A group of s servers handles n types of customers. A customer of type i is cleared from the system if on arrival he finds all servers busy or $k_i \leqslant s$ other customers of his own type in service $(i=1,2,\ldots,n)$. Customers of type i arrive in an independent Poisson stream at rate λ_i, and have exponential service times with mean μ_i^{-1}. Find the probability of blocking seen by customers of each type.

5. A group of s servers handles two types of customers on a BCC basis. Type 1 is Poisson traffic with arrival rate λ. Type 2 is quasirandom traffic generated by n sources, each bidding with rate γ when idle. Service times are exponential with means μ_1^{-1} and μ_2^{-1}, respectively.

a. Find the equilibrium probability $P_n(j_1,j_2)$ that there are j_1 type-1 customers and j_2 of the n type-2 customers simultaneously in service at an arbitrary instant.

b. Define
$$\Pi_n^i(j_1,j_2;t) = \lim_{h \to 0} P\{N_1(t)=j_1, N_2(t)=j_2 | C_i(t,t+h)\},$$
where $N_i(t)$ is the number of customers of type i that are in service at time t and $C_i(t,t+h)$ is the event $\{$a customer of type i makes a request for service in $(t,t+h)\}$; and let
$$\Pi_n^i(j_1,j_2) = \lim_{t \to \infty} \Pi_n^i(j_1,j_2;t).$$

Show that
$$\Pi_n^1(j_1,j_2) = P_n(j_1,j_2)$$
and
$$\Pi_n^2(j_1,j_2) = P_{n-1}(j_1,j_2).$$

[Therefore, the probability of blocking suffered by customers of type 1 is $\sum_{j_1+j_2=s} P_n(j_1,j_2)$, and that suffered by customers of type 2 is $\sum_{j_1+j_2=s} P_{n-1}(j_1,j_2)$.]

6. Calls arrive according to a Poisson process with rate λ at a call distributor that routes each call to one of two trunk groups, composed of s_1 and s_2 trunks, respectively. Let $n_1 \geq s_1$ and $n_2 \geq s_2$ be integers. If an arrival occurs when j_i calls are in progress on group i ($i=1,2$), then with probability $(n_i - j_i)/(n_1 + n_2 - j_1 - j_2)$ the new arrival is routed to group i. A call routed to group i that finds all s_i trunks busy is cleared from the system. Call holding times are exponential with mean μ^{-1}, and $\lambda/\mu = a$. If $P(j_1,j_2)$ is the equilibrium probability that there are j_1 calls on group 1 and j_2 calls on group 2, show that (see Appendix D of Buchner and Neal [1971])

$$P(j_1,j_2) = c \frac{\binom{n_1}{j_1}\binom{n_2}{j_2}}{\binom{n_1+n_2}{j_1+j_2}} \frac{a^{j_1+j_2}}{(j_1+j_2)!}.$$

How might one surmise a solution of this form?

7. Customers arrive according to a Poisson process with rate λ at a group of s exponential *heterogeneous* servers; that is, the servers are numbered $1,2,\ldots,s$, and the time required for server i to serve a customer is exponentially distributed with mean μ_i^{-1}. All of the servers that are idle when a customer arrives are equally likely to be selected by the customer, and blocked customers are cleared. Define the random variables $X_i = 0$ when the ith server is idle and $X_i = 1$ when the ith server is busy ($i=1,2,\ldots,s$), and let $\tilde{P}(x_1,\ldots,x_s) = P\{X_1 = x_1, \ldots, X_s = x_s\}$ be the equilibrium state probabilities ($x_i = 0,1$ for each $i = 1,2,\ldots,s$).

a. Show, with no calculation, that if $\mu_1 = \cdots = \mu_s = \mu$, then

$$\tilde{P}(x_1,\ldots,x_s) = \binom{s}{j}^{-1} P_j \quad \text{when} \quad x_1 + \cdots + x_s = j,$$

and

$$\sum_{x_1 + \cdots + x_s = j} \tilde{P}(x_1,\ldots,x_s) = P_j \quad (j = 0, 1, \ldots, s),$$

where $\{P_j\}$ is the Erlang loss distribution (3.3) of Chapter 3.

b. Write the statistical-equilibrium state equations that determine the distribution $\{\tilde{P}(x_1,\ldots,x_s)\}$ in the general case (the service rates are no longer assumed to be equal), and show that the solution is

$$\tilde{P}(x_1,\ldots,x_s) = \binom{s}{x_1 + \cdots + x_s}^{-1} \frac{(\lambda/\mu_1)^{x_1} \cdots (\lambda/\mu_s)^{x_s}}{(x_1 + \cdots + x_s)!} \tilde{P}_0,$$

where

$$\tilde{P}_0 = \left[\sum_{y_1,\ldots,y_s} \binom{s}{y_1 + \cdots + y_s}^{-1} \frac{(\lambda/\mu_1)^{y_1} \cdots (\lambda/\mu_s)^{y_s}}{(y_1 + \cdots + y_s)!} \right]^{-1}.$$

Verify that this solution can be specialized to yield the results of part a.

Queues in Tandem

Consider two sets of servers arranged in tandem, so that the output (customers completing service) from the first set of servers is the input to the second set. Assume that the arrival process at the first stage of this tandem queueing system is Poisson with rate λ, the service times in the first stage are exponential with mean μ_1^{-1}, and the queue discipline is blocked customers delayed. The customers completing service in the first stage enter the second stage, where the service times are assumed to be exponential with mean μ_2^{-1}. Customers leaving the first stage who find all servers occupied in the second stage wait in a queue in the second stage until they are served. The number of servers in stage i is s_i.

Let $P(j_1, j_2)$ be the statistical-equilibrium probability that there are j_1 customers in stage 1 and j_2 customers in stage 2. To save rewriting the state equations for each set of boundary conditions, let

$$\mu_i(j) = \begin{cases} j\mu_i & (j = 0, 1, \ldots, s_i) \\ s_i\mu_i & (j = s_i + 1, s_i + 2, \ldots) \end{cases} \quad (i = 1, 2). \tag{2.12}$$

Then the statistical-equilibrium state equations, obtained by equating the

Product Solutions (Mixed Traffic, Queues in Tandem)

rate the system leaves each state to the rate it enters that state, are

$$[\lambda + \mu_1(j_1) + \mu_2(j_2)]P(j_1,j_2)$$
$$= \lambda P(j_1-1,j_2) + \mu_1(j_1+1)P(j_1+1,j_2-1)$$
$$+ \mu_2(j_2+1)P(j_1,j_2+1)$$
$$[P(-1,j_2) = P(j_1,-1) = 0; \quad j_1 = 0,1,\ldots; \quad j_2 = 0,1,\ldots]. \quad (2.13)$$

The term $\mu_1(j_1+1)P(j_1+1,j_2-1)$ reflects the fact that a departure from stage 1 constitutes an arrival at stage 2.

In the previous two examples, it has turned out that the joint probability $P(j_1,j_2)$ could be written as the product of two factors $P_1(j_1)$ and $P_2(j_2)$:

$$P(j_1,j_2) = P_1(j_1)P_2(j_2). \quad (2.14)$$

In the example under consideration, we can evaluate without calculation the marginal distribution of the number of customers in the first stage; for, the first stage of this tandem queueing system is precisely an Erlang delay system, so that the marginal distribution of the number of customers in stage 1 is given by the Erlang delay probabilities (4.3), (4.4), and (4.5) of Chapter 3.

Therefore, because of our previous success, and because it is certainly worth a try, let us assume a product solution of the form (2.14), with the factor $P_1(j_1)$ given by the Erlang delay probabilities:

$$P_1(j_1) = \begin{cases} c_1 \dfrac{(\lambda/\mu_1)^{j_1}}{j_1!} & (j_1 = 0,1,\ldots,s_1-1), \\ c_1 \dfrac{(\lambda/\mu_1)^{j_1}}{s_1! s_1^{j_1-s_1}} & (j_1 = s_1, s_1+1,\ldots). \end{cases} \quad (2.15)$$

We shall substitute the assumed product solution (2.14) and (2.15) into the equilibrium state equations (2.13), with the hope that we will be left with a one-dimensional set of equations that we can solve for the remaining factor $P_2(j_2)$ of our assumed product solution. [If our guess turns out to be correct, then, since (2.14) holds for all j_1 and j_2, it follows that the factors on the right-hand side of (2.14) are, in fact, the marginal probabilities for each stage.]

It is readily verified that substitution of the assumed product solution (2.14) and (2.15) into the state equations (2.13) yields, after cancellation,

$$[\lambda + \mu_2(j_2)]P_2(j_2) = \lambda P_2(j_2-1) + \mu_2(j_2+1)P_2(j_2+1)$$
$$[P_2(-1) = 0; \quad j_2 = 0,1,\ldots]. \quad (2.16)$$

Inspection of the equations (2.16) shows that these are precisely the equilibrium state equations that, except for the normalization condition, define the Erlang delay probabilities. We conclude that $P_2(j_2)$ is given by

$$P_2(j_2) = \begin{cases} c_2 \dfrac{(\lambda/\mu_2)^{j_2}}{j_2!} & (j_2 = 0, 1, \ldots, s_2 - 1), \\ c_2 \dfrac{(\lambda/\mu_2)^{j_2}}{s_2! s_2^{j_2 - s_2}} & (j_2 = s_2, s_2 + 1, \ldots). \end{cases} \quad (2.17)$$

Since

$$\sum_{j=0}^{\infty} P_i(j) = 1 \quad (i = 1, 2), \quad (2.18)$$

it follows that if we define $a_i = \lambda/\mu_i$, then

$$c_i = \left(\sum_{k=0}^{s_i - 1} \frac{a_i^k}{k!} + \frac{a_i^{s_i}}{s_i!(1 - a_i/s_i)} \right)^{-1} \quad (i = 1, 2). \quad (2.19)$$

This equation implies, as expected, that a proper joint distribution exists only when $\lambda/\mu_1 < s_1$ and $\lambda/\mu_2 < s_2$.

Note that, as in the previous cases, each term on the left-hand side of Equation (2.13) corresponds to a single term on the right-hand side of (2.13) and vice versa. Specifically, we have the correspondences

$$\lambda P(j_1, j_2) = \mu_2(j_2 + 1) P(j_1, j_2 + 1),$$

$$\mu_1(j_1) P(j_1, j_2) = \lambda P(j_1 - 1, j_2),$$

$$\mu_2(j_2) P(j_1, j_2) = \mu_1(j_1 + 1) P(j_1 + 1, j_2 - 1),$$

as can be easily verified by substitution. We have thus derived three conservation-of-flow equations that hold in this particular problem. However, there was no intuitive reason to predict that these conservation laws would hold individually, but only that they would hold jointly [which, of course, is the basis for Equation (2.13)]. This must be viewed as an enlightening and unexpected result.

We have shown that the equilibrium joint probability $P(j_1, j_2)$ factors into a product of two Erlang delay probabilities, $P_1(j_1)$ and $P_2(j_2)$. We conclude that the number of customers in either stage of this tandem queue is independent of the number of customers simultaneously present in the other stage.

Product Solutions (Mixed Traffic, Queues in Tandem) 135

We have also shown that the marginal distribution of the number of customers in the second stage is the Erlang delay distribution; that is, the distribution of the number of customers in the second stage is exactly that which would obtain if the first stage were not present and the customers arrived according to a Poisson process directly at the second stage.

These remarkable results, which were first obtained by R. R. P. Jackson [1954, 1956], suggest that the output process might, in fact, be the same as the input process to the first stage, that is, a Poisson process. The truth of this conjecture was proved by Burke [1956]; Burke's theorem states that the sequence of departures from an Erlang delay system in equilibrium follows a Poisson process and, further, the state of this Erlang delay system at any arbitrary time t_0 is independent of the departure process previous to t_0. (These results were also obtained by Reich [1957].) It easily follows from Burke's theorem that the analysis given here for the case of two queues in tandem can be extended to the case of an arbitrary number of queues in tandem. A proof of Burke's theorem is outlined in Exercise 2 of Chapter 5.

As one might guess, there are many other remarkable facts about tandem queueing systems and departure processes to be uncovered. For example, in single-server tandem queues with service in order of arrival, the sojourn times (sojourn time is waiting time plus service time) of an individual customer in the various stages of the tandem queueing systems are independent random variables; however, the waiting times are not independent. Surveys of results on departure processes and tandem queues are given by Burke [1972] and Daley [1976].

The model of queues in tandem generalizes naturally into that of networks of queues (see Exercises 8 and 9). This is an area of active research, with important applications in telecommunications and computer science. (The entire September 1978 issue of *Computing Surveys*, Vol. 10, No. 3, pp. 219–362, is entitled "Queueing Network Models of Computer System Performance.") Some recent works on networks of queues are Barbour [1976]; Baskett, Chandy, Muntz, and Palacios [1975]; Beutler and Melamed [1978]; Beutler, Melamed, and Ziegler [1977]; Boxma [1979a,b]; Chandy, Howard, and Towsley [1977]; Courtois [1977]; Disney [1975]; Kelly [1976, 1979]; Lemoine [1977, 1978, 1979]; Melamed [1979a,b]; Mitrani [1979]; Noetzel [1979]; Pittel [1979]; Schassberger [1978a,b]; Simon and Foley [1979]; Stoyan [1977]; and Vantilborgh [1978]. Additional references are given in Exercises 8 and 9.

Exercises

8. *Networks of queues.*
   a. Consider a network of queues, say Q_1, Q_2, Q_3, and Q_4, with input-output structure as in Figure 4.2, where each departing customer from Q_1 is

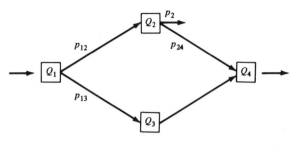

Figure 4.2

assigned, independently, to either Q_2 or Q_3, with probabilities p_{12} and p_{13} ($=1-p_{12}$), respectively; and similarly for departures from Q_2. Assume also that the input to Q_1 is Poisson, with rate λ; that Q_i is served by s_i servers, with exponential service times with mean μ_i^{-1}; and that all customers wait as long as necessary for service. Find the equilibrium distribution of the number of customers present at each queue.

  b. Consider the same network as in part a, except that now the portion of the output from Q_2 that does not go to Q_4 feeds back as input to Q_1 instead of leaving the system. Write and solve the equations that determine the equilibrium distribution of the number of customers present at each queue. (Networks of this type, with feedback, were first considered by J. R. Jackson [1957], who observed that, as in part a, the results are the same as one would obtain if the input to each queue were a Poisson process. Jackson remarked that this was "far from surprising in view of recent papers by P. J. Burke and E. Reich." However, as Burke [1972] points out, Jackson's results are, in fact, quite astonishing, because the combined input to any queue that receives feedback is *not* Poisson:

> The point is that in networks without feedback not only are the states of the component queues distributed *as if* the inputs are Poisson, these actually are Poisson; whereas when feedback is present the only statement that can be made is that these distributions are the same as those which would arise if the inputs were in fact Poisson.

For example, the results of Exercise 41 of Chapter 3 can be obtained by assuming (incorrectly) that Burke's theorem applies and therefore the combined stream of new and fed-back arrivals is a Poisson stream. Burke [1976] proves that this is not the case; nevertheless, the system behaves as if it were the case. Recently, Beutler and Melamed [1978] and Melamed [1979a] have shown that Burke's theorem can be generalized to apply to the exit arcs of Jackson-type networks with single-server nodes. For additional related results see Disney, McNickle, and Simon [1980], and the references cited therein; see also Disney, Farrell, and de Morais [1973], and Foley [1979].)

9. *Closed networks of queues.*

  a. Consider a network of m queues, $Q_1, Q_2, \ldots, Q_m$, with the structure as in Figure 4.3. Assume that Q_i is served by s_i servers, with exponential service

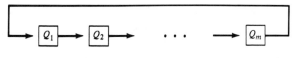

Figure 4.3

times with mean μ_i^{-1}; that all customers wait in each queue as long as necessary for service; and that the number n of customers in the network is fixed. (Thus, the n customers circulate continually among the m stages of the network.) Let $P(j_1,\ldots,j_m)$ be the equilibrium joint probability that the number N_i of customers in stage i (waiting or in service) is j_i ($i=1,2,\ldots,m$), and show that

$$P(j_1,\ldots,j_m) = \frac{\prod_{i=1}^{m} Q_i(j_i)}{\sum_S \prod_{i=1}^{m} Q_i(r_i)}, \tag{1}$$

where S denotes all combinations of the nonnegative integers $r_1,\ldots,r_m$ for which $r_1+\cdots+r_m=n$, and

$$Q_i(j_i) = \begin{cases} \dfrac{(1/\mu_i)^{j_i}}{j_i!} & (j_i < s_i), \\[6pt] \dfrac{(1/\mu_i)^{j_i}}{s_i! s_i^{j_i-s_i}} & (j_i \geq s_i). \end{cases}$$

Are the random variables $N_1,\ldots,N_m$ independent?

b. The model of part a is a specialization of one considered by Gordon and Newell [1967], in which each departure from stage i decides, independently, to enter stage j with probability p_{ij} ($i,j=1,2,\ldots,m$). Convince yourself that the conservation-of-flow equations that determine the equilibrium joint distribution $P(j_1,\ldots,j_m)$ of the number of customers present in each stage of Gordon and Newell's model are

$$\sum_{i=1}^{m} \mu_i(j_i) P(j_1,\ldots,j_m) = \sum_{i=1}^{m} \sum_{\substack{k=1 \\ k \neq i}}^{m} \mu_k(j_k+1) p_{ki} P(j_1,\ldots,j_i-1,\ldots,j_k+1,\ldots,j_m)$$

$$+ \sum_{i=1}^{m} \mu_i(j_i) p_{ii} P(j_1,\ldots,j_m), \tag{2}$$

where $\mu_i(j)$ is defined as in (2.12), and, of course, $\sum P(j_1,\ldots,j_m) = 1$. Show that if one again assumes a solution of the form (1), with $Q_i(j_i)$ now defined as

$$Q_i(j_i) = \begin{cases} \dfrac{(1/x_i)^{j_i}}{j_i!} & (j_i < s_i), \\[6pt] \dfrac{(1/x_i)^{j_i}}{s_i! s_i^{j_i-s_i}} & (j_i \geq s_i), \end{cases} \tag{3}$$

then it follows from substitution of the assumed solution (1) and (3) into the equilibrium state equations (2) that

$$\sum_{i=1}^{m} \mu_i(j_i) \left[1 - \mu_i^{-1} x_i \sum_{k=1}^{m} \mu_k x_k^{-1} p_{ki} \right] = 0. \tag{4}$$

Observe that if (4) is to hold for every j_i ($0 \leq j_i \leq n$), the term in brackets must vanish for each i; that is,

$$\sum_{k=1}^{m} p_{ki}(\mu_k x_k^{-1}) = \mu_i x_i^{-1} \qquad (i = 1, 2, \ldots, m). \tag{5}$$

It can be shown (see, for example, p. 153 of Çinlar [1975]) that if the matrix $[p_{ij}]$ is that of an *irreducible aperiodic Markov chain*, then the vector $(\mu_1 x_1^{-1}, \ldots, \mu_m x_m^{-1})$ is uniquely determined to within a multiplicative factor; furthermore, the components of this solution vector are all strictly positive. Conclude, therefore, that the assumed solution (1) and (3) is indeed the correct solution of Equation (2), provided only that the finite Markov chain defined by the matrix of transition probabilities $[p_{ij}]$ is aperiodic and irreducible (which is not a severe restriction on the kinds of transitions allowed among stages of the network).

Gordon and Newell's model subsumes as special cases the closed network models of earlier authors, such as Koenigsberg [1958] and Finch [1959]. Gordon and Newell show also that the open models of Exercise 8 ("open" in the sense that customers can enter and leave the system) and the closed models of this exercise are equivalent. For a discussion of numerical methods for computing the equilibrium distributions that describe networks of this type, see Buzen [1973].

10. The following model can be used to study talking and dialing equipment interactions in a modern private branch exchange telephone switching system. Calls arrive at random to two groups of servers, G_1 (time slots for talking) and G_2 (digit trunks for dialing). If an arrival finds all servers in G_1 busy, the call is cleared from the system. If an arrival finds at least one idle server in each of G_1 and G_2, the call holds one server in each group simultaneously for an exponentially distributed time interval with mean μ_1^{-1} (dialing time); it then releases the server in G_2 but retains the server in G_1 for an additional exponentially distributed time with mean μ_2^{-1} (conversation length). If an arrival finds all servers in G_2 busy and at least one idle server in G_1, the call holds a server (waits) in G_1 until it gains access to an idle server in G_2; it then proceeds in the same manner as a call that initially finds an idle server in both G_1 and G_2. Write and solve an appropriate set of equilibrium state equations for this system.

11. The following is a simplified version of a model that has been used to estimate the number of fire engines needed in New York City (Chaiken [1971]). Suppose that in a region of the city fire alarms occur according to a Poisson process with rate λ. With probability p_1 an alarm will request the dispatch of a

Generating Functions (Overflow Traffic)

single fire engine, which will be sufficient to handle the fire. In this case, the length of time the fire engine spends away from the firehouse is exponentially distributed with mean $\tau(1)$. With probability p_2 an alarm will again request the dispatch of only a single fire engine, but this will prove inadequate and a second fire engine will be dispatched after an exponentially distributed time interval with mean $\tau_1(2)$. The two fire engines will then work together for an exponential time with mean $\tau_2(2)$, after which one of the fire engines returns to the firehouse and the other remains at the fire for an additional exponential "cleanup" time with mean $\tau_3(2)$. Finally, with probability p_3 an alarm will request the dispatch of two fire engines simultaneously. In this case the two fire engines will work together for an exponential time interval with mean $\tau_1(3)$, after which one of the fire engines remains for an additional exponential cleanup time with mean $\tau_2(3)$. Thus there are three types of fires, each composed of a number of independent exponential phases—the type-1 fire, which occurs with probability p_1, has one phase with mean $\tau(1)$; the type-2 fire has three phases with means $\tau_1(2)$, $\tau_2(2)$, and $\tau_3(2)$; and the type-3 fire has two phases with means $\tau_1(3)$ and $\tau_2(3)$. The total number of fire engines in the region is n. Any fire engines needed beyond these n will be provided from outside the region. Let $P(i;j_1,j_2,j_3;k_1,k_2)$ be the equilibrium probability that at an arbitrary instant there are simultaneously i fires of type 1, j_ν fires of type 2 in phase ν ($\nu=1,2,3$), and k_ν fires of type 3 in phase ν ($\nu=1,2$); and let $P_m = \sum_I P(i;j_1,j_2,j_3;k_1,k_2)$, where $I=\{i,j_1,j_2,j_3,k_1,k_2: i+j_1+2j_2+j_3+2k_1+k_2 = m\}$. Then $P = \sum_{m=n}^{\infty} P_m$ is the probability that there will not be any idle fire engines in the region when an alarm occurs. Thus, calculation of $P(i;j_1,j_2,j_3;k_1,k_2)$ allows one to estimate the number of fire engines required to keep $P < \varepsilon$ for any specified ε. Show that

$$P(i;j_1,j_2,j_3;k_1,k_2) = \frac{[p_1\lambda\tau(1)]^i}{i!} \frac{[p_2\lambda\tau_1(2)]^{j_1}}{j_1!} \frac{[p_2\lambda\tau_2(2)]^{j_2}}{j_2!} \frac{[p_2\lambda\tau_3(2)]^{j_3}}{j_3!}$$
$$\times \frac{[p_3\lambda\tau_1(3)]^{k_1}}{k_1!} \frac{[p_3\lambda\tau_2(3)]^{k_2}}{k_2!}$$
$$\times \exp\{-[p_1\lambda\tau(1)+p_2\lambda\tau_1(2)+p_2\lambda\tau_2(2)+p_2\lambda\tau_3(2)$$
$$+p_3\lambda\tau_1(3)+p_3\lambda\tau_2(3)]\}.$$

(Chaiken also shows that, as one might expect, the same result holds when the phase lengths are not exponentially distributed.)

4.3. Generating Functions (Overflow Traffic)

As with state equations in one variable, generating functions are often useful in solving multidimensional state equations. It is interesting that the generating-function technique may be applicable even when some of the variables take on only finitely many distinct values.

As an example of the use of generating functions for solving such a problem, we consider the case of a queueing system composed of an s-server primary group and an infinite-server overflow group. In this

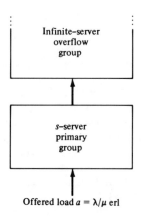

Figure 4.4

so-called *simple overflow model*, it is assumed that customers request service first from the s-server primary group, and that all customers who find all servers busy in the primary group overflow to and are handled by the infinite-server overflow group.

The system to be studied is represented by Figure 4.4. Let $P\{M=j, N=k\} = P(j,k)$ be the statistical-equilibrium probability that $M=j$ and $N=k$ customers are simultaneously in service on the primary and overflow groups, respectively. The input is Poisson with rate λ, and the service times are exponential with mean μ^{-1}.

Following Kosten [1937], we shall obtain the generating function of the distribution $\{P(j,k)\}$, from which the following important formulas for the mean $E(N) = \alpha$ and variance $V(N) = v$ are easily found:

$$\alpha = aB(s,a) \tag{3.1}$$

and

$$v = \alpha\left(1 - \alpha + \frac{a}{s+1+\alpha-a}\right), \tag{3.2}$$

where $a = \lambda/\mu$ is the offered load in erlangs and $B(s,a)$ is the Erlang loss formula.

These results form the basis for an important approximation technique, called the *equivalent random method*, which we shall discuss in Section 4.7. The derivation given here is instructive, but it is not the most elegant route leading to (3.1) (which is intuitively obvious) and (3.2). (See Brockmeyer [1954], Riordan [1956, 1962], Wallström [1966], and Descloux [1970]. The properties of overflow streams are discussed in Palm [1943] and Khintchine [1969]. Although (3.1) and (3.2) follow easily from Kosten's analysis, which we give here, according to Riordan they were first given

Generating Functions (Overflow Traffic)

explicitly in independent unpublished memoranda (1941) by H. Nyquist and E. C. Molina.) Also, despite the elementary nature of these calculations, they are involved. If the reader is willing to accept (3.1) and (3.2), he can skip the calculations without loss of continuity.

The equations for the probabilities $\{P(j,k)\}$ when $j \leq s-1$ are

$$(a+j+k)P(j,k) = aP(j-1,k) + (j+1)P(j+1,k)$$
$$+ (k+1)P(j,k+1)$$
$$[P(-1,k)=0; \quad j=0,1,\ldots,s-1; \quad k=0,1,\ldots], \quad (3.3)$$

where $a = \lambda/\mu$. When an arriving customer finds all s servers in the primary group busy, he takes a server in the overflow group. Thus, for $j = s$,

$$(a+s+k)P(s,k) = aP(s-1,k) + aP(s,k-1)$$
$$+ (k+1)P(s,k+1)$$
$$[P(s,-1)=0; \quad k=0,1,\ldots]. \quad (3.4)$$

The system is completely described by Equations (3.3), (3.4), and

$$\sum_{j=0}^{s} \sum_{k=0}^{\infty} P(j,k) = 1. \quad (3.5)$$

To solve these equations using generating functions, consider the auxiliary system of equations obtained from (3.3) by ignoring the finiteness of s:

$$(a+j+k)\hat{P}(j,k) = a\hat{P}(j-1,k) + (j+1)\hat{P}(j+1,k)$$
$$+ (k+1)\hat{P}(j,k+1)$$
$$[\hat{P}(-1,k)=0; \quad j=0,1,\ldots; \quad k=0,1,\ldots]. \quad (3.6)$$

We shall solve the set (3.6) in such a way that

$$\hat{P}(j,k) = P(j,k) \quad (j=0,1,\ldots,s; \quad k=0,1,\ldots). \quad (3.7)$$

The values $\{P(j,k)\}$ for $j > s$ yielded by the equations (3.6) are of no consequence. (Observe that this is the same strategy used in Exercise 46 of Chapter 3.)

Define the two-dimensional probability-generating function

$$\hat{\psi}(x,y) = \sum_{j=0}^{\infty} \sum_{k=0}^{\infty} \hat{P}(j,k) x^j y^k. \quad (3.8)$$

Substitution of equations (3.6) into (3.8) yields, after some straightforward manipulation, the partial differential equation

$$(1-x)\frac{\partial}{\partial x}\hat{\psi}(x,y)+(1-y)\frac{\partial}{\partial y}\hat{\psi}(x,y)=a(1-x)\hat{\psi}(x,y), \quad (3.9)$$

whose solution is

$$\hat{\psi}(x,y)=e^{-a(1-x)}G\left(\frac{1-y}{1-x}\right), \quad (3.10)$$

where G is an arbitrary function. (For discussions of partial differential equations of the form (3.9) see, for example, Garabedian [1964].)

We now determine the function G such that the coefficients $\{\hat{P}(j,k)\}$ generated by $\hat{\psi}(x,y)$, which satisfy the state equations (3.3), also satisfy the boundary condition implied by Equation (3.4). Then Equation (3.7) will be satisfied. To this end, we first expand G in powers of $(1-y)/(1-x)$. Then, from (3.10),

$$\hat{\psi}(x,y)=e^{-a(1-x)}\sum_{\nu=0}^{\infty}c_\nu\left(\frac{1-y}{1-x}\right)^\nu, \quad (3.11)$$

where the coefficients $\{c_\nu\}$ remain to be determined.

Let us expand the term $e^{-a(1-x)}(1-x)^{-\nu}$ in Equation (3.11) into a power series in x:

$$e^{-a(1-x)}(1-x)^{-\nu}=\sum_{j=0}^{\infty}\phi_\nu(j)x^j, \quad (3.12)$$

where, as is easily verified,

$$\phi_\nu(j)=\begin{cases}\dfrac{a^j}{j!}e^{-a} & (\nu=0),\\[2mm] e^{-a}\displaystyle\sum_{i=0}^{j}\binom{\nu+i-1}{i}\dfrac{a^{j-i}}{(j-i)!} & (\nu=1,2,\ldots).\end{cases} \quad (3.13)$$

Two useful identities are

$$\sum_{j=0}^{n}\phi_\nu(j)=\phi_{\nu+1}(n) \quad (3.14)$$

and

$$(j+1)\phi_\nu(j+1)=a\phi_\nu(j)+\nu\phi_{\nu+1}(j). \quad (3.15)$$

To obtain Equation (3.14), we simply equate coefficients on both sides of the identity

$$e^{-a(1-x)}(1-x)^{-\nu} = e^{-a(1-x)}(1-x)^{-(\nu+1)} - e^{-a(1-x)}(1-x)^{-(\nu+1)}x.$$

The result is

$$\phi_\nu(j) = \phi_{\nu+1}(j) - \phi_{\nu+1}(j-1),$$

from which Equation (3.14) follows directly by summing both sides of this equation. Similarly, Equation (3.15) follows after we equate coefficients on both sides of the identity

$$\frac{d}{dx} e^{-a(1-x)}(1-x)^{-\nu} = ae^{-a(1-x)}(1-x)^{-\nu} + \nu e^{-a(1-x)}(1-x)^{-(\nu+1)}.$$

Now consider Equation (3.7). When $j = s$, comparison of Equations (3.4) and (3.6) shows that if (3.7) is true, then

$$a\hat{P}(s,k-1) = (s+1)\hat{P}(s+1,k) \quad (k=1,2,\ldots) \tag{3.16}$$

and $\hat{P}(s+1,0) = 0$.

Let $\hat{\psi}_j(y)$ be the marginal generating function

$$\hat{\psi}_j(y) = \sum_{k=0}^{\infty} \hat{P}(j,k)y^k, \tag{3.17}$$

so that

$$\hat{\psi}(x,y) = \sum_{j=0}^{\infty} \hat{\psi}_j(y)x^j. \tag{3.18}$$

Then, introducing (3.16) into (3.17), we obtain

$$ay\hat{\psi}_s(y) = (s+1)\hat{\psi}_{s+1}(y). \tag{3.19}$$

From Equation (3.18) we see that

$$\hat{\psi}_j(y) = \frac{1}{j!}\left(\frac{\partial^j}{\partial x^j}\hat{\psi}(x,y)\right)_{x=0}. \tag{3.20}$$

Using (3.12) in (3.11), we have

$$\hat{\psi}(x,y) = \sum_{\nu=0}^{\infty} c_\nu(1-y)^\nu \sum_{j=0}^{\infty} \phi_\nu(j)x^j. \tag{3.21}$$

It follows from Equations (3.20) and (3.21) that

$$\hat{\psi}_j(y) = \sum_{\nu=0}^{\infty} c_\nu \phi_\nu(j)(1-y)^\nu. \tag{3.22}$$

In particular,

$$\hat{\psi}_s(y) = \sum_{\nu=0}^{\infty} c_\nu \phi_\nu(s)(1-y)^\nu. \tag{3.23}$$

If we use the identity

$$(1-y)^\nu - (1-y)^{\nu+1} = (1-y)^\nu y,$$

we can write

$$y\hat{\psi}_s(y) = \sum_{\nu=0}^{\infty} c_\nu \phi_\nu(s)(1-y)^\nu - \sum_{\nu=0}^{\infty} c_\nu \phi_\nu(s)(1-y)^{\nu+1}. \tag{3.24}$$

Therefore, equating coefficients of $(1-y)^\nu$ in Equation (3.19) we have, by virtue of Equations (3.23) and (3.24), for $\nu = 1, 2, \ldots$,

$$a c_\nu \phi_\nu(s) - a c_{\nu-1} \phi_{\nu-1}(s) = (s+1) c_\nu \phi_\nu(s+1). \tag{3.25}$$

Comparison of Equation (3.25) with (3.15) when $j = s$ shows that

$$c_\nu \nu \phi_{\nu+1}(s) = -a c_{\nu-1} \phi_{\nu-1}(s) \qquad (\nu = 1, 2, \ldots)$$

or, solving recursively,

$$c_\nu = (-1)^\nu \frac{a^\nu}{\nu!} \frac{\phi_1(s)\phi_0(s)}{\phi_{\nu+1}(s)\phi_\nu(s)} c_0 \qquad (\nu = 1, 2, \ldots). \tag{3.26}$$

It remains to find c_0.

Applying the binomial theorem to the factor $(1-y)^\nu$ in Equation (3.21), we can write

$$\hat{\psi}(x,y) = \sum_{\nu=0}^{\infty} c_\nu \sum_{k=0}^{\nu} (-1)^k \binom{\nu}{k} y^k \sum_{j=0}^{\infty} \phi_\nu(j) x^j,$$

where we define

$$\binom{0}{0} = 1,$$

and which, after interchanging the orders of summation over ν and k, becomes

$$\hat{\psi}(x,y) = \sum_{k=0}^{\infty} \left[\sum_{\nu=k}^{\infty} (-1)^k \binom{\nu}{k} c_\nu \right] y^k \sum_{j=0}^{\infty} \phi_\nu(j) x^j.$$

Finally, interchanging the orders of summation over ν and j, we have

$$\hat{\psi}(x,y) = \sum_{k=0}^{\infty} \sum_{j=0}^{\infty} \left[(-1)^k \sum_{\nu=k}^{\infty} \binom{\nu}{k} c_\nu \phi_\nu(j) \right] y^k x^j. \quad (3.27)$$

Comparison of Equations (3.8) and (3.27) shows that

$$\hat{P}(j,k) = (-1)^k \sum_{\nu=k}^{\infty} \binom{\nu}{k} c_\nu \phi_\nu(j),$$

which, in view of Equation (3.26), can be written

$$\hat{P}(j,k) = (-1)^k \sum_{\nu=k}^{\infty} (-1)^\nu \binom{\nu}{k} \frac{a^\nu}{\nu!} \frac{\phi_1(s)\phi_0(s)}{\phi_{\nu+1}(s)\phi_\nu(s)} \phi_\nu(j) c_0. \quad (3.28)$$

We now determine c_0 from Equation (3.7) and the normalization equation (3.5). Summing first with respect to the index k, we have

$$\sum_{k=0}^{\infty} P(j,k) = \sum_{k=0}^{\infty} (-1)^k \sum_{\nu=k}^{\infty} (-1)^\nu \binom{\nu}{k} \frac{a^\nu}{\nu!} \frac{\phi_1(s)\phi_0(s)}{\phi_{\nu+1}(s)\phi_\nu(s)} \phi_\nu(j) c_0,$$

which, after interchange of the order of summation, becomes

$$\sum_{k=0}^{\infty} P(j,k) = c_0 \sum_{\nu=0}^{\infty} (-1)^\nu \frac{a^\nu}{\nu!} \frac{\phi_1(s)\phi_0(s)}{\phi_{\nu+1}(s)\phi_\nu(s)} \phi_\nu(j) \sum_{k=0}^{\nu} (-1)^k \binom{\nu}{k}. \quad (3.29)$$

The last summation on the right-hand side of (3.29) is the binomial expansion of $(1-1)^\nu$; therefore

$$\sum_{k=0}^{\nu} (-1)^k \binom{\nu}{k} = 0 \quad (\nu \geq 1).$$

Since by definition $\binom{0}{0} = 1$, Equation (3.29) becomes

$$\sum_{k=0}^{\infty} P(j,k) = c_0 \phi_0(j). \quad (3.30)$$

Summing both sides of Equation (3.30) with respect to j, we have

$$\sum_{j=0}^{s} \sum_{k=0}^{\infty} P(j,k) = c_0 \sum_{j=0}^{s} \phi_0(j). \qquad (3.31)$$

Using the normalization equation (3.5), it follows that

$$c_0 = \frac{1}{\sum_{j=0}^{s} \phi_0(j)}$$

and finally, using the identity (3.14), we conclude that

$$c_0 = \frac{1}{\phi_1(s)}. \qquad (3.32)$$

Thus, in view of Equation (3.28), we have Kosten's explicit expression for the distribution $\{P(j,k)\}$:

$$P(j,k) = (-1)^k \phi_0(s) \sum_{\nu=k}^{\infty} (-1)^\nu \binom{\nu}{k} \frac{a^\nu}{\nu!} \frac{\phi_\nu(j)}{\phi_{\nu+1}(s)\phi_\nu(s)}$$

$$(j=0,1,\ldots,s; \quad k=0,1,\ldots), \qquad (3.33)$$

where $\{\phi_\nu(j)\}$ is given by Equation (3.13).

Note that, according to Equations (3.30), (3.32), and (3.13), the marginal distribution of the number of busy servers in the primary group is, of course, the Erlang loss distribution

$$\sum_{k=0}^{\infty} P(j,k) = \frac{\phi_0(j)}{\phi_1(s)} = \frac{\frac{a^j}{j!}}{\sum_{i=0}^{s} \frac{a^i}{i!}} \qquad (j=0,1,\ldots,s). \qquad (3.34)$$

We are now prepared to derive the required formulas (3.1) and (3.2) for the mean $E(N) = \alpha$ and the variance $V(N) = v$ of the overflow distribution. It follows from Equation (3.17) that

$$\left(\frac{d}{dy}\hat{\psi}_j(y)\right)_{y=1} = \sum_{k=1}^{\infty} k P(j,k) \qquad (j=0,1,\ldots,s)$$

and therefore

$$\alpha = \sum_{j=0}^{s} \left(\frac{d}{dy}\hat{\psi}_j(y)\right)_{y=1}. \qquad (3.35)$$

From Equation (3.22) we see that

$$\left(\frac{d}{dy}\hat{\psi}_j(y)\right)_{y=1} = -c_1\phi_1(j),$$

which, in view of (3.26) and (3.32), becomes

$$\left(\frac{d}{dy}\hat{\psi}_j(y)\right)_{y=1} = a\frac{\phi_0(s)}{\phi_2(s)\phi_1(s)}\phi_1(j). \tag{3.36}$$

Using (3.36) in (3.35) we have, in light of the identity (3.14),

$$\alpha = a\frac{\phi_0(s)}{\phi_1(s)}. \tag{3.37}$$

The required result (3.1), which of course could have been anticipated on other grounds, follows from (3.37).

To calculate $v = V(N)$, note that

$$\sum_{j=0}^{s}\left(\frac{d^2}{dy^2}\hat{\psi}_j(y)\right)_{y=1} = E(N^2) - \alpha.$$

Since $V(N) = E(N^2) - E^2(N)$, we have

$$\sum_{j=0}^{s}\left(\frac{d^2}{dy^2}\hat{\psi}_j(y)\right)_{y=1} = v + \alpha^2 - \alpha. \tag{3.38}$$

From Equation (3.22), we see that

$$\left(\frac{d^2}{dy^2}\hat{\psi}_j(y)\right)_{y=1} = 2c_2\phi_2(j) \quad (j=0,1,\ldots,s). \tag{3.39}$$

Therefore

$$v + \alpha^2 - \alpha = 2c_2\sum_{j=0}^{s}\phi_2(j). \tag{3.40}$$

Using the identity (3.14), we have

$$v + \alpha^2 - \alpha = 2c_2\phi_3(s). \tag{3.41}$$

Equation (3.2) now follows directly from (3.41).

The important concept to be digested here is in the use of probability-generating functions to solve multidimensional state equations which, at first glance, do not seem amenable to these techniques because of the existence of certain boundary conditions. The key steps are the extension of the original state equations so that the generating-function technique can be used, the expansion of the generating function in a power series, and the calculation of the unknown coefficients by forcing the generating function to agree with the previously ignored boundary conditions.

It is interesting to speculate whether the preceding results are valid independently of the form of the service-time distribution function. In particular, is the formula (3.2) for the variance $V(N) = v$ of the overflow distribution valid in general for systems with Poisson input? One might suspect so, since both the distribution of the number M of customers on the primary group and the distribution of the total number $M + N$ of customers in the system are known to be independent of the form of the service-time distribution function. Nevertheless, as Burke [1971] has demonstrated in a study of the simple overflow model with constant service times, this is not the case. Of course, (3.1) remains true, since it gives the mean number of customers per service time who find all primary servers busy.

Overflow models are of both theoretical and practical interest, and have received much attention throughout the past 50 years. Syski [1960] summarizes most of this work, including practical applications. Riordan [1962] and Khintchine [1969] also discuss and summarize various aspects of overflow problems. Important papers not discussed or summarized in these books include Burke [1971], Descloux [1963, 1970], Neal [1971], and Wallström [1966]. (See also Section 4.7 and Exercise 12.)

Exercise

12. *Apportioning the moments of the overflow distribution.* Suppose that customers arrive in n independent Poisson streams with rates $\lambda_1, \ldots, \lambda_n$ at a single primary group of servers and, if all servers in the primary group are busy, are served on a single infinite-server overflow group. Let μ^{-1} be the mean service time, and set $a_i = \lambda_i / \mu$ ($i = 1, 2, \ldots, n$). Let N_i be the number of customers from the ith Poisson stream present on the overflow group at an arbitrary instant in equilibrium. Then, with $a = a_1 + \cdots + a_n$ and $N = N_1 + \cdots + N_n$,

$$E(N_i) = \frac{a_i}{a} E(N), \tag{1}$$

$$V(N_i) = \left(\frac{a_i}{a}\right)^2 V(N) + \frac{a_i}{a}\left(1 - \frac{a_i}{a}\right) E(N), \tag{2}$$

$$\mathrm{Cov}(N_i, N_j) = \left(\frac{a_i}{a}\right)\left(\frac{a_j}{a}\right)[V(N) - E(N)]. \tag{3}$$

For simplicity, we shall prove the results (1)–(3) for the case $n = 2$ only.

a. Assume exponential service times. Let M be the number of customers in service on the primary group, and define $h(j, k_1, k_2) = P\{M = j, N_1 = k_1, N_2 = k_2\}$. Write the equilibrium state equations for the distribution

$\{h(j,k_1,k_2)\}$, and show that the solution is

$$h(j,k_1,k_2) = P(j,k)\binom{k}{k_1}\left(\frac{a_1}{a}\right)^{k_1}\left(\frac{a_2}{a}\right)^{k_2},$$

where $P(j,k)$ is given by Equation (3.33). [*Hint:* Substitute the above expression for $h(j,k_1,k_2)$ into the state equations, and show that Equations (3.3) and (3.4) result.]

b. Show that

$$P\{N_1 = k_1, N_2 = k_2 | N = k\} = \binom{k}{k_i}\left(\frac{a_1}{a}\right)^{k_1}\left(\frac{a_2}{a}\right)^{k_2}. \qquad (4)$$

c. Show that

$$E(N_i | N = k) = k\frac{a_i}{a}, \qquad (5)$$

$$E(N_i^2 | N = k) = k(k-1)\left(\frac{a_i}{a}\right)^2 + k\frac{a_i}{a}. \qquad (6)$$

d. Show that (5) implies (1).
e. Show that (6) implies

$$E(N_i^2) = \left(\frac{a_i}{a}\right)^2 E(N^2) + \left(\frac{a_i}{a}\right)\left(1 - \frac{a_i}{a}\right)E(N). \qquad (7)$$

f. Show that (7) implies (2). Define the *peakedness factor* for any stream as its overflow-distribution variance-to-mean ratio, and show that Equation (2) can be written

$$z_i - 1 = p_i(z - 1), \qquad (8)$$

where p_i is the proportion of customers in stream i, and z_i and z are the relevant peakedness factors.

g. Continue the argument of parts b–f to prove (3).
h. Conclude that, for exponential service times, the moments of the overflow distribution corresponding to the ith Poisson stream (at least for $n=2$) are given by Equations (1), (2), and (3) with $E(N) = \alpha$ given by (3.1) and $V(N) = v$ given by (3.2).

These results, which are useful in the application of the equivalent random method, were originally obtained by A. Descloux [1962, unpublished] and Lotze [1964]. The following argument shows that the assumption of exponential service times is not necessary for the validity of Equation (4), and therefore the results (1)–(3) are valid for arbitrary service-time distribution function: Because the original streams of traffic are Poisson, each sees the same probability of blocking on the primary group. Hence, given that a customer is blocked, the probability that this customer is from stream i is the ratio a_i/a. Further, this probability is independent of the composition of the other customers in service on the overflow group. Therefore, given the number of simultaneous customers in service on the overflow group, the distribution of the number of customers of type i is the binomial with proportion a_i/a, given by Equation (4). It is interesting that the formulas (1)–(3) are valid for an arbitrary service-time distribution function, whereas the moment $V(N)$ [but not $E(N)$] depends on the distribution of service times. [If the service times are exponentially distributed, then $V(N) = v$ is given by Equation (3.2).]

4.4. Macrostates (Priority Reservation)

In all the examples so far, the naming of the states of the system has been quite natural in the context of the problem, and our attention has been directed toward methods of solution of the corresponding equilibrium state equations. In this section we observe that it is sometimes possible and convenient to consider groupings of the "natural" states (*microstates*) into *macrostates*.

For example, consider a system in which a group of s servers handles two independent Poisson streams of traffic, where customers in stream 1 are designated high-priority customers, and those in stream 2 are designated low-priority customers. A high-priority customer will be cleared from the system if on arrival he finds no idle servers, whereas a low-priority customer will be cleared from the system if on arrival he finds less than $n+1$ idle servers. Service times for both high- and low-priority customers are assumed to be exponentially distributed with the same mean. We wish to find the probability of blocking experienced by the customers of each priority class.

One approach to this problem is to describe the system by the two-dimensional states E_{j_1,j_2}, where j_1 and j_2 are, respectively, the numbers of high-priority and low-priority customers in service; and then write and solve the "rate out = rate in" state equations for the corresponding distribution $\{P(j_1,j_2)\}$. If we define

$$P_j = \sum_{j_1+j_2=j} P(j_1,j_2) \qquad (j=0,1,\ldots,s), \tag{4.1}$$

then P_s is the probability of blocking seen by the high-priority customers, and $\sum_{j=s-n}^{s} P_j$ is the probability of blocking suffered by the low-priority customers. The indicated calculations are tedious and, as the following illustrates, unnecessary.

Consider now the grouping of all those two-dimensional microstates E_{j_1,j_2} with the property that $j_1+j_2=j$ into a single one-dimensional macrostate E_j. That is, we define the macrostate E_j as the set of all microstates E_{j_1,j_2} that correspond to a total of j customers in service. It should now be clear that the probability corresponding to the state E_j is P_j, and that the distribution $\{P_j\}$ is easily found from the one-dimensional "rate up = rate

Macrostates (Priority Reservation)

down" state equations (where λ_1 and λ_2 are the arrival rates and μ^{-1} is the mean service time):

$$(\lambda_1 + \lambda_2)P_{j-1} = j\mu P_j \qquad (j=1,2,\ldots,s-n), \qquad (4.2)$$

$$\lambda_1 P_{j-1} = j\mu P_j \qquad (j=s-n+1, s-n+2,\ldots,s). \qquad (4.3)$$

The solution, found easily by recurrence, is

$$P_j = \begin{cases} \dfrac{(a_1+a_2)^j}{j!} P_0 & (j=1,2,\ldots,s-n), \\ \dfrac{(a_1+a_2)^{s-n} a_1^{j-(s-n)}}{j!} P_0 & (j=s-n+1, s-n+2,\ldots,s), \end{cases} \qquad (4.4)$$

where $a_i = \lambda_i/\mu$ is the offered load, in erlangs, from the ith stream, and where, from the normalization requirement,

$$P_0 = \left(\sum_{k=0}^{s-n} \frac{(a_1+a_2)^k}{k!} + (a_1+a_2)^{s-n} \sum_{k=s-n+1}^{s} \frac{a_1^{k-(s-n)}}{k!} \right)^{-1}. \qquad (4.5)$$

It is worth pointing out that the example considered here is identical to Exercise 43 of Chapter 3. A reader with no knowledge of multidimensional birth-and-death processes would not make the mistake of doing this problem the hard way. Of course, if the mean service times of the high-priority and low-priority customers were unequal, the one-dimensional macrostate approach would not apply, and the advantage would lie with the more advanced reader.

In the above example, the fact that a macrostate approach is possible is almost obvious, and once the proposed macrostates are identified, the validity of the results is clear. This is not always the case. When a valid set of equilibrium macrostate equations exists, the macrostate equations can always be derived from the ordinary equilibrium microstate equations. In general, however, it is arguable whether it is easier to write down directly an appropriate set of macrostate equations, or to derive them (if indeed they exist) from the more detailed but sometimes conceptually simpler ordinary microstate equations.

As an illustration of these difficulties, we consider a model that is a generalization of the model considered above. Imagine a priority reservation system with s identical exponential servers, where customers arrive in s independent Poisson streams, $S_1, S_2, \ldots, S_s$. Customers in stream S_i may start service only when $i-1$ or fewer servers are busy. All customers wait as long as necessary to obtain service. (If we denote by λ_i the arrival rate of the customers in stream S_i, then this model reduces to the s-server Erlang delay model when $\lambda_1 = \lambda_2 = \cdots = \lambda_{s-1} = 0$.)

Let $E_{j;n_1,...,n_j}$ be the (micro-) state corresponding to the event {j servers are busy and n_i customers from stream S_i are waiting, $i = 1, 2, ..., j$, $j = 1, 2, ..., s$}, and let $P(j; n_1, ..., n_j)$ be the corresponding equilibrium probability. It is tedious but straightforward to write the conservation-of-flow equations that determine these probabilities; however, it is quite another matter to solve them.

Now consider the macrostates E_j ($j = 0, 1, ..., s$), corresponding to the events {j servers are busy}, and let {$P(j)$} be the corresponding equilibrium distribution. Clearly,

$$P(j) = \sum_{n_1,...,n_j} P(j; n_1, ..., n_j),$$

and $P(j) + P(j+1) + \cdots + P(s)$ is the blocking probability for stream S_j.

We now argue that an appropriate application of the principle of conservation of flow leads to the following equations for the macrostate equilibrium probabilities:

$$(\lambda_j + \cdots + \lambda_s)P(j-1) + \lambda_j[P(j) + \cdots + P(s)] = j\mu P(j) \qquad (j = 1, 2, ..., s). \tag{4.6}$$

Equation (4.6) and the normalization equation,

$$\sum_{j=0}^{s} P(j) = 1,$$

permit successive calculations of the distribution {$P(j)$} whenever the offered loads $a_1, ..., a_s$ ($a_i = \lambda_i/\mu$) are such that statistical equilibrium obtains.

To see the truth of (4.6), observe that the term on the right-hand side equals the equilibrium rate of transition from E_j to E_{j-1}, *including those transitions that are instantaneously reversed because a waiting customer seizes a just-released server*. On the left-hand side of (4.6), the term $(\lambda_j + \cdots + \lambda_s)P(j-1)$ equals the equilibrium rate of transition from E_{j-1} to E_j, including only those transitions that are not involved in instantaneous reversal; the term $\lambda_j[P(j) + \cdots + P(s)]$ equals the equilibrium rate of transition from E_{j-1} to E_j, including only those transitions that instantaneously reverse previous transitions from E_j to E_{j-1}. (The quantity $\lambda_j[P(j) + \cdots + P(s)]$ equals the rate at which arrivals from S_j join the queue. Eventually, each such customer will seize a server whose release caused the transition $E_j \to E_{j-1}$; the just-released server will be seized instantaneously by a waiting customer from S_j, thereby causing the transition $E_{j-1} \to E_j$.) Hence, the left-hand side of (4.6) equals the equilibrium rate of transition from E_{j-1} to E_j, and the right-hand side equals the equilibrium rate of transition from E_j to E_{j-1}; that is, rate up equals rate down.

Indirect Solution of Equations (Ordered Servers) 153

A. Descloux [1969, unpublished] has rigorously derived (4.6) directly from the conservation-of-flow equations that determine the microstate probabilities $\{P(j;n_1,\ldots,n_j)\}$. Descloux's complicated analysis further yields the conditions that must be satisfied if statistical equilibrium is to obtain (as we have tacitly assumed). Interestingly, the equilibrium conditions turn out to be polynomial functions of the offered loads, this nonlinearity being a reflection of the interaction among the different streams of traffic. For instance, when $s=3$ the equilibrium condition for stream S_1 is

$$6a_1 + 3a_2 + 2a_3 + a_1 a_3 - 6 < 0.$$

Exercise

13. Show that when $a_1 = a_2 = \cdots = a_{s-1} = 0$ and $a_s = a$, then Equation (4.6) yields

$$P(s) = C(s,a),$$

where $C(s,a)$ is the Erlang delay formula.

4.5. Indirect Solution of Equations (Load Carried by Each Server of an Ordered Group)

We have seen that in many cases of practical interest, the probability state equations are easy to specify but difficult to solve in closed form. Such problems usually call for simulation (which we shall discuss in a subsequent chapter) or numerical methods (which we shall discuss in the next section), but occasionally a closed-form solution can be obtained by indirect means.

We consider again the question of systems with *ordered hunt* as described previously in Section 3.3: The servers are numbered $1, 2, \ldots, s$, and each arriving customer takes the lowest-numbered idle server. As in Chapter 3, we assume that customers arrive according to a Poisson process. However, we replace the assumption that blocked customers are cleared by the assumption that blocked customers may wait in the queue and defect from it at any arbitrary rate; this necessitates (if the problem is not to be intractable) the further assumption of exponential service times. For this model we calculate, as before, the load carried by the jth ordered server ($j = 1, 2, \ldots, s$), which is the same as the equilibrium probability that the jth ordered server is busy.

As a first step, let us review the results of Section 3.3 for the Erlang loss model with ordered hunt. We defined the random variables X_j ($j = 1, 2, \ldots, s$), with $X_j = 0$ when the jth ordered server is idle and $X_j = 1$ when the jth ordered server is busy. Then, defining $\tilde{p}_j = P\{X_j = 1\}$, we showed [Equation (3.18) of Chapter 3] that

$$\tilde{p}_j = a[B(j-1,a) - B(j,a)], \tag{5.1}$$

where a is the offered load and $B(k,a)$ is the Erlang loss formula. Equation (5.1) was interpreted to say that the load $\tilde{p}_j$ carried by the jth ordered server in an Erlang loss system is the difference between the load that overflows the $(j-1)$th ordered server and that which overflows the jth ordered server.

This model was extended in Section 3.4 to include the case where all blocked customers wait until served. There we presented a clever intuitive argument, first advanced by Vaulot [1925], that shows that the equilibrium conditional probability $P\{X_j = 1 | N \leq s\}$ that the jth ordered server is busy, given that there are no customers waiting, is given, remarkably, by [see Equation (4.14) of Chapter 3]

$$P\{X_j = 1 | N \leq s\} = \tilde{p}_j \qquad (j = 1, 2, \ldots, s), \tag{5.2}$$

where $\tilde{p}_j$, given by (5.1), is the corresponding probability for the system in which no waiting is allowed. This result, combined with Equation (4.13) of Chapter 3, leads to the required result, namely,

$$p_j = a[B(j-1,a) - B(j,a)]P\{N \leq s\} + P\{N > s\}, \tag{5.3}$$

which we used in Exercise 19 of Chapter 3 to calculate the most economical division between flat-rate and measured-rate trunks in an Erlang delay system.

Our present objective is to give a more algebraic proof of (5.2) that is of some interest in itself. For the sake of generality, we no longer restrict ourselves to the assumption that blocked customers wait until served, but instead we allow any queue discipline that can be described in the framework of a multidimensional birth-and-death process. (That Vaulot's argument remains valid for *any* birth-and-death model with $\lambda_j = \lambda$ and $\mu_j = j\mu$ ($j = 0, 1, \ldots, s$) was observed by P. J. Burke [1963, unpublished].) To this end, we again begin by considering the Erlang loss system with ordered hunt, and we define

$$\tilde{P}(x_1, \ldots, x_s) = P\{X_1 = x_1, \ldots, X_s = x_s\} \tag{5.4}$$

and

$$\delta(x, y) = \begin{cases} 0 & \text{when } x \neq y, \\ 1 & \text{when } x = y. \end{cases} \tag{5.5}$$

Indirect Solution of Equations (Ordered Servers)

Then, using the principle of conservation of flow, we can write, for all $x_i = 0$ or 1 except $x_1 = \cdots = x_s = 1$,

$$[\lambda + (x_1 + \cdots + x_s)\mu]\tilde{P}(x_1,\ldots,x_s)$$

$$= \lambda \sum_{j=1}^{s} \delta(j, x_1 + \cdots + x_j)\tilde{P}(x_1,\ldots,x_{j-1}, x_j - 1, x_{j+1},\ldots,x_s)$$

$$+ \mu \sum_{j=1}^{s} \delta(0, x_j)\tilde{P}(x_1,\ldots,x_{j-1}, x_j + 1, x_{j+1},\ldots,x_s); \tag{5.6}$$

and when $x_1 = \cdots = x_s = 1$,

$$s\mu\tilde{P}(1,\ldots,1) = \lambda \sum_{j=1}^{s} \delta(j, x_1 + \cdots + x_j)\tilde{P}(x_1,\ldots,x_{j-1}, x_j - 1, x_{j+1},\ldots,x_s). \tag{5.7}$$

The 2^s unknown probabilities $\{\tilde{P}(x_1,\ldots,x_s)\}$ can be found from the set (5.6) of $2^s - 1$ independent equations and the normalization equation

$$\sum \tilde{P}(x_1,\ldots,x_s) = 1, \tag{5.8}$$

where the summation is carried out over all 2^s probabilities. [Equation (5.7) can be obtained by adding all the equations in (5.6), and is therefore redundant.]

Our objective is to find $\tilde{p}_j$, the load carried by the jth ordered server, which is given by

$$\tilde{p}_j = \sum \tilde{P}(x_1,\ldots,x_{j-1}, 1, x_{j+1},\ldots x_s), \tag{5.9}$$

where the sum is taken over all values of the arguments $x_1, \ldots, x_{j-1}, x_{j+1}, \ldots, x_s$. Observe, however, that it is not necessary to solve the equations (5.6) and evaluate the right-hand side of (5.9) in order to calculate $\tilde{p}_j$; we have already rigorously shown, in Chapter 3, that $\tilde{p}_j$ is given by (5.1). We now use this observation to prove that, as asserted, Equation (5.2) is valid for the more general model that allows queueing, thereby proving (5.3).

To this end, let Q be the number of customers waiting in the queue, and define

$$P(x_1,\ldots,x_s; k) = P\{X_1 = x_1,\ldots, X_s = x_s, Q = k\}. \tag{5.10}$$

Now observe that, for $k = 0$ and not all $x_i = 1$, the probabilities defined by

(5.10) satisfy the *same* conservation-of-flow equations (5.6) that are satisfied by the probabilities (5.4) for the Erlang loss system with ordered hunt. When $k=0$ and all $x_i = 1$, then the corresponding flow equation is

$$(\lambda + s\mu) P(1,\ldots,1;0)$$

$$= \lambda \sum_{j=1}^{s} \delta(j, x_1 + \cdots + x_j) P(x_1,\ldots,x_{j-1}, x_j - 1, x_{j+1},\ldots,x_s; 0)$$

$$+ \mu_{s+1} P(1,\ldots,1;1). \tag{5.11}$$

It is also true (because "rate up = rate down") that

$$\lambda P(1,\ldots,1;k) = \mu_{s+k+1} P(1,\ldots,1;k+1) \qquad (k=0,1,\ldots). \tag{5.12}$$

(Note that this model allows the departure rate μ_j to be arbitrary when $j > s$. The Erlang delay model is the special case in which $\mu_j = s\mu$ when $j > s$.) Substitution of (5.12) with $k=0$ into (5.11) yields an equation identical with (5.7). Thus we have shown that the probabilities $\{\tilde{P}(x_1,\ldots,x_s)\}$ and $\{P(x_1,\ldots,x_s;0)\}$ satisfy the same equations (for all x_i). From this important observation we can conclude that the probabilities $\{\tilde{P}\}$ and $\{P\}$ are proportional (by c, say) when $k=0$:

$$P(x_1,\ldots,x_s;0) = c\tilde{P}(x_1,\ldots,x_s). \tag{5.13}$$

Now, applying the definition of conditional probability, we can write

$$P\{X_j = 1 | N \leq s\} = \frac{P\{X_j = 1, N \leq s\}}{P\{N \leq s\}}. \tag{5.14}$$

But

$$P\{X_j = 1, N \leq s\} = \sum P(x_1,\ldots,x_{j-1}, 1, x_{j+1},\ldots,x_s; 0) \tag{5.15}$$

and

$$P\{N \leq s\} = \sum P(x_1,\ldots,x_s; 0). \tag{5.16}$$

Substitution of (5.13) into (5.15) and (5.16) yields

$$P\{X_j = 1, N \leq s\} = c \sum \tilde{P}(x_1,\ldots,x_{j-1}, 1, x_{j+1},\ldots,x_s) \tag{5.17}$$

and

$$P\{N \leq s\} = c \sum \tilde{P}(x_1,\ldots,x_s). \tag{5.18}$$

Comparison of (5.17) with (5.9) shows that

$$P\{X_j = 1, N \leq s\} = c\tilde{p}_j; \qquad (5.19)$$

and comparison of (5.18) and (5.8) shows that

$$P\{N \leq s\} = c. \qquad (5.20)$$

Substitution of (5.19) and (5.20) into (5.14) yields (5.2), and the proof is complete.

Note that nowhere did we solve any complicated set of conservation-of-flow equations. The problem was solved by recognizing that the set of equations whose solution we required had essentially the same solution as another set, namely (5.6) and (5.7); and, further, the required information could be extracted from (5.6) and (5.7) without solving these either, but rather by applying an elegant argument [that of Section 3.3 to obtain (5.1)] that is valid only for the original Erlang loss model. For an analysis of a similar model with heterogeneous exponential servers (exponential servers with different rates) see Cooper [1976].

The moral of this section is that there is more than one way to skin a cat.

Exercise

14. Consider again the premise of Exercise 7 (heterogeneous exponential servers selected in random order), but assume now that all blocked customers wait as long as necessary for service. Let Q be the number of customers waiting in the queue; define $P(x_1,\ldots,x_s; k) = P\{X_1 = x_1,\ldots,X_s = x_s, Q = k\}$ and $\rho = \lambda/(\mu_1 + \cdots + \mu_s)$; and show that if $\rho < 1$, then

$$P(x_1,\ldots,x_s; 0) = c\tilde{P}(x_1,\ldots,x_s)$$

and

$$P(1,\ldots,1; k) = c\rho^k \tilde{P}(1,\ldots,1),$$

where $\tilde{P}(x_1,\ldots,x_s)$ is given in part b of Exercise 7 and

$$c = \left(1 + \frac{\rho}{1-\rho} \tilde{P}(1,\ldots,1)\right)^{-1}.$$

These results were first obtained, in somewhat different form, by Gumbel [1960].

4.6. Numerical Solution of State Equations by Iteration (Gauss-Seidel and Overrelaxation Methods)

Thus far we have seen that queueing problems often lead to sets of multidimensional probability state equations. In each of the cases considered so far in this chapter, we have been able to solve these equations without recourse to numerical methods. In most cases of practical interest, however, product solutions, generating functions, or astute observations will not lead to neat solutions. Fortunately, equations of the type we are concerned with often yield easily to a simple numerical iteration scheme. Convergence criteria are known, but are often difficult to apply in practical cases. However, real queueing problems in which this iteration technique fails rarely, if ever, occur. Because of its simplicity, its particular suitability for digital-computer application, its apparent reliability, and the lack of anything better, we shall briefly describe an iteration procedure and its generalization and illustrate their application to the solution of multidimensional probability state equations.

Numerical analysis is a subject in itself. The present discussion is intended to provide only a starting point for one who is interested in the theory of numerical solution of large systems of linear equations. Hopefully, the following necessarily terse discussion will enable the reader to "get numbers" from multidimensional probability state equations. It is worth mentioning that the iteration procedure we shall describe is much simpler to apply than might appear from the mathematical description. The numerical example that follows the theoretical description should make this clear.

Consider a general set of linear equations

$$\mathbf{A}\mathbf{x} = \mathbf{b}, \tag{6.1}$$

where $\mathbf{A}$ is a given square matrix, $\mathbf{b}$ is a given vector, and $\mathbf{x}$ is an unknown vector. [In the case of interest here, the components of $\mathbf{x}$ are the equilibrium state probabilities that are to be found; (6.1) is the set of equilibrium state equations.] Equation (6.1) can be rewritten in the form

$$(\mathbf{I} - \mathbf{L} - \mathbf{U})\mathbf{x} = \mathbf{d}, \tag{6.2}$$

where $\mathbf{L}$ and $\mathbf{U}$ are, respectively, lower and upper triangular matrices with zeros along the main diagonal, and $\mathbf{I}$ is the unit matrix, with ones along the main diagonal and zeros everywhere else. Equation (6.2) can be written

$$\mathbf{x} = \mathbf{L}\mathbf{x} + \mathbf{U}\mathbf{x} + \mathbf{d}, \tag{6.3}$$

which suggests the iteration scheme

$$\mathbf{x}^{(n+1)} = \mathbf{L}\mathbf{x}^{(n+1)} + \mathbf{U}\mathbf{x}^{(n)} + \mathbf{d}, \tag{6.4}$$

where the vector $\mathbf{x}^{(n)}$ is the nth iterate, $n = 0, 1, \ldots$, and $\mathbf{x}^{(0)}$ is an arbitrary vector. Equation (6.4) simply says calculate each component of the vector $\mathbf{x}^{(n+1)}$ based on the latest calculated values available, which are all the components of the vector $\mathbf{x}^{(n+1)}$ calculated thus far and the remaining components of the vector $\mathbf{x}^{(n)}$ calculated at the last stage of iteration. The iteration procedure (6.4) has associated with it the names Seidel, Gauss, Liebmann, Nekrasov, and probably several others. The general technique of solving equations by iteration is also called *successive approximation* and *relaxation*. The most common name of the scheme (6.4) is *Gauss-Seidel iteration*.

It can be shown that a necessary and sufficient condition for the convergence of the iteration procedure (6.4) for arbitrary $\mathbf{x}^{(0)}$ is that all the eigenvalues of the *iteration matrix* $\mathbf{M} = (\mathbf{I} - \mathbf{L})^{-1}\mathbf{U}$ lie inside the unit circle; that is, the iteration procedure (6.4) converges independently of the initial vector $\mathbf{x}^{(0)}$ if and only if the spectral radius of the iteration matrix is less than unity.

This convergence criterion is ordinarily of theoretical interest only, since finding the spectral radius of the iteration matrix is usually of the same order of difficulty as solving the original equations (6.1). It is desirable to be able to predict behavior of the iteration scheme directly from inspection of the original matrix $\mathbf{A}$. This is possible in certain special cases. In particular, a sufficient condition for convergence of the iteration scheme is that the original matrix $\mathbf{A}$ be *irreducible* and exhibit *weak diagonal dominance*; that is, if a_{ij} is the element in row i and column j of the irreducible matrix $\mathbf{A}$, then convergence is assured if

$$|a_{ii}| \geqslant \sum_{\substack{j=1 \\ j \neq i}}^{k} |a_{ij}| \qquad (i = 1, 2, \ldots, k),$$

and for at least one i

$$|a_{ii}| > \sum_{\substack{j=1 \\ j \neq i}}^{k} |a_{ij}|,$$

where k is the number of rows (columns) of the matrix $\mathbf{A}$. This condition is not usually met by probability state equations, but it does suggest why this iteration scheme has proven useful in this context. The greater the concentration of "mass" of the matrix $\mathbf{A}$ along its main diagonal, the more likely (and faster) will the procedure converge. State equations of the type we are concerned with are characterized by a large number of zero elements off the main diagonal, making convergence likely, and simultaneously making computer programming particularly easy.

A generalization of the Gauss-Seidel method that has proven useful is the so-called method of successive *overrelaxation*, in which the overrelaxation values are obtained by weighting the most recent value by a factor $\omega \geqslant 1$:

$$\mathbf{x}^{(n+1)} = \omega[\mathbf{L}\mathbf{x}^{(n+1)} + \mathbf{U}\mathbf{x}^{(n)} + \mathbf{d}] + (1-\omega)\mathbf{x}^{(n)} \qquad (\omega \geqslant 1). \qquad (6.5)$$

The Gauss-Seidel method is thus the special case of overrelaxation for $\omega = 1$. An intuitive feel for the reason overrelaxation sometimes speeds convergence can best be gained by running through a numerical example. Mathematically, the use of a weighting factor ω will speed convergence if the corresponding iteration matrix has a smaller spectral radius.

Experience indicates that overrelaxation with $\omega \approx 1.3$ is often faster than simple Gauss-Seidel iteration. Many fundamental questions about convergence and rates of convergence remain unanswered. The interested reader should see Young [1971]. Recently, Brandwajn [1979] has proposed other iterative methods that may be superior (with respect to assurance of convergence and rate of convergence) to the Gauss-Seidel and overrelaxation methods. Related numerical methods are discussed in Buzen [1973] and Wallace [1973].

From a practical point of view, it appears that the Gauss-Seidel, overrelaxation, and similar iterative methods (such as those proposed by Brandwajn) are the most useful methods for numerical solution of large sets of equations of the type that arises from analysis of multidimensional birth-and-death processes. These methods are especially suitable for implementation on a digital computer. (Also, a point that was of utmost importance only a short time ago, when computers were relatively rare and expensive, is that iteration schemes that converge for any starting vector $\mathbf{x}^{(0)}$ are self-correcting; an error is "absorbed" by the method and results in only a larger number of iterations. Anyone who has tried to solve systems of equations on a calculator will surely appreciate this fact.)

We now give a simple illustration of the use of the Gauss-Seidel and overrelaxation methods in the solution of multidimensional statistical-equilibrium probability state equations.

Consider two server groups, each of which receives direct Poisson traffic and also overflow traffic from the other group (see Figure 4.5). Let the server groups G_1 and G_2 be of sizes s_1 and s_2, with Poisson loads of magnitudes a_1 erl and a_2 erl, and exponential service times with mean μ^{-1}. Customers who find all $s_1 + s_2$ servers busy are cleared from the system. Let $P(j_1, j_2)$ be the statistical-equilibrium probability that there are j_1

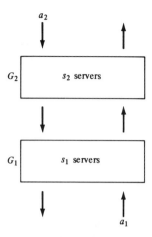

Figure 4.5

customers in service on G_1 and j_2 on G_2. Then, equating rates of flow out of and into each state, the statistical-equilibrium probability state equations are

$$(a_1+a_2+j_1+j_2)P(j_1,j_2)$$
$$= a_1 P(j_1-1,j_2) + a_2 P(j_1,j_2-1)$$
$$+ (j_1+1)P(j_1+1,j_2) + (j_2+1)P(j_1,j_2+1)$$
$$(j_1<s_1,\ j_2<s_2), \quad (6.6)$$

$$(a_1+a_2+s_1+j_2)P(s_1,j_2)$$
$$= a_1 P(s_1-1,j_2) + a_1 P(s_1,j_2-1)$$
$$+ a_2 P(s_1,j_2-1) + (j_2+1)P(s_1,j_2+1)$$
$$(j_1=s_1,\ j_2<s_2), \quad (6.7)$$

$$(a_1+a_2+j_1+s_2)P(j_1,s_2)$$
$$= a_1 P(j_1-1,s_2) + a_2 P(j_1,s_2-1)$$
$$+ a_2 P(j_1-1,s_2) + (j_1+1)P(j_1+1,s_2)$$
$$(j_1<s_1,\ j_2=s_2), \quad (6.8)$$

$$(s_1+s_2)P(s_1,s_2) = (a_1+a_2)P(s_1-1,s_2) + (a_1+a_2)P(s_1,s_2-1)$$
$$(j_1=s_1,\ j_2=s_2). \quad (6.9)$$

The normalization equation is

$$\sum_{j_1=0}^{s_1} \sum_{j_2=0}^{s_2} P(j_1,j_2) = 1. \quad (6.10)$$

Table 4.1

(4,2)	(4,1)	(4,0)	(3,2)	(3,1)	(3,0)	(2,2)	(2,1)	(2,0)	(1,2)	(1,1)	(1,0)	(0,2)	(0,1)	(0,0)
6	-6	0	-6	0	0	0	0	0	0	0	0	0	0	0
-2	11	-6	0	-5	0	0	0	0	0	0	0	0	0	0
0	-1	10	0	0	-5	0	0	0	0	0	0	0	0	0
-4	0	0	11	-1	0	-6	0	0	0	0	0	0	0	0
0	-4	0	-2	10	-1	0	-5	0	0	0	0	0	0	0
0	0	-4	0	-1	9	0	0	-5	0	0	0	0	0	0
0	0	0	-3	0	0	10	-1	0	-6	0	0	0	0	0
0	0	0	0	-3	0	-2	9	-1	0	-5	0	0	0	0
0	0	0	0	0	-3	0	-1	8	0	0	-5	0	0	0
0	0	0	0	0	0	-2	0	0	9	-1	0	-6	0	0
0	0	0	0	0	0	0	-2	0	-2	8	-1	0	-5	0
0	0	0	0	0	0	0	0	-2	0	-1	7	0	0	-5
0	0	0	0	0	0	0	0	0	-1	0	0	8	-1	0
0	0	0	0	0	0	0	0	0	0	-1	0	-2	7	-1
0	0	0	0	0	0	0	0	0	0	0	-1	0	-1	6

We shall solve these equations for the particular case $s_1 = 4$, $a_1 = 5$, $s_2 = 2$, $a_2 = 1$. The total number of states is $(s_1 + 1)(s_2 + 1) = 15$. Consider the state equations (6.6)–(6.9). For reasons that will be explained presently, we reorder these equations so that their matrix of coefficients is as shown in Table 4.1. All the entries on the main diagonal are positive, and all those off the main diagonal are negative or zero. Note how the nonzero elements cluster along the main diagonal.

Observe that this 15×15 matrix is of rank 14; that is, each column sums to zero and any one of the rows can be derived from the other 14. In other words, one of the equations is redundant, and the other 14 equations determine the solution only up to a constant factor. This factor is supplied by the normalization condition (6.10).

Although one of the equations is redundant, it appears that the convergence of the iteration procedure sometimes is accelerated by using all the state equations, as we shall do in this example. The normalization equation may be used after each complete round of iteration, as we do in this example, or it may be used only after the last stage of iteration. When all the state equations are used, normalization is accomplished by summing the unnormalized values and dividing each by their sum. Normalizing only after the last stage of iteration is more efficient as long as computer overflows and/or underflows do not occur, since the relative values are unaffected by normalization. (We normalize after each round in this example because we want to display the values at each step.)

We now apply the method of successive overrelaxation (6.5) to our matrix of coefficients, using $\omega = 1.0$ (the ordinary Gauss–Seidel iteration), 1.1, 1.2, 1.3, and 1.4. As first step, we must specify the initial values $P^{(0)}(j_1, j_2)$.

Let us first consider ordinary Gauss-Seidel iteration ($\omega = 1.0$) with the initial values all equal. Then $P^{(0)}(j_1,j_2) = \frac{1}{15} = 0.06667$. Following the order indicated by the matrix, we first calculate $P^{(1)}(4,2)$ according to

$$6P^{(1)}(4,2) = 6P^{(0)}(4,1) + 6P^{(0)}(3,2).$$

We next calculate $P^{(1)}(4,1)$ according to

$$11P^{(1)}(4,1) = 2P^{(1)}(4,2) + 6P^{(0)}(4,0) + 5P^{(0)}(3,1).$$

We continue in this manner, calculating $P^{(1)}(4,0)$, $P^{(1)}(3,2),\ldots,P^{(1)}(0,0)$. We may then normalize these values (divide each one by their sum) and start the second round of iteration, or we may start the next round of iteration immediately and normalize at completion of the iteration procedure. For this example, the normalized probabilities after $k = 0, 1, 2, 5, 11, 12,$ and 13 stages of iteration ($\omega = 1.0$) are as shown in Table 4.2.

Observe that the probabilities change only slightly from the twelfth iteration to the thirteenth. Somehow it must be decided when to stop iterating, that is, when the procedure has converged (or shown that convergence is not occurring). In the present example the convergence criterion was taken to be

$$\sum_{j_1=0}^{4} \sum_{j_2=0}^{2} |P^{(k)}(j_1,j_2) - P^{(k-1)}(j_1,j_2)| < 0.001.$$

That is, we assume that the process has converged when the sum of the absolute changes of the probabilities from one iteration to the next is less

Table 4.2

	$k=0$	$k=1$	$k=2$	$k=5$	$k=11$	$k=12$	$k=13$
$P^{(k)}(4,2)$	0.0667	0.12500	0.16304	0.23077	0.26404	0.26444	0.26466
$P^{(k)}(4,1)$	0.0667	0.08523	0.08893	0.12338	0.12464	0.12455	0.12449
$P^{(k)}(4,0)$	0.0667	0.03977	0.03866	0.04774	0.03978	0.03960	0.03949
$P^{(k)}(3,2)$	0.0667	0.08523	0.10316	0.12751	0.13995	0.14019	0.14033
$P^{(k)}(3,1)$	0.0667	0.08864	0.10358	0.12392	0.12059	0.12056	0.12054
$P^{(k)}(3,0)$	0.0667	0.06225	0.06761	0.06623	0.05431	0.05410	0.05399
$P^{(k)}(2,2)$	0.0667	0.06932	0.07597	0.05920	0.06067	0.06076	0.06082
$P^{(k)}(2,1)$	0.0667	0.08662	0.10412	0.08036	0.07458	0.07455	0.07454
$P^{(k)}(2,0)$	0.0667	0.07323	0.08478	0.05100	0.04162	0.04149	0.04141
$P^{(k)}(1,2)$	0.0667	0.06402	0.03595	0.01954	0.01876	0.01878	0.01880
$P^{(k)}(1,1)$	0.0667	0.08543	0.05956	0.03260	0.02925	0.02923	0.02922
$P^{(k)}(1,0)$	0.0667	0.07764	0.04448	0.02312	0.01901	0.01896	0.01893
$P^{(k)}(0,2)$	0.0667	0.01581	0.00755	0.00332	0.00305	0.00305	0.00305
$P^{(k)}(0,1)$	0.0667	0.02552	0.01301	0.00639	0.00564	0.00563	0.00563
$P^{(k)}(0,0)$	0.0667	0.01719	0.00958	0.00492	0.00411	0.00410	0.00409

than 0.001. This criterion, which is completely arbitrary, is satisfied in our example at $k=13$.

If we had used an ω different from the Gauss-Seidel value of unity, we might have speeded up the convergence; that is, we might have satisfied our arbitrary convergence criterion after less iterations. For this example use of an overrelaxation factor $\omega=1.3$ will produce convergence (as defined above) for the same initial values after $k=9$ iterations, a saving in computation time of about $\frac{4}{13}\approx 31\%$.

Another way to speed convergence is to start out with a set of initial values closer to the solution. That is, use some intelligence in specifying the initial values, instead of choosing them in a completely arbitrary fashion. In the present example, we know that a product solution of the form

$$P(j_1,j_2) = \frac{a_1^{j_1}}{j_1!}\frac{a_2^{j_2}}{j_2!}c$$

will satisfy the equilibrium state equations everywhere except on the boundary. Therefore it seems reasonable that the true solution should not differ numerically as much as from the product solution values as from the arbitrary value $\frac{1}{15}$.

For the example under consideration, using the product

$$P^{(0)}(j_1,j_2) = \frac{a_1^{j_1}}{j_1!}\frac{a_2^{j_2}}{j_2!}c,$$

where c is the normalization factor, leads to convergence (with respect to our criterion) in 10 iterations when $\omega=1.0$, and in 6 iterations when $\omega=1.3$ [as opposed to 13 and 9 when the initial values are $P^{(0)}(j_1,j_2)=\frac{1}{15}$]. Speeds of convergence (numbers of iterations required) for this example for various values of the overrelaxation factor ω and different initial values are summarized in Table 4.3.

The reason for ordering the equations as we have done, starting with the boundary states, should now be apparent. If, when using the product $(a_1^{j_1}/j_1!)(a_2^{j_2}/j_2!)c$ as the initial vector, we had started the iteration procedure at interior states, then no change in the calculated values would have

Table 4.3

$P^{(0)}(j_1,j_2)$ \ ω	1.0	1.1	1.2	1.3	1.4
$\dfrac{a_1^{j_1}}{j_1!}\dfrac{a_2^{j_2}}{j_2!}c$	10	8	6	6	8
$\dfrac{1}{15}$	13	11	9	9	11

occurred until a boundary equation (where the product solution does not hold) was reached. Thus the computation time of the first round of iteration would have been largely wasted.

Another useful observation is that we know, in this example, that $\Sigma_{j_1+j_2=j} P(j_1,j_2) = P_j$, where $\{P_j\}$ is the Erlang loss distribution (3.3) of Chapter 3. In particular, we can calculate $P(0,0) = P_0$ and $P(4,2) = P_6$; these can be used as starting values or, if they are not, can serve to indicate the degree of convergence achieved after a finite number of iterations. For example, $P(4,2) = P_6 = B(6,6) = 0.26492$; comparison with Table 4.2 shows that $|P(4,2) - P^{(13)}(4,2)| = 0.00026$.

These tricks may seem hardly worth the effort for the simple example considered here. But for the solution of large sets of equations of the kind likely to arise in practical problems, the computer needs all the help it can get.

Finally, it should be mentioned that when mathematical or numerical solution is impractical, simulation is often a useful tool for analysis. (This topic will be addressed in Chapter 6.) Simulation can be used to study processes that are much more complicated than those that can be studied by mathematical and numerical analysis. However, if an appropriate mathematical or numerical solution can be obtained, it is often more useful than simulation data.

4.7. The Equivalent Random Method

Consider the following important queueing system. Customers arrive in n independent Poisson streams, each stream directed at one of n different primary groups of servers. All customers who find all servers busy in their primary group are directed toward a single overflow group. The overflow group thus provides a common backup for the n primary groups. The service times on all groups are assumed to be independently and identically distributed negative-exponential random variables. Customers who are blocked on the overflow group are cleared from the system. The operations researcher might be interested in determining the numbers of servers and the particular server arrangements that ensure that a prespecified probability of loss (overflow from both primary and overflow servers) is not exceeded.

We shall describe a technique, called the equivalent random method, for the approximate analysis of such a system. The theoretical basis of the equivalent random method has already been discussed in Section 4.3. We begin with a discussion of the simplest overflow system, which consists of a single primary group with one server and an overflow group with one server, as illustrated in Figure 4.6.

Let P_1 be the proportion of time that the overflow server is busy. The load carried on the overflow server is the difference between the load $aB(1,a)$ that overflows the primary server and the load $aB(2,a)$ that

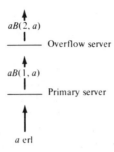

Figure 4.6

overflows both the primary server and the overflow server. Since the load carried on a single server equals the proportion of time that server is busy, it follows that

$$P_1 = a[B(1,a) - B(2,a)], \qquad (7.1)$$

where $B(s,a)$ is the Erlang loss formula.

Now consider the probability Π_1 that a customer who overflows the primary server finds the overflow server busy. Since the customers arrive at the two-server (primary and overflow) system in a Poisson stream, they see the same distribution of states on the two-server system as would an outside observer. It follows that Π_1 is the conditional probability that the overflow server is busy, given that the primary server is busy. Applying the definition of conditional probability, we have

$$\Pi_1 = \frac{B(2,a)}{B(1,a)}, \qquad (7.2)$$

which can be written

$$\Pi_1 = \frac{aB(2,a)}{aB(1,a)}. \qquad (7.3)$$

We can conclude from Equation (7.3) that the proportion of overflow customers who find the overflow server occupied, and thus are cleared from the system, is the ratio of the load overflowing the second server to the load overflowing the first server. Hence, viewing the overflow server as a system by itself, whose input stream is the overflow stream from the first server, the proportion of customers who find the overflow server busy is the ratio of the load not carried (that is, overflowing the second server) to the load offered (that is, overflowing the first server).

Comparison of Equations (7.1) and (7.2) shows that

$$P_1 \neq \Pi_1, \tag{7.4}$$

from which we conclude that the overflow stream from the first server is not a Poisson stream. This should also be apparent from the fact that the probability that a customer will request service from the second server in any interval $(t, t+h)$ is $\lambda h + o(h)$ as $h \to 0$ if the first server is busy at time t, and zero if the first server is idle at time t. If the overflow stream were Poisson, then this probability would be $\lambda h + o(h)$ as $h \to 0$, independent of any other considerations.

It should now be clear that the following analogous properties are true for overflow systems composed of a single s-server primary group and a single c-server overflow group:

1. The overflow stream is not Poisson.
2. Of those customers who overflow the primary group, the proportion Π_c who are also blocked on the overflow group is the ratio

$$\Pi_c = \frac{aB(s+c, a)}{aB(s, a)}. \tag{7.5}$$

The situation can be represented by Figure 4.7. Property 1 implies that overflow systems cannot, in general, be studied by the elementary methods developed thus far in this text (except through the use of multidimensional state equations, which is usually not practical). Property 2 is used in the equivalent random method, which was developed by Wilkinson [1956] (a

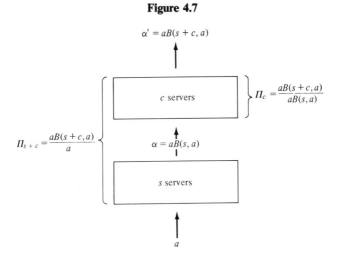

Figure 4.7

similar method is given by Bretschneider [1956]) to facilitate the approximate analysis of more complicated overflow systems, and to which we now direct our attention.

Consider the system on the left-hand side of Figure 4.8. In this system, a group of c servers acts as the common overflow group for all traffic overflowing the n primary groups. The ith primary group is assumed to consist of s_i servers, and to receive a direct Poisson load of a_i erl. The service times on all groups are assumed to be independently and identically exponentially distributed.

We shall analyze this system by means of an approximation technique widely known in teletraffic engineering as the *equivalent random method*.

For the purposes of our approximate analysis, we represent the load that overflows the ith primary group by two parameters, the mean α_i and variance v_i, which are calculated according to Equations (3.1) and (3.2):

$$\alpha_i = a_i B(s_i, a_i) \qquad (i = 1, 2, \ldots, n), \qquad (7.6)$$

$$v_i = \alpha_i \left(1 - \alpha_i + \frac{a_i}{s_i + 1 + \alpha_i - a_i} \right) \qquad (i = 1, 2, \ldots, n). \qquad (7.7)$$

That is, the load overflowing a primary group is characterized by the mean and variance of the equilibrium overflow distribution that would result if the overflowing customers were handled by an infinite-server overflow group. We denote by α' the mean value of the load that overflows the c-server overflow group.

We wish to calculate the proportion of customers who overflow the c-server overflow group, given that they overflow a primary group. That is, we wish to calculate the probability that a customer who finds all s_i servers busy in his primary group will also find all c servers busy in the overflow

Figure 4.8

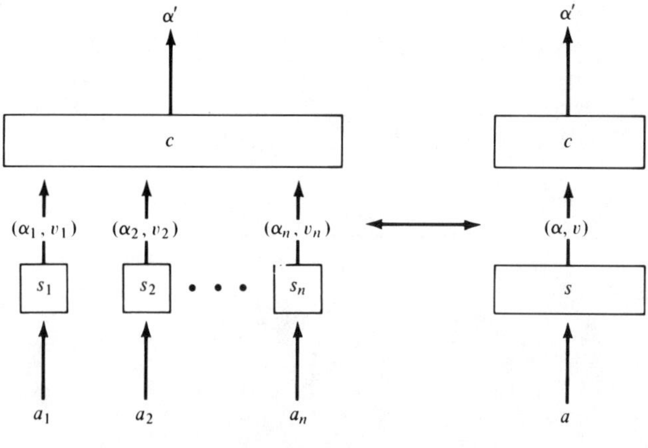

The Equivalent Random Method

group. Although in reality the value of this probability will, in general, be different for different values of index i, it seems reasonable that the value of the ratio

$$\frac{\alpha'}{\alpha_1 + \alpha_2 + \cdots + \alpha_n}$$

should provide a meaningful approximation to this probability. It remains to calculate α'.

To this end, we replace the n primary groups and their respective Poisson loads with a single "equivalent random" primary group and a single "equivalent random" load, as indicated on the right-hand side of Figure 4.8. The number s of servers in the equivalent random group and the size a (in erlangs) of the equivalent random load are chosen such that the mean α and variance v of the total overflow load offered to the overflow group remain constant. Thus, since the primary groups are independent of each other, we have

$$\alpha = \alpha_1 + \cdots + \alpha_n, \tag{7.8}$$

$$v = v_1 + \cdots + v_n. \tag{7.9}$$

Now the values of the equivalent random load a and the equivalent random group size s are chosen to satisfy the following equations for the (known) moments α and v:

$$\alpha = aB(s,a), \tag{7.10}$$

$$v = \alpha\left(1 - \alpha + \frac{a}{s+1+\alpha-a}\right). \tag{7.11}$$

Then it follows that

$$\alpha' = aB(s+c,a). \tag{7.12}$$

In summary, the equivalent random method is to replace the system on the left-hand side of Figure 4.8 with the system on the right-hand side, where the values a and s are chosen such that the resulting overflow moments α and v remain constant. Then, in this new "equivalent random" system, of those customers who overflow the s-server primary group, the proportion who also find all c servers busy on the overflow group is, according to Equation (7.5),

$$\Pi_c = \frac{aB(s+c,a)}{aB(s,a)} = \frac{\alpha'}{\alpha}. \tag{7.13}$$

We take the value Π_c, calculated according to Equation (7.13), as the probability of blocking on the overflow group suffered by overflow traffic from the primary groups. Likewise, the proportion Π of all customers who are blocked on both the primary and overflow groups is approximated by

$$\Pi = \frac{aB(s+c,a)}{a_1 + a_2 + \cdots + a_n}. \tag{7.14}$$

It remains only to discuss how to solve Equations (7.10) and (7.11) numerically for the values a and s that correspond to a given pair α and v. Curves that effect this solution are given in Wilkinson [1956]. A useful set of tables and curves, with accompanying text and examples, is given in Wilkinson [1970].

An easy and fairly accurate approximate solution to Equations (7.10) and (7.11), given by Rapp [1964], is (where $z = v/\alpha$ is the *peakedness factor*)

$$a = v + 3z(z-1) \tag{7.15}$$

and, solving Equation (7.11) for s,

$$s = \frac{a(\alpha + z)}{\alpha + z - 1} - \alpha - 1. \tag{7.16}$$

(Note that $z > 1$; see Exercise 15.) These estimates of a and s are generally on the high side of the exact values. By rounding s down to its integral part $[s]$, the corresponding value of a that produces the desired peakedness factor z is then found by solving Equation (7.16) for a:

$$a = \frac{([s] + \alpha + 1)(\alpha + z - 1)}{\alpha + z}. \tag{7.17}$$

Exercises

15. Show that, if $a > 0$, then

$$v = \alpha \quad \text{when } s = 0,$$
$$v > \alpha \quad \text{when } s \geq 1,$$

where α and v are defined by Equations (3.1) and (3.2).

16. $a_1 = 10$ erl of Poisson traffic is offered to a group of 10 servers, and $a_2 = 5$ erl of Poisson traffic is offered to a group of 5 servers. An overflow group of c servers handles the overflow from the two primary groups on a BCC basis. Service times are independent, identically distributed, exponential variables.
    a. Find the value of c that gives a loss of 10% on the overflow group.
    b. Find the loss for the system as a whole.
    c. Find the loss on the overflow group and the loss for the system as a whole if a_1 and a_2 are each increased by 50%.

The Method of Phases   171

Suppose now that an overflow system is composed of a single primary group to which two independent Poisson loads a_1 and a_2 are offered, and an overflow group that receives only the overflow from a_1. To determine the proportion of overflow customers (from a_1) who are blocked on the overflow group, one would calculate the overflow mean and variance associated with only the load a_1. These values α_1 and v_1, which can be calculated according to the formulas (1) and (2) given in Exercise 12, would then determine the equivalent random values a and s.

Exercise

17. Poisson traffic totaling 10 erl is offered to a group of 10 servers. This stream of Poisson traffic is evenly divided between two types of customers, called high- and low-priority customers. Both substreams of traffic are Poisson, and service times have a common exponential distribution. The high-priority customers who overflow the 10-server group are routed to an overflow group of c servers (where they are handled on a BCC basis). The low-priority customers who overflow the 10-server group are cleared from the system. Determine the smallest value of c such that the proportion of high-priority customers who do not receive service is no greater than 0.01.

The equivalent random method has proved to be very useful in teletraffic engineering, and doubtless will prove useful in other applications. For extensions and further discussion on the equivalent random method and its applications see, for example, Katz [1967] on the application of the equivalent random method to analysis of large communication networks with alternate routing, Neal [1971] on extending the equivalent random method to allow analysis of correlated streams of overflow traffic, Holtzman [1973] on the accuracy of the equivalent random method, Fredericks [1980] on a related method for the analysis of systems with overflow traffic, and Kuczura [1973] on a method for approximating overflow streams.

4.8. The Method of Phases

The *method of phases* (or *stages*) is an approximation procedure according to which a random variable with an arbitrary distribution is replaced by either a sum, or a mixture, or a combined sum-mixture of independent (but not necessarily identical) random variables (each being a *phase* or *stage* of the lifetime of the original random variable). This technique allows transformation of the original model into a multidimensional birth-and-death model, thus making it amenable to the powerful tools of analysis that have been described in this chapter.

Consider, for example, the s-server Erlang loss system; we assume that blocked customers are cleared, arrivals follow a Poisson process with rate λ, and service times are independent, identical, random variables with arbitrary distribution function $H(t)$. In this example of the application of the method of phases, we assume that the service time X can be approximated by a sum of n independent, but not necessarily identical, exponential random variables,

$$X = X_1 + \cdots + X_n; \tag{8.1}$$

that is, we imagine that the service time X is composed of n independent phases of service, the ith phase being exponentially distributed with distribution function $F_i(t) = P\{X_i \leq t\} = 1 - e^{-\mu_i t}$. Then $E(X) = \sum_{i=1}^{n} \mu_i^{-1}$ and $V(X) = \sum_{i=1}^{n} \mu_i^{-2}$. (In particular, when $\mu_1 = \cdots = \mu_n$, then X has the Erlangian distribution of order n, whose density function is pictured, for various values of n, in Figure 2.3. Since it is true that $(\sum_{i=1}^{n} \mu_i^{-1})^2 > \sum_{i=1}^{n} (\mu_i^{-1})^2$ (see Exercise 19), it follows that any service time described by a random variable X, where $E(X) > \sqrt{V(X)}$, can be represented as a sum of independent, exponential phases, as in (8.1), with the given mean and variance. Furthermore, by judicious choice of the values n and μ_i ($i = 1, 2, \ldots, n$), other moments might also be fitted to better approximate the given service-time distribution. Of course, the phases $X_1, \ldots, X_n$ do not necessarily correspond to any actual phases of service, but are only artifices introduced for the purpose of approximating the original process by a birth-and-death process.

Now suppose the service time X has greater variability than the exponential distribution prescribes; that is, assume $E(X) < \sqrt{V(X)}$. In this case, we can model the random variable X as a parallel arrangement of exponential phases (Morse [1958]); that is, the realization of X is obtained by choosing, with probability p_i, the realization of the random variable X_i. Thus, the distribution function of X is $F_X(t) = \sum_{i=1}^{n} p_i F_i(t)$, where, as before, $F_i(t) = 1 - e^{-\mu_i t}$; then $E(X) = \sum_{i=1}^{n} p_i \mu_i^{-1}$ and $V(X) = 2\sum_{i=1}^{n} p_i \mu_i^{-2} - (\sum_{i=1}^{n} p_i \mu_i^{-1})^2$. In this case, X is said to be a *mixture* of exponentials, and $F_X(t)$ is called the *hyperexponential* distribution function. We leave it to Exercise 19 for the reader to show that $E(X) > \sqrt{V(X)}$ when X is modeled as a serial arrangement of exponential phases, and $E(X) < \sqrt{V(X)}$ when X is modeled as a parallel arrangement of exponential phases.

Thus, as a practical matter, one may try to approximate an arbitrary random variable X by a sum [when $E(X) > \sqrt{V(X)}$], a mixture [when $E(X) < \sqrt{V(X)}$], or a combined sum-mixture (see, for example, Kleinrock [1975]). In fact, as we shall discuss shortly, such an approximation can be made with any desired degree of accuracy.

To continue with our example of the method of phases applied to the s-server Erlang loss system, suppose that a representation of the form (8.1) has been fitted to the original data or hypothesized service-time distribution. For ease of exposition let us assume $n = 2$. Now, if we let $P(j_1, j_2)$ be the equilibrium probability that simultaneously there are j_1 customers in phase 1 of service and j_2 customers in phase 2, the corresponding conservation-of-flow equations are

$$(\lambda + j_1 \mu_1 + j_2 \mu_2) P(j_1, j_2) = \lambda P(j_1 - 1, j_2)$$
$$+ (j_1 + 1) \mu_1 P(j_1 + 1, j_2 - 1) + (j_2 + 1) \mu_2 P(j_1, j_2 + 1)$$
$$(j_1 + j_2 < s) \tag{8.2}$$

and

$$(j_1 \mu_1 + j_2 \mu_2) P(j_1, j_2) = \lambda P(j_1 - 1, j_2) + (j_1 + 1) \mu_1 P(j_1 + 1, j_2 - 1)$$
$$(j_1 + j_2 = s). \tag{8.3}$$

Equations (8.2) and (8.3), together with the normalization equation,

$$\sum P(j_1, j_2) = 1, \tag{8.4}$$

can now be solved, numerically or otherwise, for the probabilities $P(j_1, j_2)$ $(j_1 + j_2 \leq s)$, from which we can calculate the equilibrium probability P_j that there are j customers in service:

$$P_j = \sum_{j_1 + j_2 = j} P(j_1, j_2) = \sum_{k=0}^{j} P(k, j-k) \qquad (j = 0, 1, \ldots, s). \tag{8.5}$$

It is interesting to note that the method of phases was originally suggested by Erlang himself. A generalization of the method, using complex-valued probabilities, has been given by Cox [1955]. Gaver [1954] has proposed an *extended method of phases*, in which the number n of terms in the sum (8.1) is itself a random variable (that is, X is a compound random variable). Recently, Neuts [1975a] has greatly extended these ideas with his introduction of the notion of *probability distributions of phase type*, of which the Erlangian and hyperexponential distributions are very special cases. Neuts has applied his theory to develop computational algorithms for the solution of a wide variety of problems in queueing theory and related areas; see, for example, Neuts [1975b, 1977, 1978a, 1978b, 1980, 1981].

A particularly interesting aspect of the method of phases is that it can sometimes be used to obtain exact results by use of a limiting process. This idea has been developed by Schassberger [1973, pp. 32–33], who proves a theorem that states, roughly speaking, that any nonnegative random variable X can be represented as accurately as desired by a compound sum (that is, the number of terms in the sum is itself a random variable) of independent, identical, exponential variables.

We now illustrate this idea by using the ordinary method of phases to prove the important theorem that was asserted in Section 3.3, namely, that the Erlang loss distribution, given by (3.3) of Chapter 3 and repeated here for convenience,

$$P_j = \frac{\frac{(\lambda/\mu)^j}{j!}}{\sum_{k=0}^{s} \frac{(\lambda/\mu)^k}{k!}} \qquad (j=0,1,\ldots,s), \tag{8.6}$$

is valid for any service-time distribution function with finite mean μ^{-1}.

To see this, observe that Equations (8.2) and (8.3), which describe an Erlang loss system in which service times are composed of a sum of two exponential phases with means μ_1^{-1} and μ_2^{-1}, are satisfied by the product solution

$$P(j_1, j_2) = \frac{(\lambda/\mu_1)^{j_1}}{j_1!} \frac{(\lambda/\mu_2)^{j_2}}{j_2!} c \qquad (j_1 + j_2 \leq s). \tag{8.7}$$

[Compare (8.2) with the equations (2.13) that describe two queues in tandem.] Substitution of (8.7) into (8.5) gives, after application of the binomial theorem,

$$P_j = \frac{\left(\frac{\lambda}{\mu_1} + \frac{\lambda}{\mu_2}\right)^j}{j!} c \qquad (j=0,1,\ldots,s); \tag{8.8}$$

if we set

$$\frac{1}{\mu} = \frac{1}{\mu_1} + \frac{1}{\mu_2}$$

in (8.8) and normalize, the result is (8.6), as promised. It is easy to see the assumption that the service time is composed of $n=2$ phases is irrelevant; the same result (8.6) would be obtained for any number n of phases and corresponding mean values μ_i^{-1} whose sum is μ^{-1}.

Thus we have proved the theorem in the special case where the service time is modeled as a serial arrangement of exponential phases. We leave it to Exercise 18 for the reader to show that the Erlang loss distribution results when the service time is modeled as a parallel arrangement of exponential phases. These results strongly suggest that our result holds for all service-time distributions. Schassberger's theorem shows that the argument made here can be sharpened to provide a rigorous proof. For other proofs, see the references given in Section 3.3.

For an example of the application of the method of phases to the study of networks of queues, see Kelly [1976] and Barbour [1976].

Exercises

18. Consider the Erlang loss system with hyperexponential service times; that is, assume that each arrival is assigned a service time chosen randomly and independently from one of (for simplicity) two different exponential distributions. (Thus, if the two distributions have means μ_1^{-1} and μ_2^{-1}, and are chosen with probabilities p_1 and $p_2 = 1 - p_1$, respectively, then the service-time density function is $p_1 \mu_1 e^{-\mu_1 t} + p_2 \mu_2 e^{-\mu_2 t}$ and the mean service time is $p_1 \mu_1^{-1} + p_2 \mu_2^{-1}$.) Show that this system is described by the Erlang loss distribution (8.6).

19. Show that if a random variable X is given by a sum of independent, exponential random variables, then $E(X) > \sqrt{V(X)}$; whereas if X is given by a mixture of independent, exponential random variables, then $E(X) < \sqrt{V(X)}$.

5

Imbedded-Markov-Chain Queueing Models

5.1. Introduction

In this chapter we investigate some important queueing models that cannot be studied in the framework of the birth-and-death process.

Recall that a queue is characterized by the input process, the service mechanism, and the queue discipline. In the preceding chapters we concentrated mainly on queues with Poisson or quasirandom input and exponential service times. These assumptions imply that the future evolution of the system from some time t depends only on the state of the system at time t, and is independent of the history of the system prior to time t. In these models, the "state" of the system could always be specified in terms of the number of customers present. (In the multidimensional case, the state was specified in terms of the number of customers of each type present at time t.)

Suppose that we are interested in a queue for which knowledge of the number of customers present at any time t is not sufficient information to permit complete analysis of the model. For example, consider the case in which the service times are assumed exponential, but the customer's arrival epochs are separated by a constant time interval. Then the future evolution of the system from some time t would depend not only on the number of customers present at time t, but also on the elapsed time since the last customer arrival epoch (because the arrival epoch of the next customer is strictly determined by the arrival epoch of the last customer).

Clearly, a new method of analysis is required. A powerful method for the analysis of certain queueing models, such as the model in the above example, is that of the *imbedded Markov chain*, introduced by Kendall [1951, 1953]. As with the birth-and-death process, there is a vast theory of Markov chains. More generally, both are subsumed under the heading

Introduction

Markov processes. As is our policy, we shall aim at as direct an approach to the analysis of our queueing models as possible, without extended excursions into the surrounding theoretical structure. Thus we shall introduce the main ideas behind the theory of the imbedded Markov chain, and show how these ideas facilitate the analysis of certain important queueing models.

We will study two important basic models, and some of their variations, which are easily analyzed by the method of the imbedded Markov chain. These two models are, in the notation introduced by Kendall [1953], the $M/G/1$ queue and the $GI/M/s$ queue. (We shall explain Kendall's notation shortly.)

The $M/G/1$ queueing model assumes that customers arrive according to a Poisson process; the distribution of service times is arbitrary; there is one server; and all blocked customers wait until served. This model differs from the 1-server Erlang delay model we studied earlier only in that the service times are no longer required to be exponentially distributed. This increase in generality requires a corresponding increase in the complexity of the mathematical apparatus needed for its analysis.

The second basic model we shall study, the $GI/M/s$ queue, assumes that the interarrival times are independent, identically distributed, nonnegative random variables; the service times are exponentially distributed; there are $s \geqslant 1$ servers; and all blocked customers wait until served. This model differs from the s-server Erlang delay model we studied earlier only in that the interarrival times are no longer required to be exponentially distributed. As with the $M/G/1$ queue, the increased generality of the model requires mathematical tools beyond those required for birth-and-death models. In both cases, the main new mathematical tool we shall use is the imbedded Markov chain.

Kendall's shorthand notation $a/b/c$ is now widely used to describe queueing models. In this notation, a specifies the arrival process, b specifies the service time, and c is the number of servers. In every case, it is assumed that all blocked customers wait until served. Of course, this notation can't characterize every model, so extended versions have been introduced to permit specification of other system characteristics, such as the order of service or the number of waiting positions if there is a maximum allowed queue length. However, these extensions are not universally accepted, and often seem to be clumsier than a simple longhand characterization of the queueing model.

Some examples of Kendall's notation are

$M/G/1$—Poisson (Markov) input, General (arbitrary) service-time distribution function, 1 server;

$GI/M/s$—General, Independently distributed interarrival times, exponential(Markov) service times, s servers;

$M/D/s$—Poisson input, constant (*D*eterministic) service times, s servers;

$E_k/M/s$—k-phase *E*rlangian interarrival-time distribution function, exponential service times, s servers;

$M/M/s$—Poisson input, exponential service times, s servers (the Erlang delay model).

Of course, each of the above models is only a special case of the model $GI/G/s$. But it is often true that the more general the analysis, the less specific (and useful) is the derived information. Such is the case here. The models that appear to be optimal with respect to the tradeoff between generality and specificity are the $M/G/1$ and $GI/M/s$ models (and some of their variations), to which we devote this chapter. References for these models will be given in the text. For a discussion of the more general model $GI/G/s$, which we shall not discuss, see De Smit [1973] and the papers referenced therein. In preparation for the analyses of these models, we start with a discussion of some important related results.

5.2. The Equation $L=\lambda W$ (Little's Theorem)

Consider a system in equilibrium in which customers arrive, remain in the system for a length of time (called the waiting time), and then depart. Let λ be the arrival rate, W be the mean waiting time, and L be the mean number of customers present. (We are adopting this notation to be consistent with common practice.) Then, if the mean values λ, W, and L exist, they are related to each other according to the equation (which holds for each realization of the process)

$$L=\lambda W. \qquad (2.1)$$

The well-known theorem embodied in Equation (2.1) (which is often referred to as *Little's theorem*, or simply as "the equation $L=\lambda W$") is one of the most general and useful results in queueing theory. It was first stated in a rigorous manner by Little [1961]. Little's original statement of (2.1) required that the system under consideration possess certain conceptually difficult mathematical properties (e.g., metric transitivity), and his proof was similarly nonintuitive. Since the publication of Little's paper, many authors have attempted to simplify the hypotheses under which (2.1) holds, or present proofs that are intuitively more appealing, or give heuristic arguments that convincingly explain (2.1) even though they lack rigor. Recently, Stidham [1974] published a relatively simple yet rigorous proof of (2.1), which is exceedingly general in that it requires only the existence and finiteness of the mean values λ and W.

In this section we will give some examples to illustrate the range and utility of (2.1), followed by a simple heuristic argument to "explain" (2.1) and then an outline of Stidham's proof.

The Equation $L = \lambda W$ (Little's Theorem)

Consider, for example, the relationship between offered load and carried load that has been used repeatedly throughout this text on heuristic grounds, namely, carried load equals the portion of the offered load that is carried. Offered load a, by definition, is the product of the arrival rate and the mean service time; hence, in the language of (2.1), the offered load equals λW. Carried load a', by definition, is the mean number of busy servers; hence, in the language of (2.1), the carried load equals L. As an application of this idea, consider the equilibrium s-server Erlang loss system, where λ is the parameter of the Poisson arrival stream [not the same quantity as defined in (2.1)], and τ is the mean service time. We define the "system" as the group of s servers. The left-hand side of (2.1) is, by definition, the carried load a'. To calculate the right-hand side of (2.1), note that the rate at which customers *enter* this system is $\lambda(1 - B(s, a))$, and the mean time spent in the system by those who enter is the mean service time τ. Therefore, (2.1) becomes

$$a' = [\lambda(1 - B(s, a))] \cdot \tau, \tag{2.2}$$

where, by definition, $a = \lambda \tau$. Hence, (2.2) becomes

$$a' = a(1 - B(s, a)), \tag{2.3}$$

which is the same as Equation (3.5) of Chapter 3.

For another example, consider again Exercise 28 of Chapter 3. In part a of that exercise, the reader is asked to verify (2.1) by direct calculation of L and W, where L is the mean number waiting in the queue and W is the mean waiting time for service. That is, in part a the "system" is defined to include only the queue, not the servers. In part b of the same exercise, the reader is asked to again verify (2.1), this time defining the system to include both the queue and the servers. In either case, the calculation is routine.

Another example is given by Exercise 37 of Chapter 3, where the reader is asked to verify Equation (8.18) of Chapter 3, which can be written

$$\sum_{j=1}^{n} j P_j[n] = a\mu [\mu^{-1} + E(W)]. \tag{2.4}$$

The left-hand side of (2.4) equals the mean number of customers present (waiting in the queue and in service) in an n-source equilibrium queueing system with quasirandom input, exponential service times (with mean μ^{-1}), and BCD queue discipline. Since a is the offered load, therefore $a\mu = \lambda$; furthermore, the term in brackets on the right-hand side of (2.4) equals the mean time a customer spends in service or waiting in the queue. Hence, Equation (2.4) is simply a special case of (2.1) and needs no further verification.

As a final example, consider Equation (8.20) of Chapter 3,

$$n = \lambda(\gamma^{-1} + E(T)),$$

which was obtained, after much calculation, for the model in which n sources submit quasirandom input to s exponential servers, with the queue discipline blocked customers delayed. We assert that this equation holds not only for that model, but also for the more general model in which each of n sources submits requests at rate γ when idle, and experiences a sojourn time T before returning to the idle state, regardless of the complexity of the network through which the sources travel, and regardless of the distributions that describe the service times and interarrival times. To prove this assertion, simply observe that we can identify n as L and $(\gamma^{-1} + E(T))$ as W.

Of course, Equation (2.1) is most useful when one wants to know W and it is difficult to calculate it directly, but L is easy, or vice versa. This is often the case, because L is a time-average of a discrete-valued random variable, whereas W is a customer-average of a (in general) continuous-valued random variable.

For example, we can use (2.1) to prove easily an important assertion made previously in this text: *In any queueing system with a nonbiased queue discipline, the mean wait for service suffered by an arbitrary customer is independent of the order of service.* (Recall that a nonbiased queue discipline is one in which customers are selected for service from among those waiting without regard to the service times of the waiting customers; that is, the next customer chosen for service has the same service-time distribution function as does an arbitrary customer.)

To prove this assertion, note that according to (2.1) the mean wait W is proportional to the mean queue length L. But the state equations from which one calculates the distribution of the number of customers in the queue do not depend on the identity of the particular customer chosen for service at any time; that is, the distribution of the number of customers waiting in the queue does not depend on the order of service if the queue discipline is nonbiased. Therefore, the mean queue length L is independent of the order of service, and hence, by Equation (2.1), so is the mean waiting time W.

Note that, in general, the distribution function of waiting times *does* depend on the order of service for nonbiased queue discipline, even though its first moment W does not.

The following heuristic argument in support of (2.1) was suggested by P. J. Burke. Assume that the mean values L and W exist, and consider a long time interval $(0, t)$ throughout which statistical equilibrium prevails. The mean number of customers who enter the system during this interval is λt. Imagine that a waiting time is associated with each arriving customer; that is, each arrival brings a waiting time with him. If the average waiting time brought in by an arrival is W, then the average amount of waiting time brought into the system during $(0,t)$ is $\lambda t W$. On the other hand, each customer present in the system uses up his waiting time linearly with time. If L is the average number of customers present throughout $(0,t)$, then Lt is the average amount of time used up in $(0,t)$. Now, as $t \to \infty$ the accumulation of waiting time must equal the amount of waiting time used up; that is, the ratio of the waiting time accumulated in $(0,t)$ to that used up in $(0,t)$ must approach unity as $t \to \infty$. Hence,

$$\lim_{t \to \infty} \frac{\lambda t W}{Lt} = 1 \quad \text{(roughly speaking),}$$

from which (2.1) follows.

We conclude this section with Stidham's [1974] statement of the theorem (2.1) and an outline of his proof. For further discussion see Brumelle [1971], Heyman and Stidham [1980], and Miyazawa [1977].

Theorem. *Let $L(x)$ be the number of customers present at time x, and define the mean number L of customers present throughout the time interval $[0, \infty)$ as*

$$L = \lim_{t \to \infty} \frac{1}{t} \int_0^t L(x) \, dx; \tag{2.5}$$

let $N(t)$ be the number of customers who arrive in $[0,t]$, and define the arrival rate λ as

$$\lambda = \lim_{t \to \infty} \frac{N(t)}{t}; \tag{2.6}$$

and let W_i be the waiting time of the ith customer, and define the mean waiting time W as

$$W = \lim_{n \to \infty} \frac{1}{n} \sum_{i=1}^n W_i. \tag{2.7}$$

If λ and W exist and are finite, then so does L, and they are related according to (2.1).

PROOF. Our strategy is first to show that for all $t \geq 0$,

$$U(t) \geq \int_0^t L(x) \, dx \geq V(t), \tag{2.8}$$

where $U(t)$ is defined as the sum of the waiting times of the customers that have arrived in $[0, t]$, and $V(t)$ is defined as the sum of the waiting times of the customers that have (arrived and) departed in $[0, t]$. We then rewrite the inequality (2.8) as

$$\frac{U(t)}{t} \geq \frac{1}{t}\int_0^t L(x)\,dx \geq \frac{V(t)}{t} \tag{2.9}$$

and establish that

$$\lim_{t \to \infty} \frac{U(t)}{t} = \lambda W \tag{2.10}$$

and

$$\lim_{t \to \infty} \frac{U(t)}{t} = \lim_{t \to \infty} \frac{V(t)}{t}. \tag{2.11}$$

Our theorem then follows from (2.9)–(2.11) and the definition (2.5).

To prove (2.8), let T_i, $i=1,2,\ldots$, be the arrival time of the ith customer $(0 \leq T_1 \leq T_2 \leq \cdots)$, and write

$$U(t) = \sum_{S_1(t)} W_i, \tag{2.12}$$

where $S_1(t)$ is the set of values of the index i for which $T_i \leq t$,

$$S_1(t) = \{i : T_i \leq t\}; \tag{2.13}$$

and similarly, we write

$$V(t) = \sum_{S_2(t)} W_i, \tag{2.14}$$

where $S_2(t)$ is the set of values of the index i for which $T_i + W_i \leq t$,

$$S_2(t) = \{i : T_i + W_i \leq t\}. \tag{2.15}$$

The truth of the inequality (2.8) is now easy to see if we use the following economic interpretation. Suppose that the system incurs a unit cost for each unit of time spent in the system by each customer. Then $U(t)$ is the total cost incurred in $[0, t]$ if each customer's cost is charged in a lump sum at the instant of his arrival; $\int_0^t L(x)\,dx$ is the total cost incurred in $[0, t]$ if each customer's cost is charged (in the usual way) continuously at unit rate while he is in the system; and $V(t)$ is the total cost incurred in $[0, t]$ if each customer's cost is charged in a lump sum at the instant of his departure. With this interpretation, (2.8) becomes obvious. (Note that this argument is

The Equation $L=\lambda W$ (Little's Theorem)

essentially the same as the heuristic argument given earlier. Indeed, this proof of $L=\lambda W$ is, in fact, a rigorous treatment of the earlier heuristic argument.)

Our next step is to establish (2.10). To this end, we calculate the product λW using the definitions (2.6) and (2.7):

$$\lambda W = \left(\lim_{t \to \infty} \frac{N(t)}{t} \right) \left(\lim_{n \to \infty} \frac{1}{n} \sum_{i=1}^{n} W_i \right). \qquad (2.16)$$

Now replace the second limit on the right-hand side of (2.16) by the equivalent representation

$$\lim_{n \to \infty} \frac{1}{n} \sum_{i=1}^{n} W_i = \lim_{t \to \infty} \frac{1}{N(t)} \sum_{i=1}^{N(t)} W_i, \qquad (2.17)$$

and use the fact that $(\lim a)(\lim b) = \lim ab$; then (2.16) becomes

$$\lambda W = \lim_{t \to \infty} \frac{1}{t} \sum_{i=1}^{N(t)} W_i. \qquad (2.18)$$

Observe that the sum on the right-hand side of (2.18) is the same as that in (2.12). Hence, we have proved (2.10).

It remains only to prove (2.11). To this end, observe that

$$\lim_{i \to \infty} \frac{W_i}{T_i} = 0. \qquad (2.19)$$

Equation (2.19) becomes obvious when one realizes that the waiting times W_i, $i=1,2,\ldots$, do not systematically increase with increasing i, whereas the arrival times T_i, $i=1,2,\ldots$, do become arbitrarily large as $i \to \infty$. (For a proof of (2.19), see Lemma 4 of Stidham [1972].) It follows from (2.19) that for any $\varepsilon > 0$, there exists a number k such that $W_i/T_i < \varepsilon$ for all $i > k$; that is,

$$W_i < T_i \varepsilon \qquad \text{for all} \quad i > k. \qquad (2.20)$$

Now suppose t is large enough so that $N(t) > k$. Observe that the index set $S_2(t)$, defined by (2.15), can be partitioned into the union of the two sets $S_3(t)$ and $S_4(t)$, defined by

$$S_3(t) = \{ i : T_i + W_i \leq t \text{ and } i \leq k \} \qquad (2.21)$$

and

$$S_4(t) = \{ i : T_i + W_i \leq t \text{ and } i > k \}. \qquad (2.22)$$

(That is, $S_3(t)$ and $S_4(t)$ partition the set $S_2(t)$ of customers who arrive and depart in $[0,t]$ into those who arrive in $[0,T_k]$ and those who arrive in $(T_k,t]$.) Hence, we can write (2.14) as

$$V(t) = \sum_{S_3(t)} W_i + \sum_{S_4(t)} W_i. \tag{2.23}$$

Now consider the index set $S_5(t)$, defined by

$$S_5(t) = \{i : T_i + T_i \varepsilon \leqslant t \text{ and } i > k\} \tag{2.24}$$

Since $T_i \varepsilon > W_i$ by (2.20), it follows that $S_4(t)$ contains at least as many points as $S_5(t)$; that is, $S_5(t) \subseteq S_4(t)$, and therefore

$$\sum_{S_5(t)} W_i \leqslant \sum_{S_4(t)} W_i. \tag{2.25}$$

It follows from (2.23) and (2.25) that

$$V(t) \geqslant \sum_{S_3(t)} W_i + \sum_{S_5(t)} W_i. \tag{2.26}$$

By adding and subtracting the quantity $\sum_{S_6(t)} W_i$, we can write (2.26) as

$$V(t) \geqslant \sum_{S_3(t)} W_i - \sum_{S_6(t)} W_i + \sum_{S_6(t)} W_i + \sum_{S_5(t)} W_i, \tag{2.27}$$

where

$$S_6(t) = \{i : T_i + T_i \varepsilon \leqslant t \text{ and } i \leqslant k\}. \tag{2.28}$$

The last two sums on the right-hand side of (2.27) can be combined, so that (2.27) can be written

$$V(t) \geqslant \sum_{S_3(t)} W_i - \sum_{S_6(t)} W_i + \sum_{S_7(t)} W_i, \tag{2.29}$$

where $S_7(t)$ is the union of $S_6(t)$ and $S_5(t)$:

$$S_7(t) = \{i : T_i + T_i \varepsilon \leqslant t\} = \left\{i : T_i \leqslant \frac{t}{1+\varepsilon}\right\}. \tag{2.30}$$

Comparison of (2.13) and (2.30) shows that

$$S_7(t) = S_1\left(\frac{t}{1+\varepsilon}\right), \tag{2.31}$$

and it then follows from (2.12) and (2.31) that (2.29) can be written

$$V(t) \geq \sum_{S_3(t)} W_i - \sum_{S_6(t)} W_i + U\left(\frac{t}{1+\varepsilon}\right); \quad (2.32)$$

that is,

$$\frac{U(t)}{t} \geq \frac{V(t)}{t} \geq \frac{1}{t}\left(\sum_{S_3(t)} W_i - \sum_{S_6(t)} W_i\right) + \frac{U(t/(1+\varepsilon))}{t}, \quad (2.33)$$

where the first inequality in (2.33) follows from (2.9).

Now let $t \to \infty$ in (2.33). The sums in parentheses in (2.33) are bounded (by $\sum_{i \leq k} W_i$), and therefore

$$\lim_{t \to \infty} \frac{1}{t}\left(\sum_{S_3(t)} W_i - \sum_{S_6(t)} W_i\right) = 0. \quad (2.34)$$

Furthermore,

$$\lim_{t \to \infty} \frac{U(t/(1+\varepsilon))}{t} = \lim_{t' \to \infty} \frac{U(t')}{(1+\varepsilon)t'} = \frac{1}{1+\varepsilon} \lim_{t \to \infty} \frac{U(t)}{t}. \quad (2.35)$$

Hence, if we take limits in (2.33) we obtain

$$\lim_{t \to \infty} \frac{U(t)}{t} \geq \lim_{t \to \infty} \frac{V(t)}{t} \geq \frac{1}{1+\varepsilon} \lim_{t \to \infty} \frac{U(t)}{t}. \quad (2.36)$$

The statement (2.36) is valid for any $\varepsilon > 0$, no matter how small; hence (2.11) follows from (2.36), and the proof is complete. □

5.3. Equality of State Distributions at Arrival and Departure Epochs

Often, queueing models are analyzed by writing equations that relate state probabilities to one another at certain well-defined points in time (epochs). The choice of the particular set of points with respect to which the equations are written is governed by the mathematical properties of the model under consideration, not by unfettered choice. For example, the $M/G/1$ model is easily analyzed by relating the system states to each other at the instants at which customers finish service and leave the system. Therefore, we define the state of the system to be the number of customers left behind by a departing customer, and the corresponding *departing customer's distribution* $\{\Pi_j^*\}$ describes the system at this special set of points. However, the quantities of interest are the state probabilities at a different set of points, namely the arrival epochs. That is, from the point of view of an arriving customer, the number of customers that he finds in the system, not the number he leaves behind, is the quantity of interest. Fortunately, for a wide range of processes that includes the important ones, the state distributions at these two sets of points are identical.

We have already encountered some similar situations. In the $M/M/s$ (Erlang delay) model, the birth-and-death formulation led directly to the outside observer's distribution $\{P_j(t)\}$. But it was shown in Chapter 3 that the arriving customer's distribution $\{\Pi_j(t)\}$ is identical to the outside observer's distribution $\{P_j(t)\}$ whenever the input is Poisson. Thus the more relevant arriving customer's distribution had been found indirectly. Likewise, the birth-and-death formulation for the case of quasirandom input in Chapter 3 yielded the outside observer's distribution, which, though not directly equal to, was easily translated into the arriving customer's distribution.

We have the following important theorem relating the equilibrium distribution $\{\Pi_j\}$ of the number of customers found by an arrival (the arriving customer's distribution) and the equilibrium distribution $\{\Pi_j^*\}$ of the number of customers left behind by a departure (the departing customer's distribution):

In any queueing system for which the realizations of the state process are step functions with only unit jumps (positive and negative), the equilibrium state distribution just prior to arrival epochs is the same as that just following departure epochs: $\{\Pi_j\} = \{\Pi_j^*\}$. (And, in addition, when the input is Poisson, $\{P_j\} = \{\Pi_j\} = \{\Pi_j^*\}$.)

Note that in this theorem, as in the theorem $L = \lambda W$, the term "queueing system" is not used in a restricted sense. For example, the "system" may be defined as the servers and the queue, or it may be defined as the queue alone. The essential characteristic is simply that the number of customers present can increase and decrease by at most one customer at a time. Thus the theorem holds for all systems (in statistical equilibrium) considered so far in this text.

The theorem becomes intuitively obvious when one observes that during any time interval the number of upward transitions $E_j \to E_{j+1}$ cannot differ by more than one from the number of downward transitions $E_{j+1} \to E_j$. That is, for every customer whose arrival causes the transition $E_j \to E_{j+1}$, there corresponds another customer whose departure causes the transition $E_{j+1} \to E_j$. Thus, the proportion of customers for whom the system is in E_j just prior to arrival approaches the same limit as the proportion of customers for whom the system is in E_j just after departure.

Because of this theorem's importance (which will become obvious) and because (strangely) it does not appear to be well known even among experts, we now give a formal statement and proof. The theorem and proof are taken from an adaptation by P. J. Burke [1968, unpublished] of a corresponding result by L. Takács for the $M/G/1$ queue.

Theorem. *Let $N(t)$ be a stochastic process whose sample functions are (almost all) step functions with unit jumps. Let the points of increase after some time $t=0$ be labeled T_α, and let the points of decrease be labeled T'_α, $\alpha = 1, 2, \ldots$. Let $N(T_\alpha -)$ be denoted A_α, and let $N(T'_\alpha +)$ be denoted D_α. (Thus, if upward jumps correspond to arrivals and downward jumps correspond to departures, then A_α is the state of the system just prior to the αth arrival epoch and D_α is the state of the system just after the αth departure epoch.) Then if either $\lim_{n \to \infty} P\{A_n \leq k\}$ or $\lim_{n \to \infty} P\{D_n \leq k\}$ exists, so does the other and these are equal. [Thus $N(t)$ has the same limiting distribution just prior to its points of increase as it does just after its points of decrease, if this limiting distribution exists.]*

PROOF. Let $N(0) = i$. We first show that $D_{n+i} \leq k \Rightarrow A_{n+k+1} \leq k$. Suppose $D_{n+i} = j \leq k$. Then there are $n+j$ of the T_α preceding T'_{n+i}. Thus $A_{n+j+1} \leq j$, and, since $k-j$ arrivals later than T_{n+j+1} the state can be at most $k-j$ greater, therefore $A_{n+k+1} \leq k$.

We will now show the converse, namely $A_{n+k+1} \leq k \Rightarrow D_{n+i} \leq k$. Suppose $A_{n+k+1} = j \leq k$. Then there are $n+i+k-j$ of the T'_α preceding T_{n+k+1}; that is, $T'_{n+i+k-j}$ is the last T'_α preceding T_{n+k+1}. Hence $D_{n+i+k-j} \leq j$ and $D_{n+i} \leq k$.

Therefore, $P\{D_{n+i} \leq k\} = P\{A_{n+k+1} \leq k\}$, and both sides must have the same limit. Thus, for any k,

$$\lim_{n \to \infty} P\{D_{n+i} \leq k\} = \lim_{n \to \infty} P\{A_{n+k+1} \leq k\};$$

that is,

$$\lim_{n \to \infty} P\{D_n \leq k\} = \lim_{n \to \infty} P\{A_n \leq k\},$$

and the proof is complete. □

Thus we have shown that, with the definitions

$$\Pi_j = \lim_{n \to \infty} P\{N(T_n -) = j\} \qquad (j = 0, 1, 2, \ldots) \tag{3.1}$$

and

$$\Pi_j^* = \lim_{n \to \infty} P\{N(T'_n +) = j\} \qquad (j = 0, 1, 2, \ldots), \tag{3.2}$$

then

$$\Pi_j = \Pi_j^* \qquad (j = 0, 1, 2, \ldots); \tag{3.3}$$

that is, when the state changes occur one at a time and statistical equilibrium prevails, then the arriving customer's distribution $\{\Pi_j\}$ and the departing customer's distribution $\{\Pi_j^*\}$ are equal.

Exercises

1. (P. J. Burke [1968, unpublished].) Consider a birth-and-death process with birth rate λ_j and death rate μ_j when the system is in state E_j ($j = 0, 1, \ldots$; $\mu_0 = 0$). Let $\{\Pi_j\}$ be the state distribution just prior to arrival epochs (births).
   a. Prove that
   $$\lambda_{j+1}\Pi_j = \mu_{j+1}\Pi_{j+1}, \qquad (1)$$
   and compare with Equation (3.15) of Chapter 2. [*Hint:* Show that
   $$\Pi_j^* = \Pi_j P\{E_{j+1} \to E_j\} + \Pi_{j+1}^* P\{E_{j+1} \to E_j\},$$
   where $\{\Pi_j^*\}$ is the state distribution just subsequent to departure epochs (deaths) and $P\{E_{j+1} \to E_j\}$ is the conditional probability that the next transition will take the system into state E_j, given that it is currently in state E_{j+1}.]
   b. Use the result (1) of this exercise to prove that in any system whose state process constitutes a (one-dimensional) birth-and-death process, the arriving customer's distribution is the same as the outside observer's distribution for a similar system with one less source. [This proves, in particular, the result (6.8) of Chapter 3 for birth-and-death systems with quasirandom input.]

2. *Burke's theorem.* For the $M/M/s$ queue in equilibrium, the sequence of service completion epochs follows a Poisson process (with the same parameter as the input process); that is, the output process is statistically the same as the input process.

   PROOF. Let T_1 be an arbitrary service completion epoch in equilibrium, and let T_2 be the next consecutive service completion epoch; and define $F_j(t)$ to be the joint probability that the number of customers in the system just after time $T_1 + t$ equals j and $T_2 > T_1 + t$.
   a. Obtain a set of differential-difference equations for $F_j(t)$, and show that the unique solution satisfying the initial condition $F_j(0) = \Pi_j^*$ is
   $$F_j(t) = \Pi_j^* e^{-\lambda t}. \qquad (1)$$
   b. Show that Equation (1) implies that the time separating two successive departures is exponentially distributed with the same mean as the interarrival times.
   c. To complete the proof of the theorem, show that the interdeparture intervals are mutually independent. [*Hint:* Because of the Markov property of the $M/M/s$ queue, it is sufficient to show that the length of an interdeparture interval and the number of customers in the system at the start of the next interval are independent. To do this use the fact that the joint probability

that the interdeparture interval length is in $(t, t+dt)$ and the number of customers in the system at the start of the next interdeparture interval is j is given by $F_{j+1}(t)\mu(j+1)dt$, where $\mu(j+1)=(j+1)\mu$ if $j+1 \leq s$ and $\mu(j+1)=s\mu$ if $j+1 > s$, and μ^{-1} is the average service time.]

This theorem was proved independently by Burke [1956] (whose proof is outlined here) and by Reich [1957]. For a related discussion and references, the reader is referred back to the subsection "Queues in Tandem" in Section 4.2.

5.4. Mean Queue Length and Mean Waiting Time in the $M/G/1$ Queue

In this section we will apply the results of the last two sections to obtain the following important results for the equilibrium $M/G/1$ queue: The expected value of the number Q of waiting customers found by an arbitrary arrival and the expected value of the time W spent by an arbitrary customer waiting in the queue for service to begin are given by

$$E(Q) = \frac{\rho^2}{2(1-\rho)}\left(1+\frac{\sigma^2}{\tau^2}\right) \tag{4.1}$$

and

$$E(W) = \frac{\rho\tau}{2(1-\rho)}\left(1+\frac{\sigma^2}{\tau^2}\right), \tag{4.2}$$

where τ is the mean and σ^2 is the variance of service times, and $\rho = \lambda\tau < 1$ (where λ is the arrival rate).

Note first that (4.2), which is known as the *Pollaczek-Khintchine formula*, follows immediately from (4.1) and the equation $L = \lambda W$. [Let "L" correspond to the mean number of customers waiting in the queue, from the viewpoint of the outside observer; and let "W" be the mean waiting time. Because the outside observer's viewpoint and the arriving customer's viewpoint are identical in systems with Poisson input, it follows that $L = E(Q)$. Hence, $E(Q) = \lambda E(W)$.] Before we establish (4.1), however, let us consider some ramifications of these results.

Suppose first that service times are exponentially distributed. Then $\sigma^2 = \tau^2$ and, if we denote by $\overline{W}_M$ the mean waiting time in the $M/M/1$ queue, it follows from (4.2) that

$$\overline{W}_M = \frac{\rho}{1-\rho}\tau. \tag{4.3}$$

Now suppose that service times are constant. Then $\sigma^2 = 0$, and if we similarly denote by $\overline{W}_D$ the mean waiting time in the $M/D/1$ queue, it follows from (4.2) that

$$\overline{W}_D = \frac{\rho}{2(1-\rho)}\tau. \tag{4.4}$$

That is,

$$\overline{W}_M = 2\overline{W}_D \tag{4.5}$$

(and, similarly, the mean queue length in $M/M/1$ is twice that in $M/D/1$). This vividly illustrates the effect of the form of the service-time distribution function on the queue length and waiting times in the $M/G/1$ queue. [It is remarkable to observe, however, that although the time spent waiting in the queue does depend on the service-time distribution function, the probability that a customer will in fact wait in the queue does not; this probability equals ρ for any equilibrium $M/G/1$ queue. To see this, recall Equation (4.11) of Chapter 3: $1 - P_0 = \lambda\tau$. Now, $P_0 = \Pi_0$ because we have Poisson input. Hence, $1 - \Pi_0 = \rho$, as asserted.]

The importance of the variance term in the formulas (4.1) and (4.2) is underscored by the following observation: An $M/G/1$ queue with finite mean service time, no matter how small, will have infinite mean queue length and infinite mean waiting time if the variance of the service-time distribution function is not finite.

We now give the promised derivation of (4.1), using a clever, though nonintuitive, argument due to Kendall [1951]. The trick is to write the equations from the viewpoint of the departing customer, and then interpret the results to describe the arriving customer's viewpoint.

Let N_k^* be the number of customers in the system (including any customer in service) just after the service completion epoch of the kth departing customer, and let X_k be the number of customers that arrive during the service time of this customer. Clearly the random variables $X_1, X_2, \ldots$ are mutually independent and identically distributed, and X_k is independent of $N_1^*, N_2^*, \ldots, N_{k-1}^*$.

If the kth departing customer does not leave the system empty, then the $(k+1)$th departure will leave behind those same customers in the queue left by the kth departure except for himself, plus all those customers who arrived during the service time of the $(k+1)$th customer. Thus

$$N_{k+1}^* = N_k^* - 1 + X_{k+1} \qquad (N_k^* > 0). \tag{4.6}$$

On the other hand, if the kth customer leaves the system empty, then the $(k+1)$th customer will leave behind just those customers who arrived during his service time. Thus

$$N_{k+1}^* = X_{k+1} \qquad (N_k^* = 0). \tag{4.7}$$

Equations (4.6) and (4.7) can be combined into a single equation

$$N^*_{k+1} = N^*_k - \delta(N^*_k) + X_{k+1}, \tag{4.8}$$

where the random variable $\delta(N^*_k)$ is defined as

$$\delta(N^*_k) = \begin{cases} 0 & \text{if } N^*_k = 0, \\ 1 & \text{if } N^*_k > 0. \end{cases} \tag{4.9}$$

We now square both sides of (4.8):

$$N^{*2}_{k+1} = N^{*2}_k + \delta^2(N^*_k) + X^2_{k+1} + 2N^*_k X_{k+1} \\ - 2X_{k+1}\delta(N^*_k) - 2N^*_k \delta(N^*_k). \tag{4.10}$$

It follows from the definition (4.9) that

$$\delta^2(N^*_k) = \delta(N^*_k) \tag{4.11}$$

and

$$N^*_k \delta(N^*_k) = N^*_k, \tag{4.12}$$

so that Equation (4.10) can be written

$$N^{*2}_{k+1} = N^{*2}_k + \delta(N^*_k) + X^2_{k+1} + 2N^*_k X_{k+1} \\ - 2X_{k+1}\delta(N^*_k) - 2N^*_k. \tag{4.13}$$

We now take expected values through (4.13). Since N^*_k and X_{k+1} are independent, we have

$$E(N^{*2}_{k+1}) = E(N^{*2}_k) + E(\delta(N^*_k)) + E(X^2_{k+1}) \\ + 2E(N^*_k)E(X_{k+1}) - 2E(X_{k+1})E(\delta(N^*_k)) \\ - 2E(N^*_k). \tag{4.14}$$

Now let $k \to \infty$ in (4.14). Assuming that a statistical-equilibrium distribution exists, then in equilibrium the distribution of the number of customers left behind by each customer is the same. Therefore $\lim_{k \to \infty} E(N^{*2}_{k+1}) = \lim_{k \to \infty} E(N^{*2}_k) = E(N^{*2})$ and (4.14) becomes

$$E(N^*) = \frac{E(\delta(N^*))[1 - 2E(X)] + E(X^2)}{2[1 - E(X)]}. \tag{4.15}$$

Similarly, taking expectations and limits through (4.8), we have

$$E(\delta(N^*)) = E(X). \tag{4.16}$$

It remains to calculate $E(X)$ and $E(X^2)$. $E(X)$ is simply the mean number of arrivals occurring during an arbitrary service time, that is, the offered load:

$$E(X) = \lambda \tau = \rho. \tag{4.17}$$

Direct calculation of $E(X^2)$ gives (see Exercise 21a of Chapter 2)

$$E(X^2) = \lambda^2(\sigma^2 + \tau^2) + \rho. \tag{4.18}$$

Substitution of Equations (4.16), (4.17), and (4.18) into (4.15) yields

$$E(N^*) = \rho + \frac{\rho^2 [1 + (\sigma^2/\tau^2)]}{2(1-\rho)}. \tag{4.19}$$

Q is the number of customers waiting in the queue just prior to an arrival epoch, and N^* is the number of customers in the system just after a departure epoch. Since the arrival and departure distributions are equal, we can equate the mean number $E(N^*)$ of customers left by an arbitrary departure to the mean number of customers found by an arbitrary arrival. The latter quantity is the sum of the mean number ρ in service and the mean number $E(Q)$ waiting for service. Thus

$$E(N^*) = \rho + E(Q). \tag{4.20}$$

Finally, comparison of (4.19) and (4.20) shows that (4.19) is equivalent to (4.1).

5.5. Riemann-Stieltjes Integrals

Before we can proceed with our study of queueing theory, we need to introduce (or review) a new tool, the Riemann-Stieltjes integral, along with the Laplace-Stieltjes transform, which will allow us to discuss discrete and continuous random variables within a single mathematical formalism. We shall continue to adhere to our policy of reviewing only the minimal theoretical background necessary to understand the applications in queueing theory. The material of this section is taken largely from Chapter 5 of Widder [1961]; see also Widder [1941].

Consider an interval (a,b) and a set of points $x_0, x_1, \ldots, x_n$ ($a = x_0 < x_1 < x_2 < \cdots < x_n = b$) that partition (a,b) into a subdivision Δ of n adjoining subintervals $(x_0, x_1), (x_1, x_2), \ldots, (x_{n-1}, x_n)$. The norm $\|\Delta\|$ of a subdivision Δ is

$$\|\Delta\| = \max(x_1 - x_0, x_2 - x_1, \ldots, x_n - x_{n-1});$$

that is, it is the length of the largest of the subintervals. We are ready to define the *Riemann-Stieltjes* (or, simply, *Stieltjes*) integral of a function $g(x)$ with respect to a function $h(x)$ from a to b:

$$\int_a^b g(x)\,dh(x) = \lim_{\|\Delta\|\to 0} \sum_{k=1}^n g(\xi_k)[h(x_k) - h(x_{k-1})], \qquad (5.1)$$

where $x_{k-1} \leqslant \xi_k \leqslant x_k$ ($k = 1, 2, \ldots, n$).

The left-hand side of (5.1) is the notation employed for the Riemann-Stieltjes integral. When $h(x) = x$ it reduces to the usual notation of the ordinary (Riemann) integral; and, of course, the right-hand side of (5.1) reduces to the definition of the Riemann integral. The Riemann-Stieltjes integral possesses many properties analogous to those of the Riemann integral, including the fact that there is a method of calculation that is much easier than the definition.

A sufficient condition for the existence of the limit on the right-hand side of (5.1) is that $g(x)$ be continuous on $[a,b]$ and $h(x)$ be nondecreasing on $[a,b]$. This will be the case for all applications in this book.

Essentially all of the elementary properties of the Riemann integral are retained by the Riemann-Stieltjes integral. In particular,

$$\int_a^b dh(x) = h(b) - h(a); \qquad (5.2)$$

also, the Riemann-Stieltjes integral is linear, that is,

$$\int_a^b [c_1 g_1(x) + c_2 g_2(x)]\,dh(x) = c_1 \int_a^b g_1(x)\,dh(x) + c_2 \int_a^b g_2(x)\,dh(x), \qquad (5.3)$$

where c_1 and c_2 are arbitrary constants; and for $a < c < b$,

$$\int_a^b g(x)\,dh(x) = \int_a^c g(x)\,dh(x) + \int_c^b g(x)\,dh(x). \qquad (5.4)$$

(A table of the elementary properties of the Riemann-Stieltjes integral is given on p. 155 of Widder [1961].)

Calculation of the Riemann-Stieltjes integral is facilitated by the following two results:

1. Suppose that $g(x)$, $h(x)$, and $h'(x) = dh(x)/dx$ are all continuous on $[a,b]$. Then

$$\int_a^b g(x)\,dh(x) = \int_a^b g(x)h'(x)\,dx. \qquad (5.5)$$

[Note that the integral on the right-hand side of (5.5) is an ordinary Riemann integral. This theorem has, in fact, been assumed from the beginning of this text.]

2. Suppose $h(x)$ is a step function with jumps at the points c_k of amounts $\Delta h_k = h(c_k +) - h(c_k -)$ $(k=1,2,\ldots,n)$, where $a < c_1 < c_2 < \ldots < c_n < b$; and suppose $g(x)$ is continuous on $[a,b]$. Then

$$\int_a^b g(x)\, dh(x) = \sum_{k=1}^n g(c_k) \Delta h_k. \tag{5.6}$$

For example, consider again Exercise 27 of Chapter 3. There the reader is asked to show that

$$P\{T>t\} = P\{W>t|W>0\} \tag{5.7}$$

for the $M/M/1$ queue with service in order of arrival, where T is the sojourn time and W is the waiting time. It follows from Equation (4.24) of Chapter 3 that

$$P\{W>t|W>0\} = e^{-(1-a)\mu t}.$$

Therefore, the problem can be restated as follows: Show that

$$P\{T>t\} = e^{-(1-a)\mu t} \quad (t \geq 0), \tag{5.8}$$

where $T = W + X$ (sojourn time equals waiting time plus service time) and

$$P\{X \leq t\} = H(t) = \begin{cases} 0 & (t<0), \\ 1 - e^{-\mu t} & (t \geq 0), \end{cases} \tag{5.9}$$

$$P\{W \leq t\} = W(t) = \begin{cases} 0 & (t<0), \\ 1 - ae^{-(1-a)\mu t} & (t \geq 0). \end{cases} \tag{5.10}$$

Note that the waiting-time distribution function (5.10), which follows from (4.25) of Chapter 3, has a jump of amount $1-a$ at the origin.

From probabilistic considerations, we can write

$$P\{T>t\} = \int_{-\infty}^{\infty} [1 - H(t-x)]\, dW(x) \tag{5.11}$$

or, equivalently,

$$P\{T>t\} = \int_{-\infty}^{\infty} [1 - W(t-x)]\, dH(x). \tag{5.12}$$

A reader with no knowledge of Riemann-Stieltjes integration could have solved the problem by rewriting (5.12) as

$$P\{T>t\} = \int_0^t [1 - W(t-x)] H'(x) dx + \int_t^\infty H'(x) dx$$
$$= \int_0^t a e^{-(1-a)\mu(t-x)} \mu e^{-\mu x} dx + e^{-\mu t},$$

from which (5.8) easily follows. However, in order to use (5.11), one must be cognizant of the theory of Riemann-Stieltjes integration to account for the effect of the jump in the function $W(x)$ at the origin. As an example, we shall evaluate the right-hand side of (5.11).

First, we break the integral on the right-hand side of (5.11) into three integrals, ranging over $(-\infty, 0-\varepsilon)$, $[0-\varepsilon, t]$, and (t, ∞), where ε is any positive number. Observe that

$$\int_{-\infty}^{0-\varepsilon} [1 - H(t-x)] dW(x) = 0 \qquad (5.13)$$

because, according to (5.10), $W(x) = 0$ for $-\infty < x < 0$, and therefore $dW(x) = W'(x) dx = 0$. Now observe that

$$\int_t^\infty [1 - H(t-x)] dW(x) = \int_t^\infty dW(x) = 1 - W(t) = ae^{-(1-a)\mu t}, \quad (5.14)$$

where the first equality in (5.14) follows from the fact (5.9) that $H(y) = 0$ when $y \leq 0$; the second equality follows from (5.2); and the third follows from (5.10). Hence, we have reduced (5.11) to

$$P\{T>t\} = \int_{0-\varepsilon}^t e^{-\mu(t-x)} dW(x) + ae^{-(1-a)\mu t}. \qquad (5.15)$$

To evaluate the integral in (5.15) we observe that $W(x)$ has a jump of amount $1-a$ at $x=0$, and $dW(x) = W'(x) dx = a(1-a)\mu e^{-(1-a)\mu x} dx$ for $x > 0$. Therefore

$$\int_{0-\varepsilon}^t e^{-\mu(t-x)} dW(x) = e^{-\mu(t-0)}(1-a) + \int_0^t e^{-\mu(t-x)} a(1-a)\mu e^{-(1-a)\mu x} dx.$$
(5.16)

It is now a simple matter to evaluate the ordinary Riemann integral on the right-hand side of (5.16), and to combine with (5.15) and simplify to get (5.8).

It is perhaps worthwhile at this point to indicate a different proof of (5.7) that requires no integration. Let N be the number of customers found by an arbitrary arrival in an equilibrium $M/M/1$ queue with offered load *a*. Then, from Equations (4.3)–(4.6) of Chapter 3, we have

$$P\{N=j\} = (1-a)a^j \quad (j=0,1,\ldots). \tag{5.17}$$

Now

$$P\{T>t\} = \sum_{j=0}^{\infty} P\{T>t|N=j\}P\{N=j\} \tag{5.18}$$

and

$$P\{W>t\} = \sum_{j=0}^{\infty} P\{W>t|N=j+1\}P\{N=j+1\}. \tag{5.19}$$

Also,

$$P\{W>t|W>0\} = \frac{P\{W>t\}}{P\{W>0\}} = \frac{1}{a}P\{W>t\}; \tag{5.20}$$

therefore, it follows from (5.19) that (5.20) can be written

$$P\{W>t|W>0\} = \frac{1}{a}\sum_{j=0}^{\infty} P\{W>t|N=j+1\}P\{N=j+1\}. \tag{5.21}$$

Substitution of (5.17) into (5.18) and (5.21) gives, respectively,

$$P\{T>t\} = \sum_{j=0}^{\infty} P\{T>t|N=j\}(1-a)a^j \tag{5.22}$$

and

$$P\{W>t|W>0\} = \sum_{j=0}^{\infty} P\{W>t|N=j+1\}(1-a)a^j. \tag{5.23}$$

But

$$P\{T>t|N=j\} = P\{W>t|N=j+1\}, \tag{5.24}$$

because when $N=j$, then T is a sum of $j+1$ service times; and when $N=j+1$, then W is a sum of $j+1$ service times. Therefore, the right-hand sides of (5.22) and (5.23) are the same; hence (5.7) is true, as asserted.

Exercise

3. Solve Exercise 18 of Chapter 2 by evaluating the integral

$$F_{I_t}(y) = \int F_{R_t}(y-x)\,dF_{A_t}(x).$$

5.6. Laplace-Stieltjes Transforms

The ordinary Laplace transform should be familiar to every student of engineering and applied mathematics. In this section we outline the properties of a generalization, called the Laplace-Stieltjes transform, which has proven to be very useful in applied probability and queueing theory. For an introduction to the theory of transforms, see Widder [1971]; and for a comprehensive discussion in the context of probability theory, see Chapter XIII of Feller [1971].

The *Laplace-Stieltjes transform* $\phi(s)$ of a distribution function $F(t)$ of a nonnegative random variable [that is, $F(t)=0$ when $t<0$] is defined as

$$\phi(s) = \int_{0-}^{\infty} e^{-st} dF(t) \qquad (s \geq 0). \tag{6.1}$$

Note that the lower limit of the integral (6.1) is defined so as to include the jump at the origin when $F(0)>0$. For convenience, throughout the remainder of the text we shall adopt the convention that $\int_0$ is to be interpreted as $\int_{0-}$; that is, we shall always assume, unless stated otherwise, that any jump at the origin is included. Thus, for example, from now on we shall write (6.1) as $\phi(s) = \int_0^\infty e^{-st} dF(t)$.

It turns out that the Laplace-Stieltjes transform of the distribution function of an arbitrary nonnegative random variable shares many of the properties of the probability-generating function of an arbitrary nonnegative integer-valued random variable, and this fact accounts for its usefulness in queueing theory. [The close analogy between the Laplace-Stieltjes transform $\phi(s)$ and the probability-generating function $g(z)$ is illuminated by Exercise 4, where the reader is asked to show that $\phi(s)$ differs from $g(z)$ only by the change of variable $z=e^{-s}$; that is, $\phi(s)=g(e^{-s})$.]

We first observe that if $F(t)$ is differentiable, $F'(t)=f(t)$, then $\phi(s)$ is the ordinary Laplace transform of $f(t)$:

$$\phi(s) = \int_0^\infty e^{-st} dF(t) = \int_0^\infty e^{-st} f(t) dt. \tag{6.2}$$

That is, the Laplace transform of a density function equals the Laplace-Stieltjes transform of the corresponding distribution function. (We remark that the names Laplace-Stieltjes transform and Laplace transform are not used universally as we use them here.)

Like the ordinary Laplace transform (and the probability-generating function), the Laplace-Stieltjes transform maps the object function, of a variable t, into another function, of a dummy variable s. In order for any transform to be useful, of course, its properties must be such that (1) manipulations in transform space are easier than in the original space, and (2) the answer in transform space determines uniquely the answer in the original space. We will now list some properties of the Laplace-Stieltjes transform that demonstrate requirement (1); but we merely assert that requirement (2) is met [if $\phi(s)$ is known for all $s \geq 0$] and refer the reader to Feller [1971] and Widder [1971] for discussion of the uniqueness of the inverse transformation. Jagerman [1978] discusses numerical inversion of the Laplace transform, and illustrates his method with examples drawn from queueing theory.

1. The moments of the distribution function $F(t)$ can be obtained directly from its Laplace-Stieltjes transform by differentiation:

$$\int_0^\infty t^n dF(t) = (-1)^n \left(\frac{d^n}{ds^n} \phi(s) \right)_{s=0}. \qquad (6.3)$$

Equation (6.3) is easily verified by formal differentiation of (6.1).

2. If $F_1(t)$ and $F_2(t)$ have Laplace-Stieltjes transforms $\phi_1(s)$ and $\phi_2(s)$, respectively, and $F(t)$ is the *convolution*

$$F(t) = \int_0^t F_1(t-x) dF_2(x) = \int_0^t F_2(t-x) dF_1(x), \qquad (6.4)$$

then the Laplace-Stieltjes transform $\phi(s)$ of $F(t)$ is the product

$$\phi(s) = \phi_1(s)\phi_2(s). \qquad (6.5)$$

Thus, if X_1 and X_2 are independent, nonnegative random variables whose distribution functions have Laplace-Stieltjes transforms $\phi_1(s)$ and $\phi_2(s)$, then the distribution function of the sum $X_1 + X_2$ has Laplace-Stieltjes transform $\phi_1(s)\phi_2(s)$. To prove this, observe that the transform (6.1) can be written as the expectation $\phi(s) = E(e^{-sX})$, where $F(t)$ is the distribution function of X. Hence, if $X = X_1 + X_2$, then $\phi(s) = E(e^{-s(X_1+X_2)}) = E(e^{-sX_1}e^{-sX_2}) = E(e^{-sX_1})E(e^{-sX_2}) = \phi_1(s)\phi_2(s)$, where the next to last equality follows from the independence of X_1 and X_2.

3. Of particular interest in queueing theory is the relationship between the generating function of the number of events that occur according to a Poisson process during a time interval whose length is a random variable and the Laplace-Stieltjes transform of the distribution function of the interval length. Thus, suppose that events occur according to a Poisson process with rate λ, and let P_j be the probability that exactly j events occur during a time interval whose length is a random variable with distribution function $F(t)$. Then

$$P_j = \int_0^\infty \frac{(\lambda t)^j}{j!} e^{-\lambda t} dF(t) \qquad (j=0,1,\ldots). \qquad (6.6)$$

If the distribution $\{P_j\}$ has the probability-generating function $g(z) = \sum_{j=0}^\infty P_j z^j$, then it follows from (6.6), after interchanging the order of summation and integration, that

$$g(z) = \int_0^\infty \sum_{j=0}^\infty \frac{(\lambda t)^j}{j!} e^{-\lambda t} z^j \, dF(t). \tag{6.7}$$

Observe that the integrand in (6.7) is the probability-generating function of a Poisson variable; hence, from Equation (4.3) of Chapter 2, Equation (6.7) can be written

$$g(z) = \int_0^\infty e^{-(\lambda - \lambda z)t} \, dF(t). \tag{6.8}$$

Now, if $F(t)$ has Laplace-Stieltjes transform $\phi(s)$,

$$\phi(s) = \int_0^\infty e^{-st} \, dF(t), \tag{6.9}$$

then it follows from comparison of (6.8) and (6.9) that

$$g(z) = \phi(\lambda - \lambda z). \tag{6.10}$$

Equation (6.10) will prove to be a most useful result.

Exercises

4. Let N be a nonnegative, integer-valued random variable with distribution function $F(t)$. If $\phi(s)$ is the Laplace-Stieltjes transform of $F(t)$ [$\phi(s) = \int_0^\infty e^{-st} dF(t)$] and $g(z)$ is the probability-generating function of N [$g(z) = \sum_{j=0}^\infty P\{N=j\} z^j$], show that $\phi(s) = g(e^{-s})$.

5. Consider again the premise of Exercise 4 of Chapter 2, but assume now that the distribution function of X_i $(i = 1, 2, \ldots)$ has Laplace-Stieltjes transform $\phi(s)$ (that is, X_i is no longer required to be discrete). Show that, in analogy with part a, S_N has Laplace-Stieltjes transform $g(\phi(s))$, and that part b remains correct.

6. We shall show in Section 5.8 that the equilibrium waiting-time distribution function $W(t)$ in the $M/G/1$ queue with service in order of arrival has Laplace-Stieltjes transform $\omega(s) = \int_0^\infty e^{-st} dW(t)$ given by

$$\omega(s) = \frac{s(1-\rho)}{s - \lambda[1 - \eta(s)]}, \tag{1}$$

where $\eta(s) = \int_0^\infty e^{-st} dH(t)$ is the Laplace-Stieltjes transform of the service-time distribution function $H(t)$, with mean $\tau = \int_0^\infty t \, dH(t)$, and $\rho = \lambda \tau < 1$, where λ is the arrival rate.

   a. Use the fact that the waiting time W has mean value $E(W)$ given by $E(W) = -\omega'(0)$ to obtain Equation (4.2), which we restate here for convenience:

$$E(W) = \frac{\rho\tau}{2(1-\rho)}\left(1 + \frac{\sigma^2}{\tau^2}\right), \qquad (2)$$

where σ^2 is the service-time variance.

b. We know that when service times are exponentially distributed with mean μ^{-1}, then the waiting-time distribution function $W(t)$ is given by [see, for example, Equation (5.10) with $a=\rho$]

$$W(t) = \begin{cases} 0 & (t<0), \\ 1 - \rho e^{-(1-\rho)\mu t} & (t \geqslant 0). \end{cases} \qquad (3)$$

Calculate $\omega(s)$ from (3) and show that it agrees with (1) when $H(t) = 1 - e^{-\mu t}$ ($t \geqslant 0$). Conclude that (3) is the inverse of (1) when service times are exponential.

The Laplace-Stieltjes transform (1) of the equilibrium waiting-time distribution function in the $M/G/1$ queue with order-of-arrival service is known as the *Pollaczek-Khintchine formula*. The corresponding result (2) for the mean waiting time (which is valid for any nonbiased queue discipline) is also often referred to by the same name.

5.7. Some Results from Renewal Theory

During our discussion of the Poisson process in Section 2.5 the following phenomenon was noted: If the mean time between adjacent arrival epochs in a Poisson process is λ^{-1}, then an outside observer who samples (observes) at a random instant will see that the forward recurrence time (the elapsed time between his sampling point and the arrival epoch of the next customer) also has mean λ^{-1}, instead of $\frac{1}{2}\lambda^{-1}$ as one might naively expect. That this mean value exceeds $\frac{1}{2}\lambda^{-1}$ was interpreted as a reflection of the fact that the observer would be more likely to sample during a long interarrival interval than during a short one. In the particular case of a Poisson process, where the interarrival times are exponentially distributed, we obtained this result about the forward recurrence time directly from consideration of the Markov property of the exponential distribution. In the more general case where the distribution of interarrival times is not restricted to the exponential, again an observer would tend to sample during a long interarrival interval, but more sophisticated mathematical tools are necessary for its analysis. In this section we review some results from renewal theory that are applicable to the analysis of this and similar questions for processes in which the interevent times are independent, identically distributed, positive random variables, with an otherwise arbitrary distribution function. Recommended references on renewal theory are Cohen [1969], Feller [1971], Neuts [1973], and Ross [1970]. We note that a rigorous treatment of the subject of this section requires some

Some Results from Renewal Theory 201

delicate arguments concerning limits that are outside the scope and spirit of this text. The reader who requires a more precise treatment of this subject should consult the references.

Suppose that $X_1, X_2,\ldots$ is a sequence of independent, identically distributed, positive random variables, with distribution function $G(x) = P\{X_i \leq x\}$ and mean value $\beta = \int_0^\infty x\, dG(x)$. (To avoid mathematical difficulties, we assume in this section that G has a density.) The sequence $\{X_i\}$ is said to be an *ordinary renewal sequence*. In a common queueing-theory application, the X_i's are interarrival times, and the corresponding arrival process is called *renewal input* or *recurrent input*.

Suppose we have a renewal process in which events or renewals (customer arrivals, say) occur at epochs $T_1, T_2,\ldots$, with $X_i = T_i - T_{i-1}$ ($i = 1, 2,\ldots;\ T_0 = 0$). Let $N(t)$ be the number of renewals (arrivals) in $(0, t]$, and let $S_n = X_1 + \cdots + X_n = T_n$. Then the events $\{N(t) \geq n\}$ and $\{S_n \leq t\}$ are equivalent; hence

$$P\{N(t) \geq n\} = P\{S_n \leq t\}. \tag{7.1}$$

Of particular interest in renewal theory is the *renewal function* $m(t)$, defined as the mean number of renewals in $(0, t]$:

$$m(t) = E(N(t)). \tag{7.2}$$

The *elementary renewal theorem* states the highly intuitive fact that the renewal function $m(t)$ satisfies

$$\lim_{t \to \infty} \frac{m(t)}{t} = \frac{1}{\beta}; \tag{7.3}$$

that is, in the language of asymptotic analysis, the mean number of renewals in $(0, t]$ is asymptotic to t/β:

$$m(t) \sim \frac{t}{\beta}. \tag{7.4}$$

We shall also need the fact that

$$m(t) = \sum_{j=1}^\infty P\{S_j \leq t\}. \tag{7.5}$$

To prove (7.5), note that

$$\sum_{j=1}^\infty P\{S_j \leq t\} = \sum_{j=1}^\infty P\{N(t) \geq j\} = \sum_{j=1}^\infty \sum_{k=j}^\infty P\{N(t) = k\}, \tag{7.6}$$

where the first equality in (7.6) follows from (7.1). Rearrangement of the terms in the right-hand sum of (7.6) yields

$$\sum_{j=1}^{\infty} \sum_{k=j}^{\infty} P\{N(t)=k\} = \sum_{i=1}^{\infty} iP\{N(t)=i\} = E(N(t)), \qquad (7.7)$$

which establishes (7.5).

Now let $P_j(t)$ be the probability that exactly j renewals occur in the interval $(T_k, T_k + t]$. Note that the point T_k is either a renewal point ($k \geq 1$) or the starting point of the process ($k=0$), and that T_k is not included in the interval. In particular, $P_j(t)$ is the probability of j renewals in $(0, t]$. Let $G^{*j}(t)$ be the j-fold convolution of G with itself, and define $G^{*0}(t)=1$ and $G^{*1}(t)=G(t)$. Then, since

$$G^{*j}(t) = P\{S_j \leq t\}$$

and

$$P\{N(t)=j\} = P\{N(t) \geq j\} - P\{N(t) \geq j+1\},$$

it follows from (7.1) that

$$P_j(t) = G^{*j}(t) - G^{*(j+1)}(t) \qquad (j=0,1,\ldots). \qquad (7.8)$$

For example, if the interarrival times are exponentially distributed, $G(t) = 1 - e^{-\lambda t}$, then $G^{*j}(t)$ is the j-phase Erlangian distribution function,

$$G^{*j}(t) = 1 - \sum_{i=0}^{j-1} \frac{(\lambda t)^i}{i!} e^{-\lambda t};$$

and (7.8) gives, of course, the Poisson probabilities,

$$P_j(t) = \frac{(\lambda t)^j}{j!} e^{-\lambda t} \qquad (j=0,1,\ldots).$$

Our main result of this section concerns the limiting distribution of the *forward recurrence time* R_t, defined as the elapsed time from an arbitrary instant t to the next renewal epoch. If we define

$$\lim_{t \to \infty} P\{R_t \leq x\} = F(x),$$

then we shall show that

$$F(x) = \frac{1}{\beta} \int_0^x [1 - G(\xi)] \, d\xi. \qquad (7.9)$$

Some Results from Renewal Theory

Thus, (7.9) gives the distribution of the remaining time from an arbitrary observation point t to the next renewal epoch in an *equilibrium renewal process* (one that has been in existence infinitely long at time t).

Exercise

7. Show that if $G(\xi) = 1 - e^{-\lambda \xi}$ in (7.9), then $F(x) = 1 - e^{-\lambda x}$; that is, if the interevent times are exponentially distributed, the forward recurrence times have the same distribution as the interevent times.

Before proceeding to a proof of (7.9), let us consider some of its ramifications. The distribution function $F(x)$ defined by (7.9) has Laplace-Stieltjes transform $\phi(s) = \int_0^\infty e^{-sx} dF(x)$ given by

$$\phi(s) = \frac{1}{\beta} \frac{1 - \gamma(s)}{s}, \tag{7.10}$$

where $\gamma(s) = \int_0^\infty e^{-sx} dG(x)$ is the Laplace-Stieltjes transform of the distribution function of times between renewals. To prove (7.10), note that from (7.9) we can write

$$\phi(s) = \int_0^\infty e^{-sx} dF(x) = \frac{1}{\beta} \int_0^\infty e^{-sx} [1 - G(x)] dx$$

$$= \frac{1}{\beta} \int_0^\infty e^{-sx} \int_x^\infty dG(y) \, dx, \tag{7.11}$$

where we have used the fact that $1 - G(x) = \int_x^\infty dG(y)$. If we interchange the order of integration on the right-hand side of (7.11), we have

$$\phi(s) = \frac{1}{\beta} \int_0^\infty \int_0^y e^{-sx} dx \, dG(y), \tag{7.12}$$

and (7.10) follows from straightforward evaluation of (7.12).

We leave it for the reader to calculate the mean $\beta^* = \int_0^\infty x \, dF(x)$ of the equilibrium forward recurrence time from the transform (7.10): $\beta^* = -\phi'(0)$, that is,

$$\beta^* = \frac{\beta}{2} + \frac{\sigma^2}{2\beta}, \tag{7.13}$$

where σ^2 is the variance of the interevent times.

Exercise

8. Verify Equation (7.13).

Equation (7.13) shows, for example, that the mean equilibrium forward recurrence time is one-half of the mean interevent time (as the naive

Figure 5.1.

person might guess) if and only if $\sigma^2 = 0$, that is, when the interevent times are constant. Likewise, it follows from (7.13) that if the interevent times are exponential (that is, if $\sigma^2 = \beta^2$), then $\beta^* = \beta$, as required by the Markov property of the exponential distribution.

To prove (7.9), note that the recurrence time R_t will not exceed the value x if and only if at least one renewal occurs between t and $t+x$. Let j be the value of the index of the last renewal to occur prior to epoch $t+x$. Then $R_t \leq x$ if and only if for some value of the index j ($j = 1, 2, \ldots$) we have $t < T_j \leq t + x$; and since T_j is the time of the last renewal prior to $t + x$, then it follows that $t + x < T_{j+1}$. That is, as Figure 5.1. illustrates, $R_t \leq x$ if and only if the point $t + x$ is bracketed by two successive renewal epochs, the first of which lies between t and $t + x$: $t < T_j \leq t + x < T_{j+1}$. Therefore,

$$P\{R_t \leq x\} = \sum_{j=1}^{\infty} \int_t^{t+x} [1 - G(t+x-y)] \, dP\{T_j \leq y\}. \tag{7.14}$$

If we interchange the order of summation and integration in (7.14), and use the fact that $T_j = X_1 + \cdots + X_j = S_j$, we can write

$$P\{R_t \leq x\} = \int_t^{t+x} [1 - G(t+x-y)] \, d\left(\sum_{j=1}^{\infty} P\{S_j \leq y\}\right); \tag{7.15}$$

and it follows from (7.5) that (7.15) can be written

$$P\{R_t \leq x\} = \int_t^{t+x} [1 - G(t+x-y)] \, dm(y). \tag{7.16}$$

It is a consequence of the *key renewal theorem* (see the references given at the beginning of this section) that

$$\lim_{t \to \infty} \int_t^{t+x} [1 - G(t+x-y)] \, dm(y) = \frac{1}{\beta} \int_0^x [1 - G(\xi)] \, d\xi, \tag{7.17}$$

and (7.9) is established. A heuristic argument for (7.17) asserts that as $t \to \infty$ so does y; therefore, it follows from (7.4) that one can replace $m(y)$ on the left-hand side of (7.17) by y/β; that is, $dm(y) \sim (1/\beta) \, dy$. Hence we can write

$$\lim_{t \to \infty} \int_t^{t+x} [1 - G(t+x-y)] \, dm(y) = \frac{1}{\beta} \int_t^{t+x} [1 - G(t+x-y)] \, dy. \quad (7.18)$$

Now, if we make the substitution $t + x - y = \xi$ in the right-hand integral of (7.18), we obtain (7.17).

Thus far we have concerned ourselves with the forward recurrence time R_t, which is the elapsed time from t until the next renewal. We now define the *covering interval* I_t as the interval defined by the renewal epochs that bracket the point t, and the *age* A_t of the covering interval at time t as the elapsed time between t and the preceding renewal epoch. The age is referred to also as the *backward recurrence time*. Clearly, $I_t = A_t + R_t$; however, the random variables A_t and R_t are not, in general, independent.

In Exercise 9 the reader is asked to verify that

$$\lim_{t \to \infty} dP\{I_t \leq x\} = \frac{1}{\beta} x \, dG(x). \quad (7.19)$$

L. Takács has given the following intuitive interpretation of the right-hand side of (7.19), which is the distribution function of the length of the interval that covers an arbitrary point t in an equilibrium renewal process: $dG(x)$ is the frequency with which intervals of length x occur in the general population of intervals defined by our renewal process. We imagine that an observer chooses an interval by observing the process at time t and selecting the interval that contains the observation point t. We have already argued that an observer sampling without regard to the underlying process is more likely to sample during a long interval than a short one; that is, the point t is more likely to be contained in a long interval than a short one. If the frequency $d\hat{G}(x)$, say, with which such an interval has length x, is proportional to the product of the frequency of occurrence of intervals of length x in the general population and the length x, then $d\hat{G}(x) = cx \, dG(x)$, where c is a constant. If we require that $\int_0^\infty d\hat{G}(x) = 1$, then $c = \beta^{-1}$ and (7.19) follows.

Other interesting facts, which the reader is asked to verify in Exercise 9, are

$$\lim_{t \to \infty} P\{R_t \leq x, I_t \leq y\} = \frac{1}{\beta} \int_0^x [G(y) - G(\xi)] \, d\xi \quad (0 \leq x \leq y), \quad (7.20)$$

$$\lim_{t \to \infty} P\{R_t \leq x | I_t = y\} = \frac{x}{y} \quad (0 \leq x \leq y), \quad (7.21)$$

$$\lim_{t \to \infty} P\{R_t > x, A_t > y\} = \frac{1}{\beta} \int_{x+y}^\infty [1 - G(\xi)] \, d\xi. \quad (7.22)$$

Equation (7.20) gives, for an equilibrium renewal process, the joint distribution function of the forward recurrence time and the length of the covering interval. Equation (7.21) shows that the location of the point t is uniformly distributed over the length of the covering interval. And (7.22) shows easily (by virtue of its symmetry in x and y) that, in an equilibrium renewal process, the forward recurrence time and the backward recurrence time are identically distributed.

Exercises

9. **a.** Show that

$$P\{R_t \leqslant x, I_t \leqslant y\} = \sum_{j=1}^{\infty} \int_{t-y}^{t-y+x} [G(y) - G(t-\xi)] dP\{T_j \leqslant \xi\}$$

$$+ \sum_{j=1}^{\infty} \int_{(t-y+x)+}^{t} [G(t-\xi+x) - G(t-\xi)] dP\{T_j \leqslant \xi\}$$

$$(0 \leqslant x \leqslant y < t).$$

[*Hint:* Let T_j be the time of occurrence of the last renewal prior to time t.]

b. Using the same kind of heuristic argument by which (7.9) was derived from (7.14), show that

$$\lim_{t \to \infty} P\{R_t \leqslant x, I_t \leqslant y\} = \frac{1}{\beta} \int_0^{y-x} [G(\xi+x) - G(\xi)] d\xi$$

$$+ \frac{1}{\beta} \int_{y-x}^{y} [G(y) - G(\xi)] d\xi,$$

and show that this implies (7.20).

c. Show that the distribution function of the length of the covering interval in an equilibrium renewal process, obtained by setting $x = y$ in (7.20), is given by

$$\lim_{t \to \infty} P\{I_t \leqslant y\} = \frac{1}{\beta} \int_0^y \xi dG(\xi),$$

and thereby deduce (7.19).

d. Prove (7.21).

e. Show that (7.20) and (7.22) are equivalent.

f. Use (7.22) to show that if the equilibrium renewal process is a Poisson process, then the forward recurrence time and the backward recurrence time are independent.

10. Customers arrive at a single server according to a renewal process with interarrival-time distribution function $G(t)$. The service-time distribution function is $H(t)$, and blocked customers are cleared.

    a. Define the *cycle time* as the time interval separating two successive epochs at which the server becomes busy, and let $F(x)$ be the cycle-time distribution function. Show that

$$F(x) = \sum_{j=1}^{\infty} \int_0^x \int_t^x [1 - G(x-y)] dG^{*j}(y) dH(t). \tag{1}$$

[*Hint:* Define R_t as the duration of time from a service completion epoch (at time t) to the next arrival epoch, so that $F(x) = \int_0^x P\{R_t \leq x - t\} dH(t)$.]

In parts b–h, assume $H(t) = 1 - e^{-\mu t}$ ($t \geq 0$).

b. Let $\phi(s) = \int_0^\infty e^{-sx} dF(x)$ be the Laplace-Stieltjes transform of $F(x)$, and show that

$$\phi(s) = \frac{\gamma(s) - \gamma(s+\mu)}{1 - \gamma(s+\mu)}, \tag{2}$$

where $\gamma(s) = \int_0^\infty e^{-st} dG(t)$ is the Laplace-Stieltjes transform of $G(t)$, and $G(0) = 0$.

c. Let α denote the mean cycle time, and show that

$$\alpha = \frac{-\gamma'(0)}{1 - \gamma(\mu)}. \tag{3}$$

d. Let P be the equilibrium probability that the server is busy, and show that

$$P = [-\gamma'(0)\mu]^{-1}[1 - \gamma(\mu)]. \tag{4}$$

e. Let Π be the equilibrium probability that an arbitrary arrival finds the server busy, and show that

$$\Pi = \gamma(\mu). \tag{5}$$

f. Equations (4) and (5) imply

$$P = [-\gamma'(0)\mu]^{-1}(1 - \Pi). \tag{6}$$

Interpret this result in terms of offered and carried loads.

g. Show that if $G(t) = 1 - e^{-\lambda t}$ ($t > 0$), then
  i. $P = \Pi$;
  ii. the cycle time is the sum of two independent exponential random variables, with parameters λ and μ, respectively.

h. Argue from first principles that if the interarrival times are of constant length τ, then

$$F(x) = 1 - e^{-j\mu\tau} \quad [j\tau \leq x < (j+1)\tau; \; j = 0, 1, 2, \ldots]. \tag{7}$$

Calculate $\phi(s) = \int_0^\infty e^{-sx} dF(x)$ from (7), and show that this agrees with (2).

11. Customers arrive according to a renewal process at a group (finite or infinite) of servers with exponential service times. The servers are numbered $1, 2, \ldots$, and each arriving customer takes the lowest-numbered idle server (ordered hunt). Thus, the input stream at the $(i+1)$th server is the *overflow* stream from the ith server. (If the number of servers is finite, then blocked customers are cleared.)

a. Show that if the arrival stream at a server is a renewal process, then the overflow stream from that server is also a renewal process.

b. Let $G_i(t)$ be the distribution function of times between successive overflows from the ith server, and show that

$$G_i(t) = \int_0^t [e^{-\mu x} + (1 - e^{-\mu x})G_i(t-x)] dG_{i-1}(x) \qquad (i=1,2,\ldots), \quad (1)$$

where $G_0(t)$ is the distribution function of interarrival times at the first server, and μ^{-1} is the mean service time.

c. If $\gamma_i(s)$ is the Laplace-Stieltjes transform of $G_i(t)$, show that

$$\gamma_i(s) = \frac{\gamma_{i-1}(s+\mu)}{1 - \gamma_{i-1}(s) + \gamma_{i-1}(s+\mu)} \qquad (i=1,2,\ldots). \quad (2)$$

Equation (2) was first derived by Palm [1943], and (1) by Takács [1959], who extended Palm's work. Çinlar and Disney [1967] have studied the more general model in which there are n waiting positions; they obtain (2) when $n=0$. Descloux [1970] has shown how systematic application of renewal theory can further simplify and extend these results.

5.8. The $M/G/1$ Queue

We have now developed sufficient mathematical machinery to derive easily many results for the equilibrium $M/G/1$ queue and some of its variants, which may well be the most widely studied class of queueing models.

A precise description of the $M/G/1$ queueing model is this: (1) Customers arrive according to a Poisson process. (2) There is a single server. (3) An arriving customer who finds the server idle begins service immediately. (4) All blocked customers wait until served. (5) The server cannot be idle when there is a waiting customer. (6) Service times are identically distributed, nonnegative random variables, independent of the arrival process and each other.

We shall denote the arrival rate by λ and the service-time distribution function by $H(t)$, with mean $\tau = \int_0^\infty t \, dH(t)$. Also, we define the *traffic intensity* (or *offered load*) ρ as the product $\rho = \lambda \tau$. We shall assume throughout that $\rho < 1$; that is, the arrival rate is less than the service completion rate. (A consequence of this assumption is that the system will, for all practical purposes, attain statistical equilibrium after a sufficiently long time.) Our objective is to obtain results that describe the statistical characteristics of the $M/G/1$ queue in equilibrium.

The Mean Waiting Time

We begin with a direct derivation of the famous Pollaczek-Khintchine formula for the expected value of the waiting time W experienced by an arbitrary customer in the equilibrium $M/G/1$ queue,

$$E(W) = \frac{\rho}{1-\rho}\left(\frac{\tau}{2} + \frac{\sigma^2}{2\tau}\right), \qquad (8.1)$$

where σ^2 is the variance of the service-time distribution function. Equation (8.1) is simply a rewriting of Equation (4.2), which was derived by a rather indirect method, and its ramifications discussed, in Section 5.4. It is the purpose of the present derivation to show that one can "explain" the presence of the mysterious and important variance term in the light of renewal theory.

We begin by expressing the expected waiting time $E(W)$ for an arbitrary customer (the test customer) in terms of the conditional expected waiting time, conditioned on whether or not the test customer is blocked:

$$E(W) = E(W|W>0)P\{W>0\} + E(W|W=0)P\{W=0\}. \qquad (8.2)$$

The second term in (8.2) is obviously zero. We have argued previously (in Section 5.4) that $1 - \Pi_0 = \rho$, where Π_0 is the equilibrium probability that an arriving customer finds the system empty. Hence $P\{W>0\} = \rho$, and Equation (8.2) reduces to

$$E(W) = \rho E(W|W>0). \qquad (8.3)$$

Now, conditional on the event $\{W>0\}$ that the test customer must wait for service, the waiting time is given by

$$W = X_1 + \cdots + X_Q + R \qquad (W>0), \qquad (8.4)$$

where Q is the number of customers (if any) waiting in the queue when the test customer arrives, given that $W>0$; X_i is the service time of the ith customer in the queue; and R is the remaining service time of the customer in service when the test customer arrives.

By assumption $E(X_i) = \tau$, and the X_i's are independent of Q. Therefore,

$$E(X_1 + \cdots + X_Q | W>0) = \tau E(Q|W>0). \qquad (8.5)$$

Hence, if we take expected values in Equation (8.4), we get

$$E(W|W>0) = \tau E(Q|W>0) + E(R|W>0). \qquad (8.6)$$

Now, by the same argument used to derive (8.3), we can also establish

$$E(Q) = \rho E(Q|W>0). \tag{8.7}$$

Applying the equation $L = \lambda W$ as in Section 5.4, we know that $E(Q) = \lambda E(W)$. Hence, it follows from (8.7) that

$$\tau E(Q|W>0) = E(W). \tag{8.8}$$

Therefore, (8.6) can be written, interestingly, as

$$E(W|W>0) = E(W) + E(R), \tag{8.9}$$

where, of course, $E(R) = E(R|W>0)$; and thus (8.3) becomes

$$E(W) = \frac{\rho}{1-\rho} E(R). \tag{8.10}$$

To evaluate $E(R)$, we note that the renewal-theory argument of Section 5.7 is applicable. That is, the sequence of service times forms a renewal process (ignoring any idle times between service times), and since the system is assumed to be in equilibrium, R is the equilibrium forward recurrence time from an arbitrary point (the arrival epoch of the test customer) to the next renewal (end of the current service time). Thus Equation (7.13) is applicable; that is, $E(R) = \beta^*$ with $\beta = \tau$:

$$E(R) = \frac{\tau}{2} + \frac{\sigma^2}{2\tau}. \tag{8.11}$$

(For a rigorous justification of this argument, see Exercise 14.) Substitution of (8.11) into (8.10) gives the Pollaczek-Khintchine formula (8.1); and the variance term is seen to reflect the fact that the test customer tends to arrive during a service time that is longer, on the average, than an arbitrary service time.

The Imbedded Markov Chain

We shall now analyze the equilibrium $M/G/1$ queue in more depth, using the concept of the *imbedded Markov chain*, an idea of great importance in queueing theory, which was introduced by Kendall [1951, 1953]. In particular, we shall derive the following expression for the probability-generating function $g(z)$ of the number of customers left behind by an arbitrary departing customer in an equilibrium $M/G/1$ system:

$$g(z) = \frac{(z-1)\eta(\lambda - \lambda z)}{z - \eta(\lambda - \lambda z)} (1-\rho), \tag{8.12}$$

The $M/G/1$ Queue

where $\eta(s)$ is the Laplace-Stieltjes transform of the service-time distribution function.

In Chapter 2 we saw that an important characteristic of a birth-and-death process is that if at any instant t the system state E_i is known, then in principle the state probabilities for all $t+x$ ($x \geq 0$) are also known, independent of whatever route the system may have followed in reaching state E_i at time t. When this is true, we may write equations relating the state probabilities at any time t to those at any other time $t+x$. We showed that a necessary and sufficient condition for a system to have this property, called the Markov property, is that the time between successive changes of state be exponentially (but not necessarily identically) distributed. A process that has this characteristic Markov property is called (naturally) a *Markov process*. An example is the number of customers in the $M/M/s$ (Erlang delay) system, which was discussed in Chapter 3.

Let us consider the birth-and-death process again, and see why the $M/G/1$ queue, unlike the $M/M/s$ queue, cannot be modeled as a birth-and-death process. Let $N(t)$ be the number of customers in the system at time t. Then, by the theorem of total probability,

$$P\{N(t+x)=j\} = \sum_{i=0}^{\infty} P\{N(t+x)=j|N(t)=i\} P\{N(t)=i\}$$

$$(t \geq 0, \quad x \geq 0; \quad i=0,1,\ldots).$$

Note that this set of equations is valid for any queueing model. All that is required is to evaluate the conditional probabilities $P\{N(t+x)=j|N(t)=i\}$, called the *transition probabilities*, and solve these equations for the state probabilities $P\{N(t)=j\}$ ($j=0,1,\ldots$).

In the analysis of those queueing models that are birth-and-death processes, we wisely chose x small and let $x \to 0$. Since all the transition probabilities except those representing a transition of step size 0 or ± 1 are $o(x)$ as $x \to 0$, we were saved from laboriously calculating the transition probabilities, only to see most of them become irrelevant as $x \to 0$. But the fact remains that in principle we could have calculated the transition probabilities first, simply because the birth-and-death process is a Markov process; that is, the time separating successive transitions is exponentially distributed, obviating the need to consider how long the system had been in state E_i at time t, or by what path state E_i at time t evolved.

In the $M/G/1$ queue this is not true; since the service-time distribution function is not assumed to be exponential, in order to specify the transition probabilities one must know not only what state the system is in at time t, but also how long it has been there. Thus, the above equations are not incorrect; they are just inapplicable to the analysis of the $M/G/1$ queue.

A *renewal point* of a stochastic process with time parameter t is a time epoch at which the Markov property holds. Thus, if t is a renewal point for a given process, then the future evolution of this process depends only on the state (appropriately defined) at time t. A *Markov chain* is a stochastic process with a discrete state space, in which every point t in the parameter space is a renewal point.

With regard to the $M/G/1$ queue, the service completion points, which are those epochs at which customers complete service and leave the system, are renewal points. For whenever a customer leaves the system, either the system becomes empty or a previously waiting customer starts service; in either case, because of the lack of memory in the (Poisson) arrival process, the transition probabilities (and therefore the future evolution of the system) depend only on the state, defined here as the number of customers in the system at the service completion point. Hence, this particular choice of time epochs (the service completion points) and definition of state (the number of customers present) permits the application of the (well-developed) theory of Markov chains. This kind of Markov chain, whose states are defined at a discrete set of renewal points that are imbedded in the continuum of all nonnegative points, is called (naturally) an *imbedded Markov chain*. Following Kendall, we now proceed to use the concept of the imbedded Markov chain to study the $M/G/1$ queue in equilibrium.

Before beginning the analysis, we note that the theory of Markov chains is widely known, and therefore we feel no need to treat it in this text. Also, the only theorem we shall need from the theory of Markov chains is intuitively reasonable, which further obviates the need for discussion here. A more formal definition of Markov chain is given in this text in Exercise 9 of Chapter 3. For further discussion, see Cohen [1969], Çinlar [1975], Feller [1968], Karlin [1968], Neuts [1973], Ross [1970], or one of the many other books that treat this topic.

Proceeding with the analysis, we define N_k^* to be the number of customers in the system (including any in service, but excluding the departing customer) at the instant the kth customer completes service. That is, if $T_1, T_2, \ldots$ are the successive service completion points, then N_k^* is the state of the system at $T_k + 0$. Then, by the theorem of total probability we can write

$$P\{N_{k+1}^* = j\} = \sum_{i=0}^{\infty} P\{N_{k+1}^* = j | N_k^* = i\} P\{N_k^* = i\}$$

$$(j = 0, 1, \ldots; \quad k = 1, 2, \ldots). \tag{8.13}$$

Consider first the transition probability $P\{N_{k+1}^* = j | N_k^* = 0\}$. If the kth customer leaves the system empty, then the $(k+1)$th customer necessarily finds the server idle. Hence the number of customers left behind by this customer will be all those who will have arrived during his service time. Since with probability $e^{-\lambda \xi}(\lambda \xi)^j / j!$ exactly j customers will arrive during any interval of length ξ, we have

$$P\{N^*_{k+1}=j|N^*_k=0\} = \int_0^\infty \frac{(\lambda\xi)^j}{j!} e^{-\lambda\xi} dH(\xi) \qquad (j \geq 0), \qquad (8.14)$$

where $H(\xi)$ is the service-time distribution function. Similarly, when $N^*_k = i > 0$, the $(k+1)$th customer will leave behind those same customers left behind by the kth customer except for himself, plus all those customers who will have arrived during the service time of the $(k+1)$th customer. Therefore

$$P\{N^*_{k+1}=j|N^*_k=i\} = \int_0^\infty \frac{(\lambda\xi)^{j-i+1}}{(j-i+1)!} e^{-\lambda\xi} dH(\xi)$$
$$(i>0,\ j \geq i-1) \qquad (8.15)$$

and, of course,

$$P\{N^*_{k+1}=j|N^*_k=i\} = 0 \qquad (i>0,\ j<i-1). \qquad (8.16)$$

Let us define

$$p_j = \int_0^\infty \frac{(\lambda\xi)^j}{j!} e^{-\lambda\xi} dH(\xi) \qquad (j=0,1,\ldots). \qquad (8.17)$$

Then the transition probabilities are

$$P\{N^*_{k+1}=j|N^*_k=i\} = \begin{cases} p_j & \text{if } i=0, \\ p_{j-i+1} & \text{if } i>0 \text{ and } j \geq i-1, \\ 0 & \text{if } i>0 \text{ and } j<i-1. \end{cases} \qquad (8.18)$$

Note that the transition probabilities are independent of the value of the index k, that is, are the same for every pair of successive customers.

With the transition probabilities thus specified, we turn our attention to Equation (8.13), which relates the distribution of the number of customers left behind by the $(k+1)$th departing customer to the distribution of the number left behind by the kth departure, for each $k=1,2,\ldots$. From a practical point of view, we are rarely interested in this distribution for any particular value of k; rather we wish to know the distribution of the number of customers left behind by an *arbitrary* departing customer. Clearly, the distribution is different for each k, but it seems reasonable that the dependence on k should diminish as $k \to \infty$. Intuitively, one would expect that $\lim_{k\to\infty} P\{N^*_k=j\} = \Pi^*_j$ should exist under the appropriate conditions for the imbedded Markov chain, just as does $\lim_{t\to\infty} P\{N(t)=$

$j\} = P_j$ for the Markov (birth-and-death) process studied in Chapter 2. The distribution $\{\Pi_j^*\}$ is thus a statistical-equilibrium distribution; Π_j^* is the probability that an arbitrary departing customer leaves behind j other customers in a system that has been operating for a sufficiently long period of time.

From a mathematical point of view, it can be shown using the theory of Markov chains that a unique proper stationary distribution

$$\Pi_j^* = \lim_{k \to \infty} P\{N_k^* = j\} \qquad (j = 0, 1, \ldots), \tag{8.19}$$

independent of the initial conditions, exists if and only if $\rho < 1$. (If $\rho \geq 1$, then $\Pi_j^* = 0$ for all finite j.) The proof of this statement is the only place where the formal theory of Markov chains is relevant. Since both the existence of the limiting distribution (8.19) and the requisite condition $\rho < 1$ are obvious in the queueing-theory context, we shall accept this intuitively plausible statement without proof. (For a formal proof, see pp. 236–238 of Cohen [1969].) Equation (8.13) becomes

$$\Pi_j^* = p_j \Pi_0^* + \sum_{i=1}^{j+1} p_{j-i+1} \Pi_i^* \qquad (j = 0, 1, \ldots). \tag{8.20}$$

The normalization equation is

$$\sum_{j=0}^{\infty} \Pi_j^* = 1. \tag{8.21}$$

We proceed to solve Equations (8.20) and (8.21) for the departing customer's distribution $\{\Pi_j^*\}$.

Define the probability-generating function

$$g(z) = \sum_{j=0}^{\infty} \Pi_j^* z^j. \tag{8.22}$$

Substituting (8.20) into (8.22), we have

$$g(z) = \Pi_0^* \sum_{j=0}^{\infty} p_j z^j + \sum_{j=0}^{\infty} \sum_{i=1}^{j+1} p_{j-i+1} \Pi_i^* z^j. \tag{8.23}$$

For simplicity let us denote by S the double sum in (8.23):

$$S = \sum_{j=0}^{\infty} \sum_{i=1}^{j+1} p_{j-i+1} \Pi_i^* z^j. \tag{8.24}$$

The $M/G/1$ Queue 215

If we reverse the order of summation in (8.24), we have

$$S = \sum_{i=1}^{\infty} \Pi_i^* \sum_{j=i-1}^{\infty} p_{j-i+1} z^j = \sum_{i=1}^{\infty} \Pi_i^* \sum_{k=0}^{\infty} p_k z^{k+i-1}, \quad (8.25)$$

where the second equality in (8.25) follows from the substitution $k = j - i + 1$. Hence, we can write (8.24) as

$$S = z^{-1} \left(\sum_{i=1}^{\infty} \Pi_i^* z^i \right) \left(\sum_{k=0}^{\infty} p_k z^k \right). \quad (8.26)$$

If we now define the probability-generating function

$$h(z) = \sum_{j=0}^{\infty} p_j z^j, \quad (8.27)$$

then, in light of (8.22) and (8.27), (8.26) can be written

$$S = z^{-1} [g(z) - \Pi_0^*] h(z); \quad (8.28)$$

and (8.23) can be written

$$g(z) = \Pi_0^* h(z) + z^{-1} [g(z) - \Pi_0^*] h(z). \quad (8.29)$$

Solving (8.29) for $g(z)$ yields

$$g(z) = \frac{(z-1)h(z)}{z - h(z)} \Pi_0^*; \quad (8.30)$$

it remains only to find Π_0^*. The normalization equation (8.21) implies

$$g(1) = 1, \quad (8.31)$$

and (8.30) and (8.31) yield, with a single application of L'Hôpital's rule,

$$\Pi_0^* = 1 - \rho. \quad (8.32)$$

[We have previously argued that $P(W > 0) = 1 - \Pi_0 = \rho$; the equality $\Pi_0 = \Pi_0^*$ is consistent with the theorem of Section 5.3.]

Finally, we note that if we define $\eta(s)$ as the Laplace-Stieltjes transform of the service-time distribution function,

$$\eta(s) = \int_0^{\infty} e^{-s\xi} dH(\xi), \quad (8.33)$$

then substitution of (8.17) into (8.27) gives [see Equation (6.10)]

$$h(z) = \eta(\lambda - \lambda z). \tag{8.34}$$

And substitution of (8.32) and (8.34) into (8.30) yields (8.12), as promised.

Thus we have derived an expression (8.12) for the probability-generating function (8.22) of the number of customers left behind by an arbitrary departure in an $M/G/1$ system in equilibrium. It follows from the theorem of Section 5.3 that (8.12) also generates the corresponding probabilities from the viewpoint of an arbitrary arrival; and it follows from the assumption of Poisson input that these are also the outside observer's probabilities:

$$\Pi_j^* = \Pi_j = P_j \qquad (j = 0, 1, 2, \ldots). \tag{8.35}$$

We emphasize that, clearly, the distribution generated by (8.12) depends on the particular choice of the service-time distribution function; nevertheless, as (8.32) shows, the probability $\Pi_0^* = \Pi_0 = P_0$ does not. Thus, the probability that an arrival will find the server busy does not depend on the service-time distribution function; but if the arrival does find the server busy, then the distribution of the number of other customers he finds waiting in the queue does depend on the distribution of service times.

The mean number of customers present in the equilibrium $M/G/1$ system (at a departure point, an arrival point, or an arbitrary point) can be found from (8.12) by a routine but tedious calculation (two applications of L'Hôpital's rule are required). The result, which we derived previously [see (4.19)] by a different argument, is

$$g'(1) = \rho + \frac{\rho^2}{2(1-\rho)}\left(1 + \frac{\sigma^2}{\tau^2}\right), \tag{8.36}$$

where σ^2 is the variance of the service-time distribution function. (The reader is referred back to Exercise 24 of Chapter 2, where we were asked to carry out these calculations.) Equation (8.36) shows, for example, that the mean number of customers waiting in the queue, [which equals $g'(1) - \rho$, the mean number in the system minus the mean number in service] is twice as great in the $M/M/1$ queue (where $\sigma^2 = \tau^2$) as in the $M/D/1$ queue (where $\sigma^2 = 0$), even though the probability that there is a queue at all is the same in both cases.

The Pollaczek-Khintchine Formula

We now turn to the waiting-time distribution function for the equilibrium $M/G/1$ queue when service is in order of arrival. Let W be the waiting time of an arbitrary customer, with distribution function $W(t) = P\{W \leq t\}$

The $M/G/1$ Queue

and Laplace-Stieltjes transform

$$\omega(s) = \int_0^\infty e^{-st} dW(t). \tag{8.37}$$

We shall show that

$$\omega(s) = \frac{s(1-\rho)}{s - \lambda[1 - \eta(s)]}. \tag{8.38}$$

The important result (8.38) is widely known as the *Pollaczek-Khintchine formula*. Using the fact that $E(W) = -\omega'(0)$, it is a straightforward but tedious task (requiring two applications of L'Hôpital's rule) to establish that

$$E(W) = \frac{\rho \tau}{2(1-\rho)} \left(1 + \frac{\sigma^2}{\tau^2}\right), \tag{8.39}$$

a result that is also often referred to as the Pollaczek-Khintchine formula. [Equation (8.39) has been derived earlier in this chapter by two independent arguments—see (4.2) and (8.1)—and its ramifications discussed in Section 5.4. The reader is also referred back to Exercise 6, where he was asked to verify (8.39) by calculating $-\omega'(0)$, and to invert the transform (8.38) when the service times are exponential.] Finally, we shall invert (8.38) to obtain the remarkable formula, first found by Beneš [1957],

$$W(t) = (1-\rho) \sum_{j=0}^\infty \rho^j \tilde{H}^{*j}(t), \tag{8.40}$$

where $\tilde{H}^{*j}(t)$ is the j-fold self-convolution (with $\tilde{H}^{*1} = \tilde{H}$ and $\tilde{H}^{*0} = 1$) of the distribution function $\tilde{H}(t)$ defined by

$$\tilde{H}(t) = \frac{1}{\tau} \int_0^t [1 - H(\xi)] d\xi. \tag{8.41}$$

We emphasize that Equations (8.38) and (8.40) require that the queue discipline be service in order of arrival, but (8.39) remains valid for any nonbiased queue discipline.

To obtain (8.38), let $\phi(s)$ be the Laplace-Stieltjes transform of the distribution function of the sojourn time (the sum of the waiting time and the service time) of an arbitrary customer (the test customer). Since the waiting time and the service time of any customer are independent random variables, the Laplace-Stieltjes transform of the distribution function of their sum is the product of the Laplace-Stieltjes transforms of the component distribution functions; that is,

$$\phi(s) = \omega(s) \eta(s). \tag{8.42}$$

Now, $\phi(s)$ is the Laplace-Stieltjes transform of the distribution function of the total length of time (sojourn time) the test customer spends in the system. The distribution of the number of customers left behind by the test customer has probability-generating function $g(z)$ given by (8.12). Since service is in arrival order, the customers left behind by the test customer must all have arrived during his sojourn time. And since the arrivals follow a Poisson process with rate λ, the generating function of the distribution of the number of arrivals during this sojourn time is related to the Laplace-Stieltjes transform of the distribution function of the length of the sojourn time according to Equation (6.10):

$$g(z) = \phi(\lambda - \lambda z). \tag{8.43}$$

Equations (8.42) and (8.43) imply $g(z) = \omega(\lambda - \lambda z)\eta(\lambda - \lambda z)$; when this expression is used in the left-hand side of (8.12) and the factor $\eta(\lambda - \lambda z)$ is canceled, we get

$$\omega(\lambda - \lambda z) = \frac{z-1}{z - \eta(\lambda - \lambda z)}(1-\rho). \tag{8.44}$$

The substitution $s = \lambda - \lambda z$ in (8.44) yields (8.38).

To derive (8.40), we rewrite (8.38) in the form

$$\omega(s) = (1-\rho)\frac{1}{1 - \rho\left(\frac{1}{\tau}\frac{1-\eta(s)}{s}\right)},$$

which expands to

$$\omega(s) = (1-\rho)\sum_{j=0}^{\infty} \rho^j \left(\frac{1}{\tau}\frac{1-\eta(s)}{s}\right)^j. \tag{8.45}$$

Referring back to Equation (7.10), we observe that the factor in parentheses in (8.45) is the Laplace-Stieltjes transform, $\tilde{\eta}(s)$ say, of the function $\tilde{H}(t)$ defined by (8.41). That is, according to (7.9) and (7.10), if $\tilde{H}(t)$ is given by (8.41) and $\tilde{\eta}(s) = \int_0^\infty e^{-st} d\tilde{H}(t)$, then

$$\tilde{\eta}(s) = \frac{1}{\tau}\frac{1-\eta(s)}{s}. \tag{8.46}$$

Therefore, Equation (8.45) can be written

$$\omega(s) = (1-\rho)\sum_{j=0}^{\infty} \rho^j [\tilde{\eta}(s)]^j. \tag{8.47}$$

Finally, term-by-term inversion of (8.47) yields (8.40), because if $\tilde{\eta}(s)$ is the transform of $\tilde{H}(t)$, then $[\tilde{\eta}(s)]^j$ is the transform of the j-fold self-convolution of $\tilde{H}(t)$.

The rather mysterious appearance of the function $\tilde{H}(t)$ in (8.40) is noteworthy. One can interpret $\tilde{H}(t)$ as the distribution function of the elapsed time from the arrival of a customer until the end of the service time of the customer (if any) in service when he arrives. In the case of exponential service times, $H(t) = 1 - e^{-\mu t}$, then $\tilde{H}(t) = H(t)$, and (8.40) yields easily to term-by-term interpretation: $(1-\rho)\rho^j$ is the probability that the test customer, who arrives at t_0, say, finds j customers present; and $H^{*j}(t)$ is the distribution function of the sum of the service times of the customers ahead of him (including the remaining service of the customer, if any, in service at t_0). But, despite its simple form, (8.40) does not seem to yield to similar term-by-term interpretation in the general case. This provides a dramatic counterexample to the folk theorem (often true) that simple results have simple explanations.

The Pollaczek-Khintchine formula (8.38) was first obtained by Pollaczek [1930] and, independently, by Khintchine [1932]. Interestingly, it was also obtained by Cramér [1930], in connection with a problem of insurance risk (to which we shall return shortly).

The elegant derivation given here is due to Kendall [1951]. In Kendall's derivation, we write equations for the state probabilities from the viewpoint of the kth customer, and let $k \to \infty$ to yield the distribution $\{\Pi_j^*\}$ of the number of customers left behind by an arbitrary departing customer, who leaves after the arrival and departure of an "infinite" number of other customers have brought the system into statistical equilibrium. Then the distribution function of the time waited by this departure is calculated. An analogous (but much more complicated) argument that takes the viewpoint of an arriving customer is outlined in Exercise 14.

Khintchine's [1932] method was to denote by W_n the waiting time of the nth arriving customer and show that for $\rho < 1$,

$$\lim_{n \to \infty} P\{W_n \leqslant t\} = W(t), \qquad (8.48)$$

where $W(t)$ has the Laplace-Stieltjes transform (8.38). Pollaczek [1930] considered the model where n customers arrive in the interval $(0, x)$, with the arrival times chosen independently from a distribution uniform over $(0, x)$. He showed that the distribution function of the waiting time of an arbitrarily selected customer from this group tends to $W(t)$, with Laplace-Stieltjes transform (8.38), when $\rho < 1$ and $n \to \infty$ and $x \to \infty$ in such a way that $n/x \to \lambda$.

Another approach is to define the *virtual waiting time* V_y as the length of time that a customer would have to wait for service to commence if he were to arrive at time y. Takács [1955, 1962a, 1967] has studied this in detail and has shown, among other things, that for $\rho<1$, as one would expect,

$$\lim_{y\to\infty} P\{V_y \leqslant t\} = W(t), \tag{8.49}$$

where $W(t)$ is the distribution function whose Laplace-Stieltjes transform is given by (8.38). A heuristic proof of (8.49) is outlined in Exercise 15.

Cohen [1969] has given a neat derivation of the Pollaczek-Khintchine formula by using the method of collective marks; this procedure is outlined in Exercise 13.

Finally, let us entertain a slight digression, namely, the insurance risk problem alluded to earlier, in connection with which Cramér [1930] derived a formula that, it turns out, is identical with the Pollaczek-Khintchine formula. Suppose that at time $u=0$ an insurance company has a reserve fund of amount t to cover claims made by its policyholders. We assume that the instants at which the claims are made follow a Poisson process with rate λ, and that the amounts of the claims are mutually independent, identically distributed, nonnegative random variables with distribution function $H(\xi)$; and we denote by $X(u)$ the total amount of claims made in $(0, u)$. We further assume that the policyholders pay their premiums into the reserve fund linearly with time. Thus, if we denote by $R(u)$ the size of the reserve fund at time u, we can write

$$R(u) = t + cu - X(u), \tag{8.50}$$

where c is the rate at which the policyholders increase the size of the reserve fund through payment of their premiums. We can, without loss of generality, define the monetary unit so that $c=1$; then (8.50) can be written

$$R(u) = t - [X(u) - u]. \tag{8.51}$$

The insurance company would like to know the probability $p(t)$ that there will always be enough money in the reserve fund to cover the claims made against it, when the initial size of the reserve fund is t. Hence, we seek the probability $p(t)$ that $R(u) \geqslant 0$ for all $u \geqslant 0$, when $R(0) = t$; that is, $p(t)$ is the probability that the right-hand side of (8.51) will never become negative:

$$p(t) = P\left\{\sup_{0 \leqslant u < \infty} [X(u) - u] \leqslant t\right\}. \tag{8.52}$$

We now give an elegant argument, due to Takács [1967, pp. 119–120], to show that the right-hand side of (8.52) and the left-hand side of (8.48) are equal; that is, the probability that an insurance company that starts with a reserve fund of amount t will never run out of money equals the probability that a customer will not wait a time in excess of t for service to commence in an equilibrium $M/G/1$ queue with service in order of arrival:

$$p(t) = W(t). \tag{8.53}$$

Denote by T_i ($i=1,2,\ldots$) the time of the ith arrival (claim), and let $X(u)$ be the sum of the service times (the total amount of claims) associated with all the arrivals in $(0,u]$. Then, clearly,

$$\sup_{0<u<\infty} [X(u)-u] = \sup_{0<i<\infty} [X(T_i+0) - T_i], \tag{8.54}$$

where $T_0=0$. If we let X_i ($i=1,2,\cdots$) be the service time of the ith customer (the amount of the ith claim), then $X(T_i+0) = X_1 + \cdots + X_i$ and (8.54) can be written

$$\sup_{0<u<\infty} [X(u)-u] = \sup_{0<i<\infty} (X_1 + \cdots + X_i - T_i), \tag{8.55}$$

where the empty sum is taken to be zero. It follows from (8.52) and (8.55) that

$$p(t) = P\left\{\sup_{0<i<\infty} (X_1 + \cdots + X_i - T_i) \leq t\right\}. \tag{8.56}$$

We now show that

$$\lim_{n\to\infty} P\{W_n \leq t\} = P\left\{\sup_{0<i<\infty} (X_1 + \cdots + X_i - T_i) \leq t\right\}, \tag{8.57}$$

from which (8.53) follows [by comparison of (8.57) with (8.48) and (8.56)].
The random variables W_n ($n=1,2,\ldots$) satisfy the recurrence relation

$$W_{n+1} = \max(0, W_n + X_n - (T_{n+1} - T_n)) \quad (n=1,2,\ldots), \tag{8.58}$$

where $W_1 = 0$ (because we assume that the server is idle at time $u=0$). Expansion of (8.58) yields

$$W_{n+1} = \max(0, X_n - (T_{n+1} - T_n), X_n + X_{n-1} - (T_{n+1} - T_{n-1}), \ldots,$$
$$X_n + \cdots + X_2 - (T_{n+1} - T_2), X_n + \cdots + X_1 - (T_{n+1} - T_1)). \tag{8.59}$$

Now, consider the right-hand side of (8.59) with the random variables $X_n, X_{n-1}, \ldots, X_1$ replaced by $X_1, X_2, \ldots, X_n$, respectively, and $(T_{n+1} - T_n)$, $(T_{n+1} - T_{n-1}), \ldots, (T_{n+1} - T_1)$ replaced by $T_1, T_2, \ldots, T_n$, respectively; the result is a new random variable that has exactly the same distribution as

(8.59). Therefore,

$$P\{W_{n+1} \leqslant t\} = P\{\max(0, X_1 - T_1, X_1 + X_2 - T_2, \ldots,$$
$$X_1 + \cdots + X_{n-1} - T_{n-1}, X_1 + \cdots + X_n - T_n) \leqslant t\}. \tag{8.60}$$

Now let $n \to \infty$ in (8.60); the result is (8.57), and the proof of (8.53) is complete. Note that the assumption of Poisson input was not used in the proof; it follows that (8.53) remains correct for the case when the times at which the arrivals (claims) occur follow an arbitrary renewal process, where $W(t)$ is now the equilibrium waiting-time distribution function for the $GI/G/1$ queue with service in order of arrival.

For more on insurance risk theory and its relationship to queueing theory, see Halmstad [1974], Seal [1969], and Takács [1967].

Exercises

12. *The $M/G/1$ queue with server vacation times.* Consider the equilibrium $M/G/1$ queue with the following variation. The server works continuously as long as there is at least one customer in the system. When the server finishes serving a customer and finds the system empty, it goes away for a length of time called a *vacation*. At the end of the vacation the server returns and begins to serve those customers, if any, who have arrived during the vacation. If the server finds no customers waiting at the end of a vacation, it immediately takes another vacation, and continues in this manner until it finds at least one waiting customer upon return from a vacation. Let λ be the arrival rate, $\eta(s)$ be the Laplace-Stieltjes transform of the service-time distribution function, and

$$f(z) = \sum_{j=0}^{\infty} P(j) z^j$$

be the probability-generating function of the number of customers waiting for service at the end of a vacation.

a. Let $\hat{\Pi}_j^*$ ($j = 0, 1, \ldots$) be the equilibrium probability that an arbitrary departing customer leaves behind j other customers, and define the probability-generating function

$$\hat{g}(z) = \sum_{j=0}^{\infty} \hat{\Pi}_j^* z^j.$$

Write the imbedded-Markov-chain equations that determine the distribution $\{\hat{\Pi}_j^*\}$, and show that

$$\hat{g}(z) = \frac{[f(z) - 1]\eta(\lambda - \lambda z)}{z - \eta(\lambda - \lambda z)} \cdot \frac{1 - \rho}{f'(1)}; \tag{1}$$

The $M/G/1$ Queue

that is,

$$\hat{g}(z) = \frac{f(z)-1}{f'(1)(z-1)} g(z), \qquad (2)$$

where $g(z)$ is given by (8.12).

b. Show that $\hat{g}(z) = g(z)$ if and only if $f(z) = P(0) + P(1)z$. Explain this result.

c. Find the equilibrium probability-generating function $\sum \hat{\Pi}_j z^j$ of the number of customers in the system found by an arbitrary arrival. Show that

$$\hat{\Pi}_0 = (1-\rho)\frac{1-P(0)}{f'(1)}.$$

d. Show that

$$\hat{g}'(1) = g'(1) + \frac{f''(1)}{2f'(1)}. \qquad (3)$$

e. Show that if the vacation lengths are independent of the arrival process, then the Laplace-Stieltjes transform $\hat{\omega}(s)$ of the waiting-time distribution function when service is in order of arrival is given by

$$\hat{\omega}(s) = \frac{1-f(1-s/\lambda)}{s-\lambda[1-\eta(s)]} \frac{(1-\rho)\lambda}{f'(1)}. \qquad (4)$$

f. If the vacation lengths are independent of the arrival process, find the probability that an arriving customer must wait for service.

g. If $\hat{W}$ is the length of time an arbitrary customer waits for his service to commence, show that

$$E(\hat{W}) = E(W) + \frac{f''(1)}{2\lambda f'(1)}, \qquad (5)$$

where $E(W)$ is given by (8.39).

h. Suppose that $f(z) = z^j$. Show that Equation (4) reduces to the Pollaczek-Khintchine formula when $j=1$. Is Equation (4) correct when $j \geq 2$? Why? What about (2), (3), and (5)?

For more on this model, see Cooper [1970] and Levy and Yechiali [1975].

13. *Derivation of the Pollaczek-Khintchine formula by the method of collective marks* (Cohen [1969]).

a. Let $g_k(z)$ be the probability-generating function of the number of customers left behind by the kth departing customer in the $M/G/1$ queue with order-of-arrival service. Show that $g_k(z)$ can be interpreted as the probability that no marked customers arrive during the sojourn time of the kth customer, where it is assumed that arriving customers are marked (or not marked) independently and each customer has probability z of not being marked.

b. Define the random variables T_k and W_k as the sojourn time and waiting time, respectively, of the kth customer; and define the events $C_k = \{$the kth customer is marked}, with complement C'_k, and $M'(X) = \{$no marked customers arrive during $X\}$, where X is a (random) time interval. Show that

$$P\{M'(T_k)\} = P\{M'(T_k), C_{k+1}\} + P\{M'(T_k), C'_{k+1}\}$$
$$= P\{W_{k+1}=0, C_{k+1}\} + P\{M'(W_{k+1}), C'_{k+1}\}$$
$$= P\{W_{k+1}=0\}(1-z) + P\{M'(W_{k+1})\}z.$$

c. Show that $P\{M'(T_k)\} = \phi_k(\lambda - \lambda z)$ and $P\{M'(W_{k+1})\} = \omega_{k+1}(\lambda - \lambda z)$, where $\phi_k(s)$ and $\omega_k(s)$ are the Laplace-Stieltjes transforms of the distribution functions of T_k and W_k, respectively.

d. Show that these results imply that

$$\omega(s) = \frac{s}{s - \lambda[1 - \eta(s)]} P\{W=0\},$$

where $\eta(s)$ is the Laplace-Stieltjes transform of the service-time distribution function, W is the equilibrium waiting time of an arbitrary customer, and $\omega(s)$ is the Laplace-Stieltjes transform of the distribution function of W.

e. Conclude that Equation (8.38), the Pollaczek-Khintchine formula, is correct.

14. *The $M/G/1$ queue from the viewpoint of arrivals.* Let N be the number of customers in the system just prior to an arbitrary arrival epoch T_c, and let R be the time needed to complete the service of the customer (if there is one) in service at T_c. Define

$$P\{N=j\} = \Pi_j \quad (j=0,1,\ldots), \tag{1}$$
$$P\{R \leqslant x, N=j\} = \Pi_j(x) \quad (x \geqslant 0; \ j=1,2,\ldots) \tag{2}$$

and

$$\psi_j(s) = \int_0^\infty e^{-sx} d\Pi_j(x) \quad (j=1,2,\ldots). \tag{3}$$

Wishart [1961] (see also Takács [1963]) has shown that for $\lambda \tau = \rho < 1$ the generating function $u(s,z)$ of the Laplace-Stieltjes transform (3),

$$u(s,z) = \sum_{j=1}^\infty \psi_j(s) z^j, \tag{4}$$

is given by

$$u(s,z) = \frac{\Pi_0 \lambda z (1-z)}{z - \eta(\lambda - \lambda z)} \frac{\eta(s) - \eta(\lambda - \lambda z)}{s - \lambda(1-z)} \tag{5}$$

with

$$\Pi_0 = 1 - \rho, \tag{6}$$

The $M/G/1$ Queue

where $\eta(s)$ is the Laplace-Stieltjes transform of the service-time distribution function $H(x)$, τ is the mean service time, and λ is the customer arrival rate. Finally, inversion of (5) gives

$$\sum_{j=1}^{\infty} \Pi_j(x) z^j = \frac{(1-\rho)\lambda z(1-z)}{\eta(\lambda-\lambda z)-z} \int_0^{\infty} e^{-\lambda(1-z)\xi}[H(\xi+x)-H(\xi)] d\xi. \quad (7)$$

a. Using the normalization equation

$$\Pi_0 + \sum_{j=1}^{\infty} \psi_j(0) = 1$$

and (5), prove Equation (6).

b. Starting with Equation (7), show that

$$\sum_{j=1}^{\infty} \Pi_j(x) = \lambda \int_0^x [1-H(\xi)] d\xi,$$

and that this in turn implies

$$P\{R \leq x | N \geq 1\} = \frac{1}{\tau} \int_0^x [1-H(\xi)] d\xi. \quad (8)$$

Observe that (8) further justifies the use of (8.11).

c. Show that

$$\sum_{j=0}^{\infty} \Pi_j z^j = \frac{(z-1)\eta(\lambda-\lambda z)}{z-\eta(\lambda-\lambda z)} (1-\rho). \quad (9)$$

Compare Equations (9) and (8.12).

d. Let $W(x)$ be the waiting-time distribution function for an arbitrary customer when service is in order of arrival. Show that (Takács [1963])

$$W(x) = \Pi_0 + \sum_{j=1}^{\infty} \int_0^x \Pi_j(x-\xi) dH^{*(j-1)}(\xi), \quad (10)$$

and that therefore $W(x)$ has Laplace-Stieltjes transform $\omega(s)$:

$$\omega(s) = \frac{s(1-\rho)}{s-\lambda[1-\eta(s)]}. \quad (11)$$

Compare Equations (11) and (8.38).
In parts e–j, the derivation leading to (5) is outlined.

e. Argue that

$$\Pi_0 = \Pi_0 \int_0^\infty e^{-\lambda t} dH(t) + \sum_{k=1}^\infty \int_0^\infty \int_0^\infty e^{-\lambda(y+z)} d\Pi_k(y) dH^{*k}(z), \quad (12)$$

$$\Pi_1(x) = \Pi_0 \int_0^\infty [H(t+x) - H(t)] \lambda e^{-\lambda t} dt$$
$$+ \sum_{k=1}^\infty \int_0^\infty \int_0^\infty e^{-\lambda(y+z)} d\Pi_k(y) dH^{*(k-1)}(z)$$
$$\times \int_0^\infty [H(t+x) - H(t)] \lambda e^{-\lambda t} dt, \quad (13)$$

$$\Pi_j(x) = \int_0^\infty [\Pi_{j-1}(t+x) - \Pi_{j-1}(t)] \lambda e^{-\lambda t} dt$$
$$+ \sum_{k=j}^\infty \int_0^\infty \int_0^\infty e^{-\lambda(y+z)} d\Pi_k(y) dH^{*(k-j)}(z)$$
$$\times \int_0^\infty [H(t+x) - H(t)] \lambda e^{-\lambda t} dt \quad (j=2,3,\ldots). \quad (14)$$

f. Show that Equations (12), (13), and (14) yield, respectively,

$$\Pi_0 = \Pi_0 \eta(\lambda) + \sum_{k=1}^\infty \psi_k(\lambda) \eta^k(\lambda), \quad (15)$$

$$\psi_1(s) = \frac{\lambda}{s-\lambda} [\eta(\lambda) - \eta(s)] \left[\Pi_0 + \sum_{k=1}^\infty \psi_k(\lambda) \eta^{k-1}(\lambda) \right], \quad (16)$$

$$\psi_j(s) = \frac{\lambda}{s-\lambda} [\psi_{j-1}(\lambda) - \psi_{j-1}(s)]$$
$$+ \frac{\lambda}{s-\lambda} [\eta(\lambda) - \eta(s)] \sum_{k=j}^\infty \psi_k(\lambda) \eta^{k-j}(\lambda) \quad (j=2,3,\ldots). \quad (17)$$

g. Show that substitution of (16) and (17) into (4) yields

$$u(s,z)[s - \lambda(1-z)] = \lambda z u(\lambda, z)$$
$$+ \lambda z [\eta(\lambda) - \eta(s)] \left[\Pi_0 + \frac{u(\lambda, \eta(\lambda)) - u(\lambda, z)}{\eta(\lambda) - z} \right]. \quad (18)$$

h. Take $s = \lambda - \lambda z$ in (18), and show

$$\Pi_0 = \frac{u(\lambda, z)}{\eta(\lambda - \lambda z) - \eta(\lambda)} + \frac{u(\lambda, \eta(\lambda)) - u(\lambda, z)}{z - \eta(\lambda)}, \quad (19)$$

so that (18) becomes

$$u(s,z) = \frac{\lambda z [\eta(s) - \eta(\lambda - \lambda z)] u(\lambda, z)}{[s - (\lambda - \lambda z)][\eta(\lambda) - \eta(\lambda - \lambda z)]}. \quad (20)$$

i. Show that (15) implies

$$u(\lambda, \eta(\lambda)) = \Pi_0 [1 - \eta(\lambda)], \quad (21)$$

and deduce from (21) and (19) that

$$u(\lambda,z) = \frac{\Pi_0(1-z)[\eta(\lambda)-\eta(\lambda-\lambda z)]}{z-\eta(\lambda-\lambda z)}. \tag{22}$$

j. Show that substitution of (22) into (20) yields (5).
k. Verify by taking Laplace-Stieltjes transforms in (7) that (7) gives the inverse of (5).

15. *The integrodifferential equation of Takács.* Let V_t be the virtual waiting time in the $M/G/1$ queue with order-of-arrival service, and define the distribution function $V(t,x) = P\{V_t \leq x\}$.
   a. Show that, for $h > 0$,

   $$V(t+h,x) = (1-\lambda h)V(t,x+h)$$
   $$+ \lambda h \int_0^{x+h} H(x+h-y)d_y V(t,y) + o(h) \quad (h \to 0).$$

   b. Show that

   $$\frac{\partial V(t,x)}{\partial t} = \frac{\partial V(t,x)}{\partial x} - \lambda V(t,x) + \lambda \int_0^x H(x-y)d_y V(t,y). \tag{1}$$

   This is known as the *integrodifferential equation of Takács* (see Takács [1955] and also Takács [1962a] and Hasofer [1963]).
   c. It is not unreasonable to assume that in equilibrium Equation (1) becomes

   $$\frac{dV(x)}{dx} = \lambda V(x) - \lambda \int_0^x H(x-y)dV(y) \quad (x > 0), \tag{2}$$

   where we have assumed that

   $$\lim_{t \to \infty} \frac{\partial V(t,x)}{\partial t} = 0$$

   and

   $$\lim_{t \to \infty} V(t,x) = V(x).$$

   Define the Laplace-Stieltjes transform $\theta(s) = \int_0^\infty e^{-sx} dV(x)$, and show that Equation (2) transforms to

   $$\theta(s) - V(0) = \frac{\lambda}{s}\theta(s) - \frac{\lambda}{s}\eta(s)\theta(s). \tag{3}$$

   d. Solve (3) for $\theta(s)$ and compare with (8.38).

The Busy Period

A notion that will prove useful and instructive is that of the *busy period*, defined as the length of time from the instant a customer enters a previously empty system until the next instant at which the system is completely empty.

A typical realization of a busy period is illustrated in Figure 5.2. The height of the graph at any time t is the duration of time from t until all those customers present in the system at time t are finished with service. Arrivals are represented by arrows pointing up, and departures by arrows pointing down. At each arrival epoch the graph jumps up by an amount equal to the service time of the customer who just arrived. The graph decreases linearly with time until either another customer arrives or the graph decreases to zero, signifying the end of the busy period. Observe that the height of the graph at any time t equals the virtual waiting time V_t, if service is in order of arrival. The realization pictured in Figure 5.2 is a busy period composed of six service times.

We shall now derive an expression for the distribution function $B(t)$ of the busy period for the $M/G/1$ queue. Consider the system in which the server does not start to serve any customers until there are $j \geqslant 1$ customers in the system, and then serves these j customers one at a time, and all subsequent arrivals, until for the first time the system becomes empty. Call this length of time a *j-busy period*. Hence a j-busy period is the length of continuous busy time of a server that starts serving when j customers are in the system.

Let $B(t)$ be the distribution function of the busy period, and $B_j(t)$ be the distribution function of the j-busy period, where $B_1(t) = B(t)$. Following Takács [1962a], we now show that the distribution function $B_j(t)$ of the j-busy period is the j-fold convolution with itself of the distribution function $B(t)$ of the ordinary 1-busy period,

$$B_j(t) = B^{*j}(t) \qquad (j = 0, 1, 2, \ldots; \quad B^{*0}(t) = 1) \qquad (8.61)$$

(where the case $j = 0$ is included to simplify forthcoming calculations).

A little thought should convince the reader that the length of a j-busy period is independent of the order in which the waiting customers are

Figure 5.2.

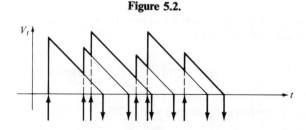

The $M/G/1$ Queue

served. This being so, we choose to consider the following particular order of service. Suppose the server is about to begin service on one of $j \geq 1$ waiting customers. When it finishes serving the first of these j customers, it then serves all those customers, if any, who arrived during the service time of the first customer, and then those customers, if any, who arrived during these service periods, and so on. That is, it serves first all those customers who arrive during the 1-busy period generated by the first of the original j customers. When the 1-busy period generated by the first of the original j customers terminates, the server starts on the next of the original j customers, and serves all those customers who arrive during the 1-busy period generated by him. The server continues in this manner, so that the length of time during which it is continuously busy is the sum of j 1-busy periods. These j 1-busy periods are clearly mutually independent and identically distributed. Hence Equation (8.61) is true.

We shall now use (8.61) to obtain the Laplace-Stieltjes transform $\beta(s)$,

$$\beta(s) = \int_0^\infty e^{-st} dB(t), \tag{8.62}$$

of the 1-busy-period distribution function $B(t)$. Note that a 1-busy period will not exceed a value t if and only if the first service time lasts a length $\xi \leq t$ and the $j \geq 0$ customers who arrive during the first service time generate a j-busy period that does not exceed $t - \xi$. Thus we may write

$$B(t) = \sum_{j=0}^\infty \int_0^t \frac{(\lambda \xi)^j}{j!} e^{-\lambda \xi} B_j(t-\xi) dH(\xi), \tag{8.63}$$

where $H(\xi)$ is the service-time distribution function.

Taking Laplace-Stieltjes transforms on both sides of (8.63), and noting that (8.61) implies that the Laplace-Stieltjes transform $\beta_j(s)$ of the j-busy-period distribution function $B_j(t)$ is

$$\beta_j(s) = [\beta(s)]^j \quad (j = 0, 1, \ldots), \tag{8.64}$$

we obtain

$$\beta(s) = \sum_{j=0}^\infty \int_0^\infty \frac{(\lambda \xi)^j}{j!} e^{-\lambda \xi} e^{-s\xi} [\beta(s)]^j dH(\xi). \tag{8.65}$$

Equation (8.65) reduces to

$$\beta(s) = \sum_{j=0}^\infty \int_0^\infty \frac{[\lambda \xi \beta(s)]^j}{j!} e^{-(\lambda + s)\xi} dH(\xi)$$

$$= \int_0^\infty e^{-[s + \lambda - \lambda \beta(s)]\xi} dH(\xi). \tag{8.66}$$

The integral on the right-hand side of (8.66) is the Laplace-Stieltjes transform $\eta(x)$ of the service-time distribution function, with argument $x = s + \lambda - \lambda\beta(s)$. Thus we have the functional equation

$$\beta(s) = \eta(s + \lambda - \lambda\beta(s)). \tag{8.67}$$

We can obtain the mean b of the busy period by differentiating Equation (8.67) and using the fact that $-\beta'(0) = b$ and $-\eta'(0) = \tau$, where τ is the mean service time:

$$\beta'(s) = [1 - \lambda\beta'(s)]\eta'(s + \lambda - \lambda\beta(s)) \tag{8.68}$$

and therefore

$$-b = -(1 + \lambda b)\tau,$$

or

$$b = \frac{\tau}{1 - \lambda\tau}. \tag{8.69}$$

[This result was derived earlier by a heuristic argument—see Equation (4.12) of Chapter 3.] Similarly, a second differentiation of (8.67) gives

$$\beta''(0) = \frac{\eta''(0)}{(1-\rho)^3}, \tag{8.70}$$

from which one can calculate the variance of the busy-period distribution function.

It is interesting to consider directly the special case of the $M/D/1$ queue. Let N_j be the number of customers served during a j-busy period, and set

$$f_n^{(j)} = P\{N_j = n\}.$$

Following Prabhu [1965], we can write the following recurrence:

$$f_n^{(j)} = \begin{cases} e^{-\lambda\tau j} & (n = j), \\ \sum_{k=1}^{n-j} \frac{(\lambda\tau j)^k}{k!} e^{-\lambda\tau j} f_{n-j}^{(k)} & (n \geq j+1), \end{cases} \tag{8.71}$$

where τ is the (constant) service time. It is easy to show (see Exercise 16) that the equations (8.71) have solution

$$f_n^{(j)} = \frac{j}{n}\left[\frac{(\lambda\tau n)^{n-j}}{(n-j)!}e^{-\lambda\tau n}\right] \quad (n \geq j). \tag{8.72}$$

The distribution (8.72) is often referred to as the *Borel-Tanner* distribution or, when $j=1$, the *Borel* distribution (Borel [1942], Tanner [1953, 1961]).

It can be shown (see Prabhu [1965]) that $\sum_{n=j}^{\infty} f_n^{(j)} = 1$ when $\rho \leq 1$, and $\sum_{n=j}^{\infty} f_n^{(j)} < 1$ when $\rho > 1$; also $E(N_j) = j(1-\rho)^{-1}$ when $\rho < 1$, and $E(N_j) = \infty$ when $\rho \geq 1$. Thus if $\rho > 1$ there is a positive probability that the busy period will continue indefinitely, whereas if $\rho \leq 1$ it will terminate eventually; however, the expected number of customers served during the j-busy period will be finite only if ρ is strictly less than one.

Note that the expression in brackets on the right-hand side of Equation (8.72) is the probability that during a time interval of length $n\tau$ (which is the time required to serve the n customers whose service times comprise this j-busy period) exactly $n-j$ customers arrive. Evidently, the factor j/n equals the probability that these $n-j$ arrivals occur in such a way as to make this a busy period.

In an exposition of the use of combinatorial methods and ballot theorems in the theory of queues and related subjects, Takács showed that, remarkably, the above interpretation of the factor j/n is valid in the general case of the $M/G/1$ queue (see Theorem 5, p. 102, of Takács [1967]). It follows, therefore, for the $M/G/1$ queue that if $B_j(n,t)$ is the joint probability that the number of customers served during a j-busy period is n and the length of the j-busy period does not exceed t, then

$$B_j(n,t) = \frac{j}{n}\int_0^t \frac{(\lambda\xi)^{n-j}}{(n-j)!}e^{-\lambda\xi}dH^{*n}(\xi) \quad (n \geq j), \tag{8.73}$$

where $H^{*n}(\xi)$ is the n-fold convolution with itself of the service-time distribution function. Since

$$B_j(t) = \sum_{n=j}^{\infty} B_j(n,t),$$

it follows that

$$B_j(t) = \sum_{n=j}^{\infty} \frac{j}{n}\int_0^t \frac{(\lambda\xi)^{n-j}}{(n-j)!}e^{-\lambda\xi}dH^{*n}(\xi). \tag{8.74}$$

In particular, Equation (8.74) with $j=1$ provides us with the inversion of the Laplace-Stieltjes transform $\beta(s)$ defined by the functional equation

(8.67). Thus,

$$B(t) = \sum_{n=1}^{\infty} \frac{1}{n} \int_0^t \frac{(\lambda \xi)^{n-1}}{(n-1)!} e^{-\lambda \xi} dH^{*n}(\xi). \qquad (8.75)$$

[Prior to his derivation of Equation (8.75) by combinatorial methods, Takács [1962a] obtained (8.75) by inversion directly from the functional equation (8.67).]

We have observed that the transform $\beta(s)$ can be inverted to obtain an infinite-series representation for the distribution function $B(t)$, as given by Equation (8.75). With regard to numerical inversion of the transform $\beta(s)$, it can be shown by using the monotonicity of the Laplace-Stieltjes transform that the iteration scheme

$$x_{i+1}(s) = \eta(s + \lambda - \lambda x_i(s)) \qquad [i = 0, 1, \ldots; \quad 0 \leq x_0(s) \leq 1] \qquad (8.76)$$

converges to $\beta(s)$, $x_i(s) \to \beta(s)$, for $s \geq 0$ whenever $\lambda \tau < 1$. This is illustrated in Figure 5.3. Therefore one may calculate, for each $s \geq 0$, the value of the Laplace-Stieltjes transform $\beta(s)$, and, at least in theory, invert numerically. In practice, as we shall see, the quantity of interest may well be the Laplace-Stieltjes transform $\beta(s)$ itself, not the original distribution function $B(t)$.

The interesting point here is not so much the mathematics of solving functional equations (although this is important if one wants to get an answer), but the characteristic combination of mathematical and heuristic reasoning that leads to the functional equation (8.67) and the explicit solution (8.75).

Figure 5.3.

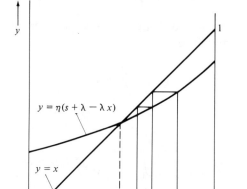

Exercises

16. Verify that (8.72) is the solution of (8.71).

17. a. Let N_k be the number of customers served during a k-busy period in the $M/G/1$ queue. Give an argument, similar to that following Equation (8.61), to show that

$$N_{i+j} = N_i + N_j, \qquad (1)$$

where N_i and N_j are independent.

b. Show that

$$E(N_k) = \frac{k}{1-\rho}.$$

c. Show that the truth of Equation (1) for the $M/D/1$ queue implies the identity

$$\sum_{k=i}^{n-j} \frac{k^{k-i-1}}{(k-i)!} \frac{(n-k)^{n-k-j-1}}{(n-k-j)!} = \frac{(i+j)}{ij} \frac{n^{n-i-j-1}}{(n-i-j)!}$$

$$(i \geqslant 1, \ j \geqslant 1, \ n \geqslant i+j). \qquad (2)$$

In particular, when $i=j=1$ in Equation (2), we obtain the interesting identity

$$\sum_{k=1}^{n-1} \binom{n}{k} k^{k-1}(n-k)^{n-k-1} = 2(n-1)n^{n-2} \qquad (n \geqslant 2). \qquad (3)$$

Equation (3), which is difficult to derive from first principles, is a member of a class of formulas related to Abel's generalization of the binomial theorem (see, for example, pp. 18–23 of Riordan [1968]).

18. Let N_j be the number of customers served during a j-busy period in the $M/M/1$ queue, and show that

$$P(N_j = n) = \frac{j}{n}\binom{2n-j-1}{n-j}\frac{\rho^{n-j}}{(1+\rho)^{2n-j}} \qquad (n \geqslant j).$$

[*Hint:* Use (8.73).] Observe that the distribution of N_j depends on the form of the service-time distribution function [as comparison with Equation (8.72) shows], whereas $E(N_j)$ does not (see Exercise 17b).

19. *The "polite" customer.* Let us call a customer *polite* if she declines to enter service whenever anyone else is waiting for service, and denote by W_p her equilibrium waiting time. (Clearly, no queue is big enough for more than one polite customer at a time.)

a. Argue that

$$P\{W_p \leqslant x | W_p > 0\} = \frac{1}{b} \int_0^x [1 - B(\xi)] d\xi, \quad (1)$$

where $B(t)$ is the distribution function of the busy period, and b is its mean. [*Hint:* Use Equation (7.9).]

b. Conclude from (1) that

$$P\{W_p \leqslant x\} = (1 - \rho) + \frac{\rho(1-\rho)}{\tau} \int_0^x [1 - B(\xi)] d\xi.$$

(See Takács [1963].)

c. Show that

$$E(W_p | W_p > 0) = \frac{\tau}{2(1-\rho)} + \frac{(1-\rho)\sigma_B^2}{2\tau},$$

where σ_B^2 is the variance of the busy-period distribution function.

d. Conclude that

$$E(W_p | W_p > 0) = \frac{\tau^2 + \sigma^2}{2\tau(1-\rho)^2},$$

where σ^2 is the variance of the service-time distribution function.

e. Show that for the $M/M/s$ queue,

$$E(W_p | W_p > 0) = \frac{1}{(1-\rho)^2 s\mu}.$$

20. *Service in reverse order of arrival.*

a. Let T_c be the arrival epoch of an arbitrary customer who arrives when the server is busy, and let T_1 be the first subsequent service completion epoch. Let $\tilde{P}_j(x)$ be the joint probability that j customers arrive in (T_c, T_1) and that $T_1 - T_c \leqslant x$. Argue that

$$\tilde{P}_j(x) = \int_0^x \frac{(\lambda \xi)^j}{j!} e^{-\lambda \xi} d\tilde{H}(\xi),$$

where $\tilde{H}(t)$ is given by (8.41).

b. Let $W(t)$ be the equilibrium waiting-time distribution function, and argue that

$$W(t) = (1-\rho) + \rho \sum_{j=0}^{\infty} \int_0^t \tilde{P}_j(t-x) dB_j(x), \quad (1)$$

where $B_j(x)$ is the distribution function of the j-busy period.

c. Let $\omega(s)$ be the Laplace-Stieltjes transform of $W(t)$, and show by taking transforms in (1) that

$$\omega(s) = (1-\rho) + \lambda \frac{1 - \eta(s + \lambda - \lambda\beta(s))}{s + \lambda[1 - \beta(s)]}, \qquad (2)$$

where $\eta(s)$ and $\beta(s)$ are the Laplace-Stieltjes transforms of the service-time and busy-period distribution functions, respectively.

d. Show that Equation (2) reduces to

$$\omega(s) = (1-\rho) + \frac{\lambda[1 - \beta(s)]}{s + \lambda[1 - \beta(s)]}. \qquad (3)$$

For more on this model see Riordan [1961], Wishart [1960], and, especially, Takács [1963], who obtains formulas for the moments of the waiting-time distribution function, and gives the inverse of (3) as

$$W(t) = (1-\rho) + \lambda \sum_{j=1}^{\infty} e^{-\lambda t} \frac{(\lambda t)^{j-1}}{j!} \int_0^t [1 - H^{*j}(x)] dx.$$

5.9. The $M/G/1$ Queue with Finite Waiting Room

In this section we discuss the single-server queue with Poisson input, general service times, and a finite number n of waiting positions. [Hence there are $n+2$ states E_j $(j=0,1,\ldots,n,n+1)$, where the value of the index j is the number of customers in the system.] In the analysis of this important model we overcome a difficulty that has not appeared in any previous model.

When the number of waiting positions is finite, a customer will depart from the system either after he has completed service or, if he arrives when all waiting positions are occupied, immediately upon arriving. Note that no customer departing after having been served can leave behind him a completely full system; at least one waiting position must be empty.

We consider first those states $E_0, E_1, \ldots, E_n$ in which a customer who has just completed service can leave the system. Let Π_j^* $(j=0,1,\ldots,n)$ be the equilibrium probability that a departing customer (after completing service) leaves behind j other customers. Then, following the previous analysis for the infinite-waiting-room model, we obtain again Equation (8.20), this time valid only for $j=0,1,\ldots,n-1$, which can be written

$$\Pi_{j+1}^* = p_0^{-1}\left(\Pi_j^* - p_j \Pi_0^* - \sum_{i=1}^{j} p_{j-i+1} \Pi_i^*\right) \qquad (j=0,1,\ldots,n-1). \quad (9.1)$$

Equation (9.1) can be solved numerically by recurrence; one can successively determine $\Pi_1^*, \Pi_2^*, \ldots, \Pi_n^*$ in terms of Π_0^*.

If we are interested only in the distribution of the number of customers left behind by a customer who gets served, then Π_0^* is determined from the normalization equation

$$\Pi_0^* + \Pi_1^* + \cdots + \Pi_n^* = 1. \tag{9.2}$$

Observe that since this distribution $\{\Pi_0^*, \Pi_1^*, \ldots, \Pi_n^*\}$ satisfies the same equations (9.1) as does that of the infinite-waiting-room model, the finite-waiting-room probabilities $\Pi_0^*, \Pi_1^*, \ldots, \Pi_n^*$ are proportional to the corresponding infinite-waiting-room probabilities. We assume in what follows that the probabilities $\Pi_0^*, \Pi_1^*, \ldots, \Pi_n^*$ are now known.

We wish to find the distribution of the number of customers found by an arbitrary arrival (not just those arrivals that are served). That is, we wish to find the arriving customer's distribution $\{\Pi_0, \Pi_1, \ldots, \Pi_n, \Pi_{n+1}\}$. We have found the distribution $\{\Pi_0^*, \Pi_1^*, \ldots, \Pi_n^*\}$ of the number of customers left in the system by a departing customer who is completing service. By the theorem of Section 5.3, $\{\Pi_0^*, \Pi_1^*, \ldots, \Pi_n^*\}$ is also the distribution that describes the viewpoint of the arrivals, excluding those who find the system in state E_{n+1}; that is, Π_j^* is the conditional probability that an arbitrary arrival finds the system in state E_j, given that $j < n+1$. Therefore, the probabilities $\{\Pi_j\}$ and $\{\Pi_j^*\}$ both satisfy the same equation (9.1) when $j < n+1$, and it follows that they are proportional when $j < n+1$:

$$\Pi_j = c\Pi_j^* \qquad (j = 0, 1, \ldots, n). \tag{9.3}$$

It remains to find a new equation that permits calculation of the probability Π_{n+1} that an arriving customer finds all n waiting positions occupied. The constant of proportionality c is then calculated from the normalization equation

$$\Pi_0 + \Pi_1 + \cdots + \Pi_n + \Pi_{n+1} = 1. \tag{9.4}$$

Observe that Π_{n+1} is the proportion of customers who are not served. If one wished to design a system with sufficient number of waiting positions so that only a given proportion of the arriving customers are turned away (cleared), then he could determine the number n of waiting positions required from a plot of Π_{n+1} versus n.

We return to consideration of the new equation required to calculate Π_{n+1}. The carried load a' (the mean number of busy servers) is

$$a' = P_1 + \cdots + P_{n+1} = 1 - P_0; \tag{9.5}$$

therefore, for systems with Poisson input, it follows from the equality of the outside observer's distribution $\{P_0, P_1, \ldots, P_n, P_{n+1}\}$ and the arriving

The $M/G/1$ Queue with Finite Waiting Room

customer's distribution $\{\Pi_0, \Pi_1, \ldots, \Pi_n, \Pi_{n+1}\}$ that a' is given by

$$a' = 1 - \Pi_0. \tag{9.6}$$

The probability that an arbitrary customer is served is $1 - \Pi_{n+1}$; therefore, since a' is that portion of the offered load $a(=\lambda\tau=\rho)$ that is carried (served), we have

$$a' = a(1 - \Pi_{n+1}). \tag{9.7}$$

(In what follows, we use this symbol a instead of ρ to emphasize its interpretation as offered load.) Combination of (9.6) and (9.7) yields the required equation:

$$\Pi_{n+1} = \frac{a - (1 - \Pi_0)}{a}. \tag{9.8}$$

Substitution of (9.3) and (9.8) into (9.4) yields

$$c = \frac{1}{\Pi_0^* + a\sum_{j=0}^{n} \Pi_j^*}. \tag{9.9}$$

Substitution of (9.3) into (9.4) and use of (9.2) yields

$$c = 1 - \Pi_{n+1}; \tag{9.10}$$

also, (9.9) becomes

$$c = \frac{1}{\Pi_0^* + a}. \tag{9.11}$$

Hence, we conclude from (9.3) and (9.11) that

$$\Pi_j = \frac{\Pi_j^*}{\Pi_0^* + a} \quad (j = 0, 1, \ldots, n) \tag{9.12}$$

and, from (9.10) and (9.11),

$$\Pi_{n+1} = 1 - \frac{1}{\Pi_0^* + a}. \tag{9.13}$$

Thus we see that although the finiteness of the waiting room precludes the writing of an equation for the probability Π_{n+1}^* from the point of view of a customer completing service, we can obtain the necessary equation (9.8) independently from consideration of the carried load. (For a much more detailed study of this model, see Chapter III.6 of Cohen [1969]; see also Bhat [1973] and Hokstad [1977] for variations of this model.)

Exercise

21. It is required to calculate the arriving customer's equilibrium state distribution $\{\Pi_j\}$ for the $M/G/1$ queue with n waiting positions.

a. Show that $\{\Pi_j\}$ can be calculated from the following algorithm:

1. Assume an arbitrary value for Π_0^*, say $\Pi_0^* = 1$, and calculate $\Pi_1^*, \ldots, \Pi_n^*$ from Equation (9.1).
2. Set $d = \Pi_0^* + \Pi_1^* + \cdots + \Pi_n^*$ and calculate

$$\Pi_j = \frac{\Pi_j^*}{\Pi_0^* + ad} \quad (j=0,1,\ldots,n),$$

$$\Pi_{n+1} = \frac{\Pi_0^* + (a-1)d}{\Pi_0^* + ad}.$$

b. Suppose that the state distribution, say $\{\hat{\Pi}_j\}$, for the corresponding infinite-waiting-room queue has been previously calculated. Show that

$$\Pi_j = \frac{\hat{\Pi}_j}{\hat{\Pi}_0 + a \sum_{\nu=0}^{n} \hat{\Pi}_\nu} \quad (j=0,1,\ldots,n),$$

$$\Pi_{n+1} = \frac{\hat{\Pi}_0 + (a-1) \sum_{\nu=0}^{n} \hat{\Pi}_\nu}{\hat{\Pi}_0 + a \sum_{\nu=0}^{n} \hat{\Pi}_\nu}.$$

c. Using the results of part b, calculate the distribution $\{\Pi_j\}$ for the particular case of exponential service times. Compare with the corresponding probabilities $\{P_j\}$ computed from the "rate up = rate down" equations for a birth-and-death process.

Another approach to the analysis of the $M/G/1$ queue with finite waiting room is through use of the mean busy period. Let $b(n)$ denote the mean busy period in the $M/G/1$ queue with n waiting positions. Then the probability $1 - P_0$ that the server is occupied is the ratio $b(n)/[\lambda^{-1} + b(n)]$ of the mean busy period to the mean cycle time. But this probability also equals the carried load [see (9.5)]; hence

$$a' = \frac{b(n)}{\lambda^{-1} + b(n)}. \tag{9.14}$$

It follows from (9.14) and (9.7) that

$$\Pi_{n+1} = 1 - \frac{b(n)}{\tau + ab(n)}; \tag{9.15}$$

therefore, one can calculate the loss probability Π_{n+1} if one can calculate

The $M/G/1$ Queue with Finite Waiting Room

the mean busy period $b(n)$. We now give a recurrence for the calculation of $b(n)$.

We begin by writing

$$b(n) = \tau + \sum_{j=1}^{n-1} p_j \sum_{k=n-j+1}^{n} b(k) + \sum_{j=n}^{\infty} p_j \sum_{k=1}^{n} b(k) \qquad (n=0,1,\ldots), \tag{9.16}$$

where p_j is defined by (8.17), and any undefined sum is taken to equal zero. To prove (9.16) observe first that, clearly, $b(0) = \tau$. Now assume $n \geq 1$. Observe that the busy period is composed of the service time of the first customer plus some additional time if there are any new arrivals during the first service time. Suppose that exactly j ($1 \leq j \leq n-1$) arrivals occur during the first service time. Then, as the second service time begins, there will be $j-1$ customers waiting in the queue. Since the length of the busy period does not depend on the order of service of waiting customers, we can imagine that none of these $j-1$ waiting customers will enter service until any and all new customers are served who enter the waiting room after the start of the second service time. Thus, the mean time until the next (if there is one) of the original j customers enters service is $b(n-j+1)$ (because only $n-j+1$ waiting positions were available to new arrivals during this time). Hence, with this queue discipline, the mean time required for the system to become completely empty is $b(n-j+1) + \cdots + b(n)$; this explains the second term on the right-hand side of (9.16). Finally, if $j \geq n$ customers arrive during the first service time, then the mean time until the system becomes completely empty is $b(1) + \cdots + b(n)$; this explains the third term on the right-hand side of (9.16).

Equation (9.16) can be rewritten

$$b(n) = \tau + \sum_{k=0}^{n-1} \left(1 - \sum_{j=0}^{k} p_j\right) b(n-k) \qquad (n=0,1,\ldots). \tag{9.17}$$

If we subtract the $(n-1)$th equation from the nth in (9.17), we get the following system, which appears most suitable for calculation by recurrence of $b(1), b(2), \ldots$:

$$b(n) = \begin{cases} p_0^{-1} \tau & (n=1), \\ p_0^{-1} \left[(1-p_1) b(n-1) - \sum_{j=1}^{n-2} p_{n-j} b(j) \right] & (n=2,3,\ldots). \end{cases} \tag{9.18}$$

Equation (9.18) thus provides an easy algorithm for the calculation of the loss probability Π_{n+1}, given by (9.15). An attractive feature of this algorithm is that in the process of calculating the loss probability Π_{n+1} for the

$M/G/1$ queue with n waiting positions, one obtains as a by-product the corresponding loss probabilities for all the systems with smaller waiting rooms.

We have given two methods for the calculation of the probability of loss in the $M/G/1$ queue with finite waiting room. For this model, the first method is probably easier to implement. However, the busy-period method is conceptually easier, and may be easier in practice if the number of waiting positions is small and the model is not standard. An example is provided by Exercise 22.

Exercises

22. A particle-counting device consists of three one-particle buffers and a recorder that visits the buffers and counts the particles. Particles arrive at each of the buffers in three independent Poisson streams, each with rate λ. Any particle that finds its particular buffer occupied is cleared from the system, that is, not counted. A buffer is held by a particle both while waiting for the counter and while being counted. The time required for the recorder to count a particle (and empty a buffer) is a constant of length unity. Show that the proportion p of particles not counted by this device is

$$p = 1 - \frac{e^{3\lambda} + e^{\lambda} - e^{2\lambda}}{1 + 3\lambda(e^{3\lambda} + e^{\lambda} - e^{2\lambda})}.$$

23. Let $B(j,k)$ be the duration of the j-busy period in the $M/G/1$ queue with $j+k-1$ waiting positions, with $E(B(j,k)) = b(j,k)$; and let $P(j,k)$ be the probability that throughout the busy period there will always be at least one unoccupied waiting position.

a. Show that

$$B(j,k) = B(1,k) + B(j-1, k+1),$$

and thus

$$B(j,k) = B(1,k) + B(1, k+1) + \cdots + B(1, j+k-1).$$

b. Show that

$$b(j,k) = b(j, k-1) + [1 - P(j,k)]b(1, j+k-1) \qquad (k \geq 1).$$

[*Hint:* Imagine that the customer (if there is such a customer) whose arrival causes the waiting room to be fully occupied for the first time will not enter service until there are no other waiting customers.]

c. Conclude that

$$P(j,k) = \frac{b(1, k-1)}{b(1, j+k-1)} \qquad (k \geq 1).$$

d. Show that when service times are exponential, then

$$P(j,k) = P(j,k,a) = \frac{1 + a + a^2 + \cdots + a^{k-1}}{1 + a + a^2 + \cdots + a^{j+k-1}},$$

where a is the offered load.

e. The *classical ruin problem* can be formulated as follows: A gambler with initial capital j ($1 \leq j < \infty$) dollars plays against an adversary whose initial capital is k ($1 \leq k < \infty$) dollars. The game consists of independent trials, at each of which the gambler wins a dollar from his adversary with probability p or loses a dollar to his adversary with probability $q = 1 - p$. The game continues until the gambler's capital either is reduced to zero or is increased to $j + k$, that is, until one of the two players is ruined. Let $P'(j,k,p)$ denote the probability of the gambler's ruin, and show that $P'(j,k,p) = P(j,k,a)$, where $p = a/(1+a)$; that is,

$$P'(j,k,p) = P\left(j, k, \frac{p}{q}\right).$$

The relationship between combinatorial problems and stochastic processes, as in part e, has been systematically studied by Takács [1967]. This exercise is adapted from a paper by Cooper and Tilt [1976], whose main theme concerns the result of part c.

5.10. The $M/G/1$ Queue with Batch Arrivals

An interesting variant of the $M/G/1$ queueing model occurs when one assumes *batch arrivals*. That is, each arrival epoch now corresponds to the arrival of a batch of customers, where the batch sizes are independent, identically distributed, random variables. The customers within a batch are served one at a time, and, as before, the service times of the customers are independent, identically distributed, random variables, with distribution function $H(x)$. (Thus, when the batch size is always one, this model reduces to the ordinary $M/G/1$ queue.) Let W be the equilibrium waiting time for an arbitrary (test) customer. Then $W = W_1 + W_2$, where W_1 is the time from arrival to start of service of the test customer's batch, and W_2 is the remaining time until start of service of the test customer. We shall show that

$$E(W_1) = \frac{\lambda m^2 \tau^2}{2(1 - \lambda m \tau)} \left(1 + \frac{m\sigma^2 + \tau^2 \hat{\sigma}^2}{m^2 \tau^2}\right) \tag{10.1}$$

and

$$E(W_2) = \frac{(m-1)\tau}{2} + \frac{\hat{\sigma}^2 \tau}{2m}, \tag{10.2}$$

where τ and σ^2 are, as before, the mean and variance of the service-time distribution function; and m and $\hat{\sigma}^2$ are the mean and variance of the distribution of the number of customers in an arbitrary batch; and $\lambda m\tau < 1$.

This model is worthy of our attention here, not only because it is an interesting variant of the ordinary $M/G/1$ model, but also because, as pointed out by Burke [1975], several treatments of this model that appear in the literature give incorrect expressions for $E(W_2)$.

Let N be the size of an arbitrary batch, and let S_N be the sum of the service times of the customers in an arbitrary batch. Then $S_N = X_1 + \cdots + X_N$, where X_i is the service time of the ith customer in the batch, where $E(X_i) = \tau$ and $V(X_i) = \sigma^2$. Now, if we view a batch as if it were a single customer with service time S_N, then, according to (8.39), we can write

$$E(W_1) = \frac{[\lambda E(S_N)] E(S_N)}{2[1 - \lambda E(S_N)]} \left(1 + \frac{V(S_N)}{E^2(S_N)}\right). \tag{10.3}$$

According to Exercise 5, the compound sum S_N has mean

$$E(S_N) = m\tau \tag{10.4}$$

and variance

$$V(S_N) = m\sigma^2 + \tau^2 \hat{\sigma}^2. \tag{10.5}$$

Substitution of (10.4) and (10.5) into (10.3) yields (10.1).

To obtain (10.2) observe that the discussion of Section 2.1 applies: Let N' be the number of customers (balls) who arrive in the batch (urn) that contains the test customer. Then, according to Equation (1.5) of Chapter 2,

$$P\{N' = j\} = \frac{jP\{N = j\}}{m}, \tag{10.6}$$

(where the "cycle" of that discussion corresponds to the "batch" in our present discussion). Now,

$$E(W_2) = \sum_{j=1}^{\infty} E(W_2 | N' = j) P\{N' = j\}, \tag{10.7}$$

where, clearly,

$$E(W_2 | N' = j) = \frac{(j-1)}{2} \tau. \tag{10.8}$$

Substitution of (10.6) and (10.8) into (10.7) yields (10.2). (The error alluded to earlier resulted from ignoring the difference between N, the size of an arbitrary batch, and N', the size of the batch to which the test customer belongs.)

Exercise

24. Consider the equilibrium $M/G/1$ queue with batch arrivals, and assume that the batches are served in order of arrival. Let $\hat{\omega}(s)$ be the Laplace-Stieltjes transform of the waiting-time distribution function for an arbitrary customer, and show that

$$\hat{\omega}(s) = \frac{s(1-\lambda m\tau)}{s-\lambda[1-g(\eta(s))]} \frac{1-g(\eta(s))}{m[1-\eta(s)]},$$

where $\eta(s) = \int_0^\infty e^{-sx} dH(x)$, and $g(z)$ is the probability-generating function of the number of customers in a batch.

5.11. Optimal Design and Control of Queues: The N-Policy and the T-Policy

So far in this text we have discussed only *descriptive* queueing models, that is, models that are completely specified with regard to method of operation. One reason for studying descriptive models is that they permit one to determine the values of the model parameters that result in the most efficient (that is, *optimal*) operation. For example, if it is required to design an Erlang loss system that is optimal according to the criterion that, for a given offered load a, no more than $p\%$ of the arriving customers will be blocked, then one simply chooses the number s of servers to be the minimum value such that $B(s, a) \leq p/100$. (Here we are making the tacit and reasonable assumption that the cost of operating the system is an increasing function of the number of servers, and our objective is to meet the stated blocking criterion at lowest cost.) Recently, much attention has been directed to the problem of *prescriptive* queueing models, that is, models in which the optimization is carried out over a class of operating policies, rather than over a set of parameters for a single operating policy that is fixed as part of the model. Of course, one might view a set of possible operating policies as nothing more than a set of parameters for a single (more general) operating policy; that is, one might claim that the distinction between the *static optimization* of descriptive models and the *dynamic optimization* of prescriptive models is merely semantic obfuscation. It turns out, however, that the concept of dynamic optimization of queues is a fruitful one.

There is now available a sufficient body of research in this area so that one can talk about a general theory of optimal design and control (or operation) of queues. This theory is an application and extension of other theories and concepts such as dynamic programming, sequential decision theory, and Markov and semi-Markov decision processes. In particular, the theory of Markov and semi-Markov decision processes is discussed in Ross [1970]. Four papers that complement each other and among them provide a comprehensive explication and bibliography of research into optimal design and control of queueing systems are Stidham [1977], Stidham and Prabhu [1973], Sobel [1973], and Crabill, Gross, and Magazine [1977].

In this section we shall illustrate these ideas by considering several different types of optimal control policy for the $M/G/1$ queue. However, our methodology will not be in the framework of the (rather abstract) general theory presented in Stidham and Prabhu [1973]; instead, we shall exploit the particular structure of the model and rely on certain intuitive arguments to obtain the optimal control policies.

We begin by considering the $M/G/1$ queue with the following cost structure. The server can be in either of two states, "off" and "on," and can be turned off or on at any time. The server can provide service only when it is on. Customers who arrive when the server is off wait in the queue. The *running cost* of keeping the server on is c_0 dollars per unit time, whether or not any customers are being served. To save money, the server can be turned off when it is idle; however, a *switching cost* of c_1 dollars is incurred each time the server is turned back on. Also, a (linear) *holding cost* of c_2 dollars per unit time is incurred for each customer present in the system, whether waiting or being served. Our objective is to find the *optimal operating policy*, that is, the rules for turning the server on and off that result in the lowest long-run cost.

A *stationary policy* is defined as one that always prescribes the same action whenever the system is in a given state. For obvious reasons, attention is usually restricted to the class of stationary policies, and the following three questions are asked: (1) when does an optimal stationary policy exist, and, if it exists, (2) what is its structure, and (3) what are the values of its parameters? These are difficult questions to answer in general (see Stidham and Prabhu [1973]), but we can often make some reasonable guesses for specific problems. In particular, if we assume that an optimal stationary policy exists, and can make a reasonable guess as to its structure, then it may not be difficult to obtain the values of the parameters of the putative optimal policy.

For the particular problem at hand, let us assume that an optimal stationary policy exists, and try to infer what the structure of such a policy might be. It seems obvious that if the server is on and there are customers present, there would be no advantage in turning it off, because to do so would incur unnecessary holding costs. Likewise, if the server is off and there are no customers present, there would be no advantage in turning it on, because to do so would incur unnecessary running costs. Therefore, we conjecture that there is an optimal stationary policy and that it is one of the following:

1. *Turn the server off when there are no customers present, and turn it on when there are n customers present.* Such a policy is called an *N-policy*.
2. *Leave the server on permanently.* Such a policy is called a *do-nothing policy*.

If there is indeed an optimal stationary policy, and if it is in fact of the form conjectured, then our optimization problem reduces to finding the value of the parameter n that produces the optimal N-policy, and comparing this policy with the do-nothing policy. A proof that an N-policy or a do-nothing policy is, in fact, the optimal stationary policy (when the control information available is the number of customers present at each point in time) is given in Heyman [1968], using the same kind of dynamic-programming argument that is formalized in Stidham and Prabhu [1973].

Of course, it makes sense to consider the N-policy only if such a policy can be implemented. For example, suppose that the server of our model represents an electronic telephone switching machine that cannot directly "see" if a customer has arrived, but must periodically "look" to determine whether a customer is present. If we assume that each look corresponds to turning the server on, then we might conjecture that there is an optimal deterministic stationary policy that is one of the following:

1. *Turn the server off at the end of a busy period (when there are no customers present), and turn it on (that is, look) after an interval of length t. If there are no customers present when the server looks, turn it off and look again after another interval of length t. Repeat this procedure until the server finds a customer waiting. Then, leave the server on until the end of the busy period.* Such a policy is called a *T-policy*.
2. *Leave the server on permanently.*

Hence, optimization of this system reduces to finding the value of the parameter t that yields the optimal T-policy, and comparing this policy with the do-nothing policy. (The T-policy, motivated by the application described here, has been studied by Heyman [1977].)

It is intuitively clear (and we will show it) that, all other things being equal, the optimal T-policy is inferior to the optimal N-policy (because the T-policy uses less information about the status of the queue than the N-policy). The point to be made here is that the structure of the optimal stationary policy depends upon the constraints built into the original model.

We shall consider each of these policies in turn. Our approach is to study the general case of the P-policy; specialize our results to obtain formulas for the optimal N-policy and the optimal T-policy; and prove that, all other things being equal, the former is superior to the latter.

To this end, we denote by a_P the expected value of the cost per customer to turn the server on under policy P; and we denote by b_P the expected value of the waiting cost per customer under policy P in excess of that which would be incurred under any policy according to which the server can never be off when a customer is waiting. Let

$$c_P = a_P + b_P. \tag{11.1}$$

Define the random variable M_P as the number of customers served, under policy P, during the busy period of a particular customer (the test customer) whose viewpoint we will adopt. If we let Y be the number of times during a cycle that the server looks at the system and finds no customers waiting (which may be always zero, as in the N-policy), then $c_1(1+Y)$ is the total cost per cycle of activating the server. (We are assuming that the server must be turned on to look.) Therefore, $c_1(1+Y)/M_P$ is the test customer's share of the cost of turning on the server (to implement the test customer's cycle). Now

$$a_P = E\left(\frac{c_1(1+Y)}{M_P}\right) = c_1[1+E(Y)]E\left(\frac{1}{M_P}\right), \tag{11.2}$$

where the first equality in (11.2) follows by definition, and the second follows from the properties of the expected-value operator and the independence of Y and M_P. To calculate $E(1/M_P)$, which is given by

$$E\left(\frac{1}{M_P}\right) = \sum_{i=1}^{\infty} \frac{1}{i} P\{M_P = i\}, \tag{11.3}$$

observe that the discussion of Section 2.1 of Chapter 2 applies: Let N_P be the number of customers (balls) who are served during an arbitrary busy period (urn) under policy P. Then, according to Equation (1.5) of Chapter 2, it follows that

$$P\{M_P = i\} = \frac{iP\{N_P = i\}}{E(N_P)}. \tag{11.4}$$

Substitution of (11.4) into (11.3) yields

$$E\left(\frac{1}{M_P}\right) = \frac{1}{E(N_P)}. \tag{11.5}$$

To calculate $E(N_P)$, we write

$$E(N_P) = \sum_{j=1}^{\infty} E(N_P|A_P=j)P\{A_P=j\}, \qquad (11.6)$$

where the random variable A_P is the number of arrivals waiting when the busy period begins, under policy P. It follows from Exercise 17b that

$$E(N_P|A_P=j) = \frac{j}{1-\rho}. \qquad (11.7)$$

Hence, it follows from substitution of (11.7) into (11.6) that

$$E(N_P) = \frac{E(A_P)}{1-\rho}. \qquad (11.8)$$

To calculate $E(A_P)$, observe that the vacation-time model of Exercise 12 is applicable: If we say that when the server is off it is on vacation, then

$$E(A_P) = \frac{f_P'(1)}{1-P(0)}, \qquad (11.9)$$

where $f_P(z)$ is the probability-generating function of the number of customers present at the end of a vacation, and $P(0)$ is the probability that no customers arrive during a vacation. It now follows from (11.5), (11.8), and (11.9) that (11.2) can be written

$$a_P = \frac{c_1(1-\rho)[1-P(0)]}{f_P'(1)}[1+E(Y)]. \qquad (11.10)$$

To calculate b_P, let the random variable X_P be the test customer's sojourn time under policy P, and let X be the sojourn time that the test customer would experience if the server can never be off when a customer is waiting. Then the expected value of the test customer's excess waiting cost is

$$b_P = E[c_2(X_P - X)] = c_2[E(X_P) - E(X)], \qquad (11.11)$$

where the first equality in (11.11) follows from the definition of b_P and the assumption of a constant linear holding cost c_2, and the second equality follows from the properties of the expected-value operator. Now, using $L = \lambda W$, it follows that $E(X) = g'(1)/\lambda$, where $g(z)$ is the equilibrium probability-generating function of the number of customers present in the ordinary $M/G/1$ queue; similarly, we can write $E(X_P) = g_P'(1)/\lambda$, where

$g_P(z)$ is the corresponding probability-generating function in the $M/G/1$ queue operating under policy P. If we again adopt the viewpoint that the server's off periods correspond to vacations, then it follows from Equation (3) of Exercise 12 that

$$g_P''(1) - g'(1) = \frac{f_P''(1)}{2f_P'(1)}; \qquad (11.12)$$

therefore,

$$b_P = \frac{c_2 f_P''(1)}{2\lambda f_P'(1)}. \qquad (11.13)$$

Finally, it follows from (11.10) and (11.13) that (11.1) becomes

$$c_P = \frac{c_1(1-\rho)[1-P(0)]}{f_P'(1)}[1+E(Y)] + \frac{c_2 f_P''(1)}{2\lambda f_P'(1)}. \qquad (11.14)$$

Equation (11.14) is the key result; to find the optimal policy of type P, it suffices to find the value of the parameter associated with policy P that minimizes the right-hand side of (11.14).

For example, consider the N-policy. Then, clearly, $E(Y)=0$, $P(0)=0$, and

$$f_N(z) = z^n, \qquad (11.15)$$

and hence

$$f_N'(1) = n \qquad (11.16)$$

and

$$f_N''(1) = n(n-1). \qquad (11.17)$$

Thus, for the N-policy Equation (11.14) becomes

$$c_N = c_N(n) = \frac{c_1(1-\rho)}{n} + \frac{c_2(n-1)}{2\lambda} \quad (n=1,2,\ldots). \qquad (11.18)$$

Let n^* be the value of the parameter n that minimizes the right-hand side of (11.18):

$$c_N(n^*) = \min_n c_N(n). \qquad (11.19)$$

If we treat n as a continuous variable, call it x, and denote by x^* the

Optimal Design and Control of Queues: N-Policy and T-Policy

minimizing value of x, so that

$$c_N(x^*) = \min_{x>0} c_N(x), \quad (11.20)$$

then x^* can be obtained by the well-known method of calculus: If $c_N''(x) > 0$, then x^* is the root of the equation

$$\frac{d}{dx} c_N(x) = 0 \quad (x > 0).$$

It easily follows from (11.18) that

$$\frac{d}{dx} c_N(x) = -\frac{c_1(1-\rho)}{x^2} + \frac{c_2}{2\lambda} \quad (11.21)$$

and

$$\frac{d^2}{dx^2} c_N(x) > 0 \quad (x > 0).$$

Hence x^*, which is the value of x for which the right-hand side of (11.21) vanishes, is

$$x^* = \sqrt{\frac{2\lambda(1-\rho)c_1}{c_2}}. \quad (11.22)$$

n^* is one of the two integers adjacent to x^*; it is found by substituting these two integers into (11.18), and taking as n^* the integer that gives the smaller value of c_N.

Thus, we have found the value n^* of the parameter n that gives the optimal N-policy. However, the optimal N-policy is not necessarily better than the do-nothing policy, which is the policy of never turning the server off. To analyze this policy, we calculate the cost of running the server during the idle period that precedes the test customer's busy period, and distribute this cost equally among the customers served during the test customer's busy period. Let $c(0)$ denote the expected value of this cost; then

$$c(0) = E\left(\frac{c_0 X_0}{M_0}\right),$$

where X_0 is the interarrival time and M_0 is the number served during the test customer's (ordinary 1-) busy period. Then

$$E\left(\frac{c_0 X_0}{M_0}\right) = c_0 E(X_0) E\left(\frac{1}{M_0}\right) = \frac{c_0}{\lambda} E\left(\frac{1}{M_0}\right),$$

and, by the same reasoning that led to (11.5),

$$E\left(\frac{1}{M_0}\right) = \frac{1}{E(N_0)},$$

where N_0 is the number of customers served during an ordinary 1-busy period. Now, $E(N_0) = 1/(1-\rho)$, and therefore,

$$c(0) = \frac{c_0(1-\rho)}{\lambda}. \tag{11.23}$$

If we set $c(0) = c_N(n^*)$, we obtain $c_0 = \lambda c_N(n^*)/(1-\rho)$. We conclude that if $c_0 > \lambda c_N(n^*)/(1-\rho)$ the optimal policy is the N-policy with parameter n^*, while if $c_0 < \lambda c_N(n^*)/(1-\rho)$ the optimal policy is never to turn the server off. Note that although the long-run operating cost depends, in general, on the service-time distribution function, the optimal policy depends only on the mean service time τ (and the parameters c_1, c_2, and λ).

For example, suppose $\lambda = \frac{1}{2}$, $\tau = 1$, $c_1 = 12$, and $c_2 = 2$. Then it follows from (11.22) that $x^* = \sqrt{3}$, and thus

$$n^* = \begin{cases} 1 & \text{if } c_N(1) < c_N(2), \\ 2 & \text{if } c_N(1) > c_N(2). \end{cases}$$

We have from (11.18) that $c_N(1) = 6$ and $c_N(2) = 5$. Therefore, $n^* = 2$ is the parameter of the optimal N-policy. If $c_0 > \lambda c_N(n^*)/(1-\rho) = 5$, then the cheapest long-term operation is obtained with the N-policy with $n^* = 2$; whereas if $c_0 < 5$, then the best operating policy is to leave the server on permanently.

Exercise

25. Let c be the minimum mean operating cost per unit time in equilibrium for the $M/G/1$ queue operating either continuously or under an N-policy, and show that

$$c = \begin{cases} c_0 + c_2\left[\rho + \frac{\rho^2}{2(1-\rho)}\left(1 + \frac{\sigma^2}{\tau^2}\right)\right] & \text{when } c_0 < \frac{\lambda c_N(n^*)}{1-\rho}, \\ c_0\rho + \lambda c_N(n^*) + c_2\left[\rho + \frac{\rho^2}{2(1-\rho)}\left(1 + \frac{\sigma^2}{\tau^2}\right)\right] & \text{when } c_0 > \frac{\lambda c_N(n^*)}{1-\rho}, \end{cases}$$

where σ^2 is the variance of the service-time distribution function. Evaluate c for the above example when service times are exponentially distributed and (a) $c_0 = 4$, and (b) $c_0 = 6$. Repeat the calculations for the case of constant service times.

Let us now turn to consideration of the T-policy. We leave it to Exercise 26 for the reader to show that Equation (11.14) specializes to

$$c_T = c_T(t) = \frac{c_1(1-\rho)}{\lambda t} + \frac{c_2 t}{2}. \tag{11.24}$$

Let us define t^* as the value of t that minimizes the right-hand side of (11.24):

$$c_T(t^*) = \min_{t>0} c_T(t). \tag{11.25}$$

Then, exactly the same arithmetic that led from (11.18) to (11.22) will lead from (11.24) to

$$t^* = \sqrt{\frac{2(1-\rho)c_1}{\lambda c_2}}. \tag{11.26}$$

Inspection of (11.22) and (11.26) shows that

$$x^* = \lambda t^*. \tag{11.27}$$

Equation (11.27) shows that the mean number of customers present when the server is activated (turned on to look) in the optimal T-policy is precisely the value of the parameter n^* that would define the optimal N-policy if that value were not required to be an integer. Observe also that, as with the N-policy, the optimal T-policy does not depend on the form of the service-time distribution function.

We stated earlier that it is intuitively clear that, all other things being equal, the optimal T-policy is inferior to the optimal N-policy, because the T-policy uses less information about the status of the queue. To prove this, suppose first that x^* happens to be an integer. Then $c_T(t^*) - c_N(n^*) = c_T(t^*) - c_N(x^*) = c_T(t^*) - c_N(\lambda t^*)$, where the last equality follows from (11.27). Now application of (11.24) and (11.18) shows that $c_T(t^*) - c_N(\lambda t^*) = c_2/2\lambda > 0$; that is,

$$c_T(t^*) > c_N(n^*). \tag{11.28}$$

Now suppose that x^* is not an integer. We first show that

$$c_T(t^*) > c_N(x^* + 1), \tag{11.29}$$

and then that

$$c_N(x^*+1) > c_N(n^*); \qquad (11.30)$$

these two inequalities show that (11.28) is correct also when x^* is not an integer, which will prove our assertion.

To prove (11.29), we write

$$c_N(x^*+1) = \frac{c_1(1-\rho)}{x^*+1} + \frac{c_2 x^*}{2\lambda} < \frac{c_1(1-\rho)}{x^*} + \frac{c_2 x^*}{2\lambda}, \qquad (11.31)$$

where the equality follows from (11.18), and the inequality follows from the fact that

$$\frac{1}{x^*+1} < \frac{1}{x^*}.$$

But it follows from (11.27) and (11.24) that the right-hand side of the inequality (11.31) equals $c_T(t^*)$; we have established (11.29).

To prove (11.30), let us denote by n_1 and n_2 the integers adjacent to x^* on the left and right, respectively. Then, $n_2 < x^* + 1$, which implies that

$$c_N(n_2) < c_N(x^*+1) \qquad (11.32)$$

[because, according to (11.21), $c_N(x)$ is an increasing function of x when $x > x^*$]. Thus, (11.30) is true when $n^* = n_2$. If $n^* = n_1$, then, of course, $c_N(n_1) < c_N(n_2)$, and it then follows from (11.32) that $c_N(n_1) < c_N(x^*+1)$. Thus, (11.30) is true also when $n^* = n_1$. The proof is complete.

Exercise

26. Consider the $M/G/1$ queue operating under a T-policy, with parameter t.
   a. Show that

$$E(Y) = \frac{e^{-\lambda t}}{1 - e^{-\lambda t}}.$$

   b. Show that

$$f_T(z) = e^{-(1-z)\lambda t}.$$

   c. Conclude that (11.24) is correct.

The results x^*, n^*, and $c_N(n)$ for the N-policy were first derived by Yadin and Naor [1963]. This model was extended and studied further by Heyman [1968]. Other studies of this model are referenced in the three survey papers mentioned at the start of this section. All of the results given here for the T-policy were first derived by Heyman [1977]. An interesting and related model is the *D-policy*, according to which the server is turned off at the end of each busy period and turned back on only when the sum of the service times of the waiting customers exceeds some fixed value d. This model is more difficult to analyze, and cannot be handled by the method given here. The interested reader is referred to Balachandran [1973] and Balachandran and Tijms [1975].

5.12. The $M/G/1$ Queue with Service in Random Order

In this section we consider the equilibrium $M/G/1$ queue in which whenever a customer completes service the next customer to begin service is selected at random from among all the customers waiting in the queue. Our objective is to derive algorithms for the numerical computation of the complementary conditional waiting-time distribution function $P\{W>t|W>0\}$, which is the probability that an arbitrary customer who is forced to wait in the queue will wait more than t (for any $t>0$) for service to commence. The unconditional waiting-time distribution function can then be calculated from the relation $P\{W>t\} = P\{W>t|W>0\} \times P\{W>0\}$; for the $M/G/1$ queue, $P\{W>0\} = \rho$. This model has been studied by LeGall [1962] and Kingman [1962], who used transform methods. These yield formulas for the moments of the waiting-time distribution function (see Takács [1963]), but numerical calculation of the waiting-time distribution function requires the inversion (numerical or otherwise) of transforms.

The algorithms given here, which are taken from Carter and Cooper [1972], do not use transform methods. Nevertheless, numerical calculations may be impractical except in certain special (but important) cases. We will present the results of numerical calculation for the special case $M/E_k/1$ with $k=1, 2,$ and 3. These results are in excellent agreement with those of Neuts [1977], who used this model to compare with his method of using phase-type approximations for numerical calculations.

The special case of the $M/D/1$ queue with service in random order has been studied by Burke [1959], who, by utilizing certain simplifying results that apply only to the case of constant service times, obtained an expression suitable for the calculation of the waiting-time distribution function. Burke's paper includes a useful set of curves. Additional discussion on the numerical properties of the waiting-time distribution function for the $M/D/1$ random-service queue is given in Lee [1966].

Suppose that an arbitrary customer (the test customer) arrives at time T_c and finds $j \geq 1$ other customers in the system (either in service or waiting for service). Then the length of time that the test customer waits for service to begin is the sum of

1. the elapsed time between the test customer's arrival epoch T_c and the first subsequent service completion epoch T_1, and

2. the elapsed time between T_1 and the instant at which the test customer commences service.

The analysis of the total waiting time is complicated by two facts. First, the time intervals (1) and (2) are not, in general, statistically independent. Second, as we have pointed out, any customer in service at epoch T_c when the test customer arrives does not, in general, have the same service-time distribution function as does an arbitrary customer. Specifically, the relationship of the distribution function $\hat{H}(x)$ of the length of the service interval containing the arrival epoch T_c to the distribution function $H(x)$ of the length of an arbitrary service interval is obtained by application of Equation (7.19), namely,

$$d\hat{H}(x) = \frac{1}{\tau} x\, dH(x); \qquad (12.1)$$

likewise, the distribution function $\tilde{H}(x)$ of the length of the remainder of this service interval, from T_c to T_1, is obtained by application of (7.9) [see also Equation (8) of Exercise 14], namely,

$$\tilde{H}(x) = \frac{1}{\tau} \int_0^x [1 - H(\xi)]\, d\xi. \qquad (12.2)$$

Note that in the special case of constant service times these difficulties disappear. When the service times are constant, the delay suffered by a blocked customer between the instant of his arrival and the first postarrival departure epoch is independent of the delay subsequent to this epoch, and the distribution of the service interval containing T_c is the same as that of an arbitrary service interval (that is, constant).

We are now ready to calculate for the case of general service times the conditional probability $P\{W > t | W > 0\}$, where W is the waiting time of the test customer. We denote by T_c the test customer's arrival epoch, and by T_1 the first departure epoch subsequent to T_c. We also define $N(T)$ as the number of customers in the system (including service) at any time T, and $W(T)$ as the remaining waiting time for service to commence for any customer who is waiting at time T. Then we may write

$$P\{W > t | W > 0\} = P\{T_1 - T_c > t | W > 0\}$$
$$+ \sum_{j=2}^{\infty} \frac{j-1}{j} \int_0^t P\{W(T_1 + 0) > t - \xi | W > 0, N(T_1 + 0) = j\}$$
$$\times d_\xi P\{N(T_1 + 0) = j, T_1 - T_c \leq \xi | W > 0\}.$$
$$(12.3)$$

Let us define

$$Q_j(\xi) = P\{N(T_1 + 0) = j, T_1 - T_c \leq \xi | W > 0\} \qquad (j = 1, 2, \ldots). \quad (12.4)$$

The $M/G/1$ Queue with Service in Random Order

Now consider the service interval containing the test customer's arrival epoch T_c. The probability that this interval has length between x and $x + dx$ is $d\hat{H}(x)$, given by Equation (12.1). The probability $p_i(x)$ that, in addition to the test customer, exactly i other customers arrive during such an interval of length x is, by assumption, the Poisson probability

$$p_i(x) = \frac{(\lambda x)^i}{i!} e^{-\lambda x} \qquad (i = 0, 1, \ldots). \tag{12.5}$$

Since the arrival process is Poisson, then for fixed x the length of the interval (T_c, T_1) is uniformly distributed over $(0, x)$. It follows that $Q_j(\xi)$ has density

$$\frac{d}{d\xi} Q_j(\xi) = Q_j'(\xi)$$

given by

$$Q_j'(\xi) = \int_\xi^\infty \left(\Pi_0^* p_{j-1}(x) + \sum_{i=1}^j \Pi_i^* p_{j-i}(x) \right) \frac{1}{x} d\hat{H}(x) \qquad (j = 1, 2, \ldots), \tag{12.6}$$

(where $\{\Pi_i^*\}$ is the equilibrium state distribution as seen by a departure), which, in view of Equation (12.1), can be written

$$Q_j'(\xi) = \frac{1}{\tau} \int_\xi^\infty \left(\Pi_0^* p_{j-1}(x) + \sum_{i=1}^j \Pi_i^* p_{j-i}(x) \right) dH(x) \qquad (j = 1, 2, \ldots). \tag{12.7}$$

Now let us define

$$\check{W}_j(x) = P\{ W(T_1 + 0) > x \mid W > 0, N(T_1 + 0) = j \} \qquad (j = 2, 3, \ldots) \tag{12.8}$$

and $\check{W}_1(x) = 0$. Then Equation (12.3) can be written

$$P\{W > t \mid W > 0\} = 1 - \tilde{H}(t) + \sum_{j=2}^\infty \frac{j-1}{j} \int_0^t \check{W}_j(t - \xi) Q_j'(\xi) d\xi, \tag{12.9}$$

where $\tilde{H}(x)$ is given by Equation (12.2), $Q_j'(\xi)$ is given by Equation (12.7), and $\check{W}_j(x)$ satisfies the following set of integral equations:

$$\check{W}_j(x) = 1 - H(x) + \sum_{i=0}^\infty \frac{j+i-2}{j+i-1} \int_0^x p_i(\xi) \check{W}_{j+i-1}(x - \xi) dH(\xi)$$

$$(j = 2, 3, \ldots). \tag{12.10}$$

It remains to solve Equation (12.10) before (12.9) can be used. One formal approach is to assume that each function $\check{W}_j(x)$ has a convergent Maclaurin series expansion, and then use (12.10) to determine the coefficients in the expansion. This method is outlined in Exercise 27. Although numerical calculation by this algorithm may appear hopelessly complicated, it has in fact been demonstrated for the $M/E_k/1$ queue. The results are reproduced in Table 5.1.

Another method for obtaining the functions $\check{W}_j(x)$, suggested by P. J. Burke [1967, unpublished], involves the introduction of an additional conditioning variable (sometimes called a *supplementary variable*). Let S_k be the sum of the first k service times commencing at epoch T_1, and let the additional conditioning variable $\check{X}(x)$ be the value of the largest integer k ($k=0,1,...$) such that $S_k < x$. We define $\check{W}_{j,k}(x)$ as follows:

$$\check{W}_{j,k}(x) = P\{ W(T_1+0) > x | W > 0, N(T_1+0) = j, \check{X}(x) = k \}$$

$$(j=2,3,...; \quad k=0,1,...) \quad (12.11)$$

and $\check{W}_{1,k}(x) = 0$, where, as before, T_1 is the first service completion epoch after the arrival of the test customer, $N(T_1+0)$ is the number of customers in the system (including service) just after epoch T_1, and $W(T_1+0)$ is the remaining waiting time for service to commence for a customer who is waiting just after time T_1.

Table 5.1. Sample Calculations by the Maclaurin-Series Method for $P\{W>t|W>0\}$ for the $M/E_k/1$ Random-Service Queue

t[a]	$\rho=0.3$			$\rho=0.8$		
	$k=1$	$k=2$	$k=3$	$k=1$	$k=2$	$k=3$
0.0	1.0000	1.0000	1.0000	1.0000	1.0000	1.0000
0.2	0.8487	0.8364	0.8336	0.9254	0.9166	0.9137
0.4	0.7237	0.6889	0.6766	0.8609	0.8395	0.8312
0.6	0.6199	0.5365	0.5394	0.8045	0.7709	0.7565
0.8	0.5331	0.4599	0.4261	0.7547	0.7107	0.6908
1.0	0.4602	0.3758	0.3361	0.7102	0.6578	0.6338
1.5	0.3236	0.2305	0.1894	0.617	0.551	0.521
2.0	0.2319	0.1455	0.112	0.544	0.470	0.440
2.5	0.1689	0.0945	0.072	0.483	0.407	
3.0	0.1248	0.063		0.433	0.358	
3.5	0.093	0.043		0.39	0.33	
4.0	0.070			0.35		
4.5	0.054			0.33		

[a] In units of mean service time.

We have that [see Equation (7.8)]

$$P\{\check{X}(x)=k\} = H^{*k}(x) - H^{*(k+1)}(x)$$

$$(k=0,1,\ldots; \quad H^{*0}(x)=1; \quad x \geq 0), \quad (12.12)$$

where $H^{*k}(x)$ is the k-fold convolution with itself of the service-time distribution function. From the theorem of total probability it follows that

$$\check{W}_j(x) = \sum_{k=0}^{\infty} \check{W}_{j,k}(x) P\{\check{X}(x)=k\} \quad (j=2,3,\ldots), \quad (12.13)$$

and hence our problem will be solved if we can determine the conditional probabilities $\check{W}_{j,k}(x)$ ($j=2,3,\ldots;\ k=0,1,\ldots$).
Clearly

$$\check{W}_{j,0}(x) = 1 \quad (j=2,3,\ldots). \quad (12.14)$$

Let us define for $k>0$

$$H(\xi|k,x) = P\{S_1 \leq \xi | \check{X}(x)=k\}. \quad (12.15)$$

It follows from the definition of conditional probability that

$$d_\xi H(\xi|k,x) = \frac{P\{\check{X}(x-\xi)=k-1\}\, dH(\xi)}{P\{\check{X}(x)=k\}} \quad (k=1,2,\ldots). \quad (12.16)$$

We now have the following recurrence for $k \geq 1$:

$$\check{W}_{j,k}(x) = \sum_{i=0}^{\infty} \frac{j+i-2}{j+i-1} \int_0^x p_i(\xi) \check{W}_{j+i-1,k-1}(x-\xi)\, d_\xi H(\xi|k,x)$$

$$(j=2,3,\ldots;\ k=1,2,\ldots). \quad (12.17)$$

The required function $P\{W>t|W>0\}$ can now be determined in principle from Equation (12.9), complemented by Equations (12.7), (12.13), and (12.14) and (12.17). This method has the advantage over the MacLaurin-series method that it does not require any assumptions about existence or convergence of a Maclaurin series expansion of $\check{W}_j(x)$, but whether it offers any computational advantages is problematical. For an outline of the application of this method to the $M/D/1$ queue, see Exercise 29.

Clearly, reduction to practice of these algorithms is difficult except in special cases. One may well argue that obtaining the kind of numerical results that engineers need from the formulae produced by mathematicians is often more an art than a science. Recent progress in computational probability has been made by Neuts [1977] in utilizing phase-type approximations of arbitrary distributions to obtain numerical results, and by Jagerman [1978] in methods for the numerical inversion of Laplace transforms. Of course, one can always resort to digital-computer simulation, which we shall discuss in Chapter 6.

Exercises

27. *The Maclaurin-series method for the $M/G/1$ random-service queue* (Carter and Cooper [1972]). Assume that $\check{W}_j(x)$, defined by Equation (12.10), has the Maclaurin series expansion

$$\check{W}_j(x) = \sum_{\nu=0}^{\infty} \frac{x^\nu}{\nu!} \check{W}_j^{(\nu)} \qquad (j=2,3,\ldots;\quad \check{W}_j^{(0)}=1). \tag{1}$$

Set

$$h(x) = \frac{d}{dx} H(x) \tag{2}$$

and

$$\check{b}_i(x) = h(x) p_i(x). \tag{3}$$

Also, for any function $f(x)$, define

$$f^{(k)} = \left(\frac{d^k}{dx^k} f(x) \right)_{x=0}.$$

a. Show that

$$\check{W}_j^{(\nu)} = -H^{(\nu)} + \sum_{i=0}^{\nu-1} \frac{j+i-2}{j+i-1} \sum_{k=i}^{\nu-1} \check{b}_i^{(k)} \check{W}_{j+i-1}^{(\nu-1-k)}$$

$$(j=2,3,\ldots;\quad \nu=1,2,\ldots). \tag{4}$$

[Equation (4) permits, in principle, evaluation of the sum on the right-hand side of (1).]

b. Let $F(t) = P\{W > t | W > 0\}$, and assume the representation

$$F(t) = \sum_{\nu=0}^{\infty} \frac{t^\nu}{\nu!} F^{(\nu)} \qquad (F^{(0)}=1). \tag{5}$$

Show that Equation (12.9) yields the following formula for the coefficients $\{F^{(\nu)}\}$ in the expansion (5):

$$F^{(\nu)} = -\tilde{H}^{(\nu)} + \sum_{j=2}^{\infty} \frac{j-1}{j} \sum_{k=0}^{\nu-1} Q_j^{(k+1)} \check{W}_j^{(\nu-1-k)} \qquad (\nu=1,2,\ldots). \tag{6}$$

c. Show that Equation (12.7) yields

$$Q_j^{(1)} = Q_j'(0) = \frac{1}{\tau} \Pi_{j-1}^* \quad (j=1,2,\ldots) \tag{7}$$

and

$$Q_j^{(k+1)} = -\frac{1}{\tau} \sum_{m=0}^{k-1} \binom{k-1}{m} H^{(k-m)} \left(\Pi_0^* p_{j-1}^{(m)} + \sum_{i=1}^{j} \Pi_i^* p_{j-1}^{(m)} \right)$$
$$(k=1,2,\ldots;\ j=1,2,\ldots). \tag{8}$$

Thus Equations (4), (6), (7), and (8) permit calculation of the Maclaurin series representation (5) of the conditional waiting-time probability $P\{W > t | W > 0\}$. Sample calculations according to this algorithm for the $M/E_k/1$ random-service queue are given in Table 5.1.

28. (Carter and Cooper [1972].) Suppose that

$$H(x) = 1 - e^{-x/\tau}. \tag{1}$$

a. Show that for $\nu = 1$, Equation (6) of Exercise 27 becomes

$$F^{(1)} = -\frac{1}{\tau} + \sum_{j=2}^{\infty} \frac{j-1}{j} Q_j^{(1)}. \tag{2}$$

b. Show that Equation (2) reduces to

$$F^{(1)} = -\frac{1}{\tau} \frac{1-\rho}{\rho} \ln \frac{1}{1-\rho}. \tag{3}$$

c. Show that for $\nu = 2$, Equation (6) of Exercise 27 becomes

$$F^{(2)} = \frac{1}{\tau^2} + \sum_{j=2}^{\infty} \frac{j-1}{j} \left(Q_j^{(1)} \check{W}_j^{(1)} + Q_j^{(2)} \right). \tag{4}$$

d. Show that

$$Q_j^{(2)} = -\frac{1}{\tau} H^{(1)} \left(\Pi_0^* p_{j-1}(0) + \sum_{i=1}^{j} \Pi_i^* p_{j-i}(0) \right) \tag{5}$$

and that this reduces to

$$Q_j^{(2)} = -\frac{1}{\tau^2} (1-\rho)\rho^j \quad (j=2,3,\ldots). \tag{6}$$

e. Show that Equation (4) of Exercise 27 yields

$$\check{W}_j^{(1)} = -\frac{1}{\tau} \frac{1}{j-1} \quad (j=2,3,\ldots). \tag{7}$$

f. Show that Equation (4) reduces to

$$F^{(2)} = \frac{1}{\tau^2}(1-\rho)\left[2 - \frac{1-\rho}{\rho}\ln\frac{1}{1-\rho}\right]. \qquad (8)$$

g. Compare the results (3) and (8) with the expansion (6) of Exercise 32 of Chapter 3.

29. *The additional-conditioning-variable method for the $M/D/1$ random-service queue* (Carter and Cooper [1972]). Suppose that

$$H(x) = \begin{cases} 0 & \text{when } x < \tau, \\ 1 & \text{when } x \geqslant \tau. \end{cases} \qquad (1)$$

a. Show that Equation (12.7) reduces to

$$Q_j'(\xi) = \begin{cases} \dfrac{1}{\tau}\Pi_{j-1}^* & \text{when } \xi \leqslant \tau, \\ 0 & \text{when } \xi > \tau. \end{cases} \qquad (2)$$

b. Show that for $t < \tau$, Equation (12.9) reduces to

$$P\{W > t | W > 0\} = 1 - \frac{t}{\tau} + \frac{1}{\tau}\sum_{j=2}^{\infty} \frac{j-1}{j}\Pi_{j-1}^* \int_0^t \check{W}_j(t-\xi)\, d\xi$$

$$(0 \leqslant t < \tau) \qquad (3)$$

and, for $t \geqslant \tau$,

$$P\{W > t | W > 0\} = \frac{1}{\tau}\sum_{j=2}^{\infty} \frac{j-1}{j}\Pi_{j-1}^* \int_0^\tau \check{W}_j(t-\xi)\, d\xi$$

$$(\tau \leqslant t < \infty). \qquad (4)$$

c. Show that, since $\check{W}_j(x)$ is a step function, the integrals on the right-hand side of Equations (3) and (4) can be evaluated, giving

$$P\{W > t | W > 0\} = 1 - \frac{t}{\tau} + \frac{t}{\tau}\sum_{j=2}^{\infty} \frac{j-1}{j}\Pi_{j-1}^*$$

$$(0 \leqslant t < \tau) \qquad (5)$$

and

$$P\{W > t | W > 0\} = \frac{1}{\tau}\sum_{j=2}^{\infty} \frac{j-1}{j}\Pi_{j-1}^* \left\{ \left(t - \left[\frac{t}{\tau}\right]\tau\right)\check{W}_j(t) \right.$$

$$\left. + \left(\tau - t + \left[\frac{t}{\tau}\right]\tau\right)\check{W}_j(t-\tau)\right\}$$

$$(\tau \leqslant t < \infty), \qquad (6)$$

where $[x]$ is defined as the largest integer not exceeding x.

d. Show that

$$H(\xi|k,x) = H(\xi). \tag{7}$$

e. Show that Equation (12.17) reduces to

$$\check{W}_{j,k}(x) = \sum_{i=0}^{\infty} \frac{j+i-2}{j+i-1} p_i(\tau) \check{W}_{j+i-1,k-1}(x-\tau)$$

$$(j=2,3,\ldots; \quad k=1,2,\ldots). \tag{8}$$

f. Show that Equation (12.12) reduces to

$$P\{\check{X}(x) = k\} = \begin{cases} 1 & \text{when} \quad k = \left[\dfrac{x}{\tau}\right] \\ 0 & \text{otherwise.} \end{cases} \tag{9}$$

g. Show that Equations (9) and (12.13) imply

$$\check{W}_j(x) = \check{W}_{j,[x/\tau]}(x) \quad (j=2,3,\ldots). \tag{10}$$

h. Show that if we set $k=[x/\tau]$ in Equation (8), we obtain

$$\check{W}_j(x) = \sum_{i=0}^{\infty} \frac{j+i-2}{j+i-1} p_i(\tau) \check{W}_{j+i-1}(x-\tau)$$

$$(j=2,3,\ldots; \quad x \geq \tau). \tag{11}$$

i. Show that Equation (11) is complemented by

$$\check{W}_j(x) = 1 \quad (j=2,3,\ldots; \quad x < \tau). \tag{12}$$

j. Show that the recurrence (11) follows immediately from Equation (12.10).

Equations (11) and (12) permit calculation of Equation (6). It can be shown in a straightforward manner that these results are equivalent to those of Burke [1959], who gives several useful sets of curves for the $M/D/1$ random-service queue.

5.13. Queues Served in Cyclic Order

So far in this text we have been concerned with well-defined models. Quite often, however, applications require models that do not fit into the standard molds and are not available in the literature in packaged form. Then the reader must develop his own model if possible. We give here an example of a model suggested by a practical problem that has received some attention in the literature, but is not one that would usually be called "standard." The point is that even though this model cannot be equated directly to a standard queueing model, it can be analyzed by the same techniques.

We shall first state the general problem, and then solve it for a particular special case. In so doing we hope to illustrate the underlying concepts while simultaneously avoiding inessential detail. The success of the method in the special case then suggests its own generalization. It is interesting to observe that in the literature, for reasons of economy and mathematical "elegance," such problems are usually presented in the most general form that the authors can handle. Such presentation often obscures the process by which the solution was reached, leaving the reader with only the answers and undue respect for the intelligence of the author.

We consider a system of queues served in cyclic order by a single server. The ith queue is characterized by general service-time distribution function $H_i(x)$ and Poisson input with parameter λ_i. The process begins with the arrival of a customer at some queue, say A, when the system is otherwise empty. The server starts on this customer immediately, and continues to serve queue A until for the first time the server becomes idle and there are no customers waiting in queue A. The server then looks at the next queue in the cyclic order, queue $A+1$, and serves those customers, if any, that have accumulated during the serving period of queue A. The server continues to serve $A+1$ until for the first time the server becomes idle and there are no customers waiting in queue $A+1$. The process continues in this manner, with the queues being served in cyclic order, until for the first time the system becomes completely empty. The process is then reinitiated by the arrival of the next customer. No time is required to switch from one queue to the next.

It should be pointed out that this description is quite precise, and, in fact, itself constitutes a large chunk of the solution. The original question is simply, how does one analyze a system in which queues are served in cyclic order?

Systems in which a single server is shared among several queues are common. For example, in a time-shared computer system the users have access through teletypewriters to a central computer that is shared among them. Similarly, in some electronic telephone switching systems, the central control spends much of its time polling various hoppers and performing work requests that if finds in these hoppers. Another possible application of this model is in the design of a dynamically controlled traffic light at an intersection.

These systems, of course, do not fit our model exactly. In particular, the assumptions (or lack of assumptions) made with regard to the various distributions, and the exact way in which the server moves among the queues, are the result of a tradeoff between the model's closeness of fit to reality and its mathematical tractability.

We shall solve the problem posed above for the very special case of two queues, each characterized by the same general service-time distribution function $H(x)$ and Poisson input with the same parameter λ. We start by writing probability state equations for the Markov chain imbedded at the instants at which the server finishes serving a queue. We call these time instants *switch points*. At a switch point, all the customers in the system are waiting in the queue other than that on which the server has just been working (because the server switches from a queue at the first instant the queue becomes empty). Since the two queues are identical, the "state" of the system at a switch point can be defined as the number of waiting customers, without regard for the identity of the queue in which they are waiting.

Note that we have chosen to study this system at a particular set of renewal points, the switch points, which is a "smaller" set than that of the usual choice of renewal points, the set of all service completion points. It is not clear, a priori, that this choice will yield as much information as would a solution based on the set of all service completion points. (It turns out that it does.) But if one were to imbed the chain at the latter set of points, the definition of state would require two variables—one for the number of customers in each queue at a service completion point.

Let $P(k)$ be the probability that at an arbitrary switch point the number of waiting customers is k ($k=0,1,\ldots$). This event can occur through the following exhaustive and mutually exclusive contingencies:

1. The server leaves a queue and finds $j \geq 1$ customers waiting for service in the other queue, and during the time that the server spends at the latter queue, exactly k customers arrive at the first queue. This length of time is precisely a j-busy period, with distribution function $B_j(t) = B^{*j}(t)$ and corresponding Laplace-Stieltjes transform $[\beta(s)]^j$.
2. The server leaves a queue and finds no customers waiting in the other queue. When the next arrival occurs (at either queue) the server spends a 1-busy period there, and during that time k customers arrive at the other queue.

These considerations lead directly to the imbedded-Markov-chain equations for the equilibrium state probabilities:

$$P(k) = \sum_{j=1}^{\infty} P(j) \int_0^{\infty} \frac{(\lambda t)^k}{k!} e^{-\lambda t} dB^{*j}(t)$$

$$+ P(0) \int_0^{\infty} \frac{(\lambda t)^k}{k!} e^{-\lambda t} dB(t) \qquad (k=0,1,\ldots). \qquad (13.1)$$

We now solve the equations (13.1) by the use of a probability-generating function $g(x)$:

$$g(x) = \sum_{k=0}^{\infty} P(k) x^k. \qquad (13.2)$$

Substitution of (13.1) into (13.2) yields

$$g(x) = \sum_{k=0}^{\infty} \left\{ \sum_{j=1}^{\infty} P(j) \int_0^{\infty} \frac{(\lambda t)^k}{k!} e^{-\lambda t} dB^{*j}(t) \right.$$

$$\left. + P(0) \int_0^{\infty} \frac{(\lambda t)^k}{k!} e^{-\lambda t} dB(t) \right\} x^k$$

$$= \sum_{j=1}^{\infty} P(j) \int_0^{\infty} \sum_{k=0}^{\infty} \frac{(\lambda t x)^k}{k!} e^{-\lambda t} dB^{*j}(t)$$

$$+ P(0) \int_0^{\infty} \sum_{k=0}^{\infty} \frac{(\lambda t x)^k}{k!} e^{-\lambda t} dB(t),$$

which becomes, after summing over k,

$$g(x) = \sum_{j=1}^{\infty} P(j) \int_0^{\infty} e^{-\lambda(1-x)t} dB^{*j}(t) + P(0) \int_0^{\infty} e^{-\lambda(1-x)t} dB(t). \quad (13.3)$$

The second integral on the right-hand side of (13.3) is simply the Laplace-Stieltjes transform $\beta(s)$ of the busy-period distribution function, with argument $s = \lambda(1-x)$; similarly, the first integral on the right of (13.3) is the Laplace-Stieltjes transform with argument $s = \lambda(1-x)$ of the j-fold convolution with itself of the busy-period distribution function. Since the transform of a j-fold self-convolution is the original transform raised to the jth power, Equation (13.3) becomes

$$g(x) = \sum_{j=1}^{\infty} P(j) [\beta(\lambda - \lambda x)]^j + P(0) \beta(\lambda - \lambda x). \quad (13.4)$$

But, with the addition of the term $P(0)$, the sum in (13.4) is simply the generating function $g(z)$ with argument $z = \beta(\lambda - \lambda x)$. Hence, Equation (13.4) may be written

$$g(\beta(\lambda - \lambda x)) - g(x) = P(0)[1 - \beta(\lambda - \lambda x)]. \quad (13.5)$$

Equation (13.5) is a functional equation for the probability-generating function (13.2). It remains to solve Equation (13.5) and find $P(0)$.

Let us now define the iteration procedure

$$z_{\nu+1} = \beta(\lambda - \lambda z_\nu) \quad (\nu = 0, 1, \ldots; \quad z_0 = x). \quad (13.6)$$

[Notice that since $z_0 = x$, the νth iterate z_ν is a function of x; that is,

$z_\nu = z_\nu(x)$.] Successive use of (13.6) in (13.5) gives

$$g(z_1) - g(x) = P(0)(1 - z_1),$$
$$g(z_2) - g(z_1) = P(0)(1 - z_2),$$
$$\vdots$$
$$g(z_\nu) - g(z_{\nu-1}) = P(0)(1 - z_\nu),$$

which, when added together, give

$$g(z_\nu) - g(x) = P(0) \sum_{j=1}^{\nu} (1 - z_j). \tag{13.7}$$

We shall now show that

$$\lim_{\nu \to \infty} z_\nu(x) = 1 \quad (x \leq 1). \tag{13.8}$$

From the definition (13.6) of z_ν as a Laplace-Stieltjes transform, we see that for any $z_0 = x \leq 1$, the numbers $z_1, z_2, \ldots$ all lie in $(0, 1]$ and thus are bounded (by 1). Since $g(x)$ is a probability-generating function, the sequence $g(z_1), g(z_2), \ldots$ is also bounded (by 1). Since $z_1, z_2, \ldots$ all lie in $(0, 1]$, the sum on the right-hand side of (13.7) increases monotonically with ν. Therefore the left-hand side of (13.7), and in particular the sequence $g(z_1), g(z_2), \ldots$, increases monotonically with ν. Thus the sequence $g(z_1), g(z_2), \ldots$ is both bounded and monotonically increasing, and therefore has a finite limit. Hence, the left-hand side of (13.7) has a limit, which implies that the series of nonnegative terms on the right-hand side of (13.7) converges. But a necessary condition for the convergence of a series of nonnegative terms is that the νth term must go to zero as $\nu \to \infty$. Hence $\lim_{\nu \to \infty} (1 - z_\nu) = 0$; that is, we have proved the assertion (13.8).

Returning from our digression, we see that taking limits as $\nu \to \infty$ in Equation (13.7) gives, by virtue of (13.8),

$$g(1) - g(x) = P(0) \sum_{j=1}^{\infty} (1 - z_j). \tag{13.9}$$

Using the normalization requirement $g(1) = 1$ in Equation (13.9), we have

$$g(x) = 1 - P(0) \sum_{j=1}^{\infty} [1 - z_j(x)]. \tag{13.10}$$

Setting $x = 0$ in Equation (13.10) and noting that $g(0) = P(0)$, we obtain

$$P(0) = \left\{ 1 + \sum_{j=1}^{\infty} [1 - z_j(0)] \right\}^{-1}. \tag{13.11}$$

The functional equation (13.5) has now been solved in the sense that $g(x)$ can be evaluated [using (13.10) and (13.11)] for any $x \leq 1$. We can now use this fact to obtain some important characteristics of our queueing system.

To obtain the mean number $\bar{n} = g'(1)$ of customers in the system at a switch point, we differentiate through the functional equation (13.5):

$$-\lambda \beta'(\lambda - \lambda x) g'(\beta(\lambda - \lambda x)) - g'(x) = \lambda P(0) \beta'(\lambda - \lambda x). \quad (13.12)$$

Setting $x = 1$ in Equation (13.12) gives

$$-\lambda \beta'(0) \bar{n} - \bar{n} = \lambda P(0) \beta'(0). \quad (13.13)$$

Recall that $\beta(s)$ is the Laplace-Stieltjes transform of the $M/G/1$ busy-period distribution function, which has mean $\tau/(1 - \lambda \tau)$. Hence, $-\beta'(0) = \tau/(1 - \lambda \tau)$, and Equation (13.13) yields

$$\bar{n} = P(0) \frac{\lambda \tau}{1 - 2\lambda \tau}. \quad (13.14)$$

Note that (13.14) is meaningful only when $2\lambda \tau < 1$. This is exactly what one would intuitively expect: A statistical-equilibrium distribution for this single-server system (of two queues) will exist if and only if the total load offered to the server is less than unity. Numerical evaluation of Equation (13.14) requires calculation of only one state probability, $P(0)$, which can be computed to any desired degree of accuracy from Equations (13.11), (13.6), and (8.76). Observe that although numerical values of the Laplace-Stieltjes transform $\beta(s)$ are required for this calculation, inversion of the transform is not. (Note also that, in general, $P(0) \neq 1 - 2\lambda \tau$. Why?)

Because of the symmetry in the special case we have considered here, the mean wait $E(W)$ for service to commence for a customer arriving at a given queue is given by Equation (8.39) with $\rho = 2\lambda \tau$. In the general case, however, the required formula is more complicated. It is worth mentioning that this general formula does not require any iteration for its numerical evaluation.

Models similar to this have been studied by many authors. In particular, this model was used in a nontechnical article explaining queueing theory for the layman (Leibowitz [1968]) as an example of an interesting and important queueing model. For further discussion and references, see Cooper and Murray [1969], Cooper [1970], Eisenberg [1972], Halfin [1975], Kuehn [1979], and Stidham [1972]. (There is a subtle error in Cooper [1970]; for the record, we give the correct result in Exercise 31.)

Exercises

30. *The $M/G/1$ queue with gating.* An arriving customer who finds the server idle causes a gate to close. When this customer's service is complete, the gate opens and admits into a waiting room all those customers who arrived during this service time, and then closes. When all the customers in the waiting room have been served, the gate opens and admits into the waiting room all those customers who arrived during the collective service times of the preceding group of customers, after which it closes. The process continues in this manner. When the gate opens and finds no waiting customers, it remains open until it closes behind the next arrival. Show that the mean $\bar{n}$ of the number of customers who enter the waiting room when the gate opens is given by

$$\bar{n} = \frac{\rho}{1-\rho}\left\{1 + \sum_{j=1}^{\infty}(1-x_j)\right\}^{-1}$$

with

$$x_{j+1} = \eta(\lambda - \lambda x_j) \quad (j=0,1,\ldots; \quad x_0=0),$$

where ρ is the utilization factor, λ is the arrival rate, and $\eta(s)$ is the Laplace-Stieltjes transform of the service-time distribution function.

31. Show that Equation (44) of Cooper [1970] is incorrect, and should be replaced by

$$\omega_i(s) = (1-\rho) + \frac{\rho \lambda_i}{\overline{m}_{i-1}[s - \lambda_i + \lambda_i \eta_i(s)]}$$
$$\times \left[g_{i-1}(\eta_i(s), 1, \ldots, 1) - g_{i-1}(1 - s\lambda^{-1}, 1, \ldots, 1)\right]$$
$$(i = 0, 1, \ldots, N-1),$$

where

$$\overline{m}_{i-1} = \left(\frac{\partial}{\partial x}g_{i-1}(x, 1, \ldots, 1)\right)_{x=1}.$$

[*Hint*: The error is introduced in Equation (42).]

5.14. The $GI/M/s$ Queue

In Section 5.8 we studied the $M/G/1$ queue by writing equations relating the state probabilities to each other at successive renewal points, where, by definition, a renewal point is an epoch at which the future evolution of the system depends on the present state only, and not on the path by which that state was reached. We showed that in the $M/G/1$ queue the service completion epochs, which are the instants at which customers leave the server, constitute a set of renewal points. We then wrote an equation relating the state probabilities to each other at a pair of successive renewal points, namely the service completion points of the kth and $(k+1)$th departing customers [Equation (8.13)]. Using a statistical-equilibrium argument, we obtained an equation [Equation (8.20)] defining the statistical-equilibrium distribution $\{\Pi_j^*\}$ of the number of customers left behind by an arbitrary departing customer, from which blocking and waiting-time information was subsequently obtained.

We shall analyze the $GI/M/s$ queue by the same technique: Identify a set of renewal points, relate the state probabilities at successive renewal points to each other, assume the existence of a limiting stationary distribution, and solve the resulting system of equations.

Suppose that customers arrive at epochs $T_1, T_2, \ldots$, and assume that the interarrival times $T_{k+1} - T_k$ ($k = 0, 1, \ldots$; $T_0 = 0$) are mutually independent, identically distributed, random variables, with common distribution function $G(x) = P\{T_{k+1} - T_k \leq x\}$ ($k = 0, 1, \ldots$), and mean interarrival time λ^{-1}. All customers wait as long as necessary for service. Let N_k be the number of customers in the system just prior to the arrival of the kth customer; N_k is the number of customers present at $T_k - 0$. Since the input is recurrent, and since the service times are by assumption identically distributed exponential variables (with mean μ^{-1}), independent of the arrival epochs and each other, the arrival epochs $T_1, T_2, \ldots$ are renewal points. Hence, by the theorem of total probability,

$$P\{N_{k+1} = j\} = \sum_{i=0}^{\infty} P\{N_{k+1} = j | N_k = i\} P\{N_k = i\}$$

$$(j = 0, 1, \ldots; k = 1, 2, \ldots). \quad (14.1)$$

As was true with the $M/G/1$ queue, the transition probabilities

$$p_{ij} = P\{N_{k+1} = j | N_k = i\}$$

depend on the indices i and j, but not on the index k. It can be shown using the theory of Markov chains that a unique stationary distribution

$$\Pi_j = \lim_{k \to \infty} P\{N_k = j\} \quad (j = 0, 1, \ldots) \quad (14.2)$$

exists if and only if $\rho = \lambda/s\mu < 1$. [Equation (14.2), for the $GI/M/s$ queue, is the analogue of (8.19) for the $M/G/1$ queue.]

Therefore, after taking limits on both sides of Equation (14.1), it follows that the equilibrium probability state equations that determine the distribution (14.2) are

$$\Pi_j = \sum_{i=0}^{\infty} p_{ij} \Pi_i \quad (j = 0, 1, \ldots), \quad (14.3)$$

together with the normalization equation,

$$\sum_{j=0}^{\infty} \Pi_j = 1. \tag{14.4}$$

We now proceed to specify the transition probabilities $\{p_{ij}\}$ and solve the equations (14.3) and (14.4) that determine the equilibrium arriving customer's distribution $\{\Pi_j\}$. (Note that our analysis will yield the arriving customer's distribution directly. This is in contrast to the models previously considered, in which the distribution originally obtained required further analysis before it could be interpreted as or translated into the distribution that corresponds to the viewpoint of the arriving customer.)

To obtain the transition probabilities, note first that any arrival can find at most one more customer present than was found by the preceding arrival; hence

$$p_{ij} = 0 \qquad (i+1-j<0). \tag{14.5}$$

Now consider the case where the $(k+1)$th arrival finds all servers busy. Then the kth arrival could have found at most one server idle, and all s servers must have been continuously busy during the interarrival time $T_{k+1} - T_k$. Since the service times are assumed to be exponentially distributed, with mean $\tau = \mu^{-1}$, it follows that

$$P\{N_{k+1} = j | N_k = i, T_{k+1} - T_k = x\} = \frac{(s\mu x)^{i+1-j}}{(i+1-j)!} e^{-s\mu x}$$

$$(i \geq s-1, \quad j \geq s, \quad i+1-j \geq 0);$$

and, since the interarrival time $T_{k+1} - T_k$ has distribution function $G(x)$, therefore

$$p_{ij} = \int_0^{\infty} \frac{(s\mu x)^{i+1-j}}{(i+1-j)!} e^{-s\mu x} dG(x) \qquad (i \geq s-1, \quad j \geq s, \quad i+1-j \geq 0).$$

$$\tag{14.6}$$

It will turn out that for our purposes we need only the transition probabilities (14.5) and (14.6); we leave it to Exercise 34 for the reader to show that the remaining transition probabilities are given by

$$p_{ij} = \int_0^{\infty} \binom{i+1}{j} e^{-j\mu x} (1 - e^{-\mu x})^{i+1-j} dG(x) \qquad (i \leq s-1, \quad i+1-j \geq 0)$$

$$\tag{14.7}$$

and

$$P_{ij} = \int_0^\infty \int_0^x \binom{s}{j} e^{-j\mu(x-y)}(1-e^{-\mu(x-y)})^{s-j} \frac{(s\mu y)^{i-s}}{(i-s)!} e^{-s\mu y} s\mu \, dy \, dG(x)$$

$$(i \geqslant s, \quad j < s, \quad i+1-j \geqslant 0). \quad (14.8)$$

Let us now consider the equilibrium probability state equations (14.3) for the values of the index $j \geqslant s$; then substitution of (14.6) into (14.3) yields

$$\Pi_j = \sum_{i=j-1}^\infty \Pi_i \int_0^\infty \frac{(s\mu x)^{i+1-j}}{(i+1-j)!} e^{-s\mu x} dG(x) \qquad (j = s, s+1, \ldots),$$

which, with a change of variable, becomes

$$\Pi_j = \sum_{i=0}^\infty \Pi_{i+j-1} \int_0^\infty \frac{(s\mu x)^i}{i!} e^{-s\mu x} dG(x) \qquad (j = s, s+1, \ldots). \quad (14.9)$$

Equation (14.9) is most easily solved by a serendipitous guess; we assume a solution of the simplest form that could conceivably be correct, namely, a geometric distribution:

$$\Pi_j = A\omega^{j-s} \qquad (j \geqslant s-1). \quad (14.10)$$

Substitution of (14.10) into (14.9) yields the equation

$$\omega = \int_0^\infty e^{-(1-\omega)s\mu x} dG(x). \quad (14.11)$$

Observe that the right-hand side of (14.11) is the Laplace-Stieltjes transform $\gamma(z)$ of the interarrival-time distribution function, evaluated at $z = (1-\omega)s\mu$. Hence, Equation (14.11) can be written

$$\omega = \gamma((1-\omega)s\mu). \quad (14.12)$$

It can be shown (see, for example, Takács [1962a]) that (14.12) has a unique root in $(0, 1)$ whenever $\rho = \lambda/s\mu < 1$, and in this case the numerical solution can be found using the iteration procedure

$$\omega_{i+1} = \gamma((1-\omega_i)s\mu) \qquad (i = 0, 1, \ldots; \quad 0 \leqslant \omega_0 < 1). \quad (14.13)$$

Note that for Poisson input, the root ω of Equation (14.12) equals the server occupancy; $\omega = \lambda/s\mu = \rho$. For this reason, ω is sometimes referred to as the *generalized occupancy*, although the true occupancy, defined to be the carried load per server in erlangs, is still $\rho = \lambda/s\mu$, and in general $\omega \neq \rho$.

The GI/M/s Queue

We conclude that our guess (14.10) is indeed correct. It remains for us to calculate the constant A and the remaining state probabilities $\Pi_{s-2}, \Pi_{s-3}, \ldots, \Pi_0$. The determination of these probabilities, described now by a finite number of equations, requires much labor. Since these probabilities are not required explicitly for most applications, we content ourselves here with the statement of a formula for the calculation of the constant A:

$$A = \left\{ \frac{1}{1-\omega} + \sum_{j=1}^{s} \frac{1}{C_j(1-\gamma_j)} \binom{s}{j} \frac{s(1-\gamma_j)-j}{s(1-\omega)-j} \right\}^{-1}, \qquad (14.14)$$

where

$$\gamma_j = \gamma(j\mu) \qquad (j=0,1,\ldots,s) \qquad (14.15)$$

and

$$C_j = \prod_{i=1}^{j} \left(\frac{\gamma_i}{1-\gamma_i} \right) \qquad (j=1,2,\ldots,s). \qquad (14.16)$$

A derivation following Takács [1962a] of (14.14) and the probabilities $\Pi_0, \Pi_1, \ldots, \Pi_{s-2}$ is outlined in Exercise 35.

We now turn to calculation of the blocking probability and, for service in order of arrival, the length of time W that an arbitrary customer spends waiting in the queue for service to commence. An arriving customer will be forced to wait for service if on arrival he finds at least s other customers in the system, and this event has probability

$$P\{W>0\} = \sum_{j=s}^{\infty} \Pi_j.$$

Hence, it follows from (14.10) that the probability of blocking is given by

$$P\{W>0\} = \frac{A}{1-\omega}. \qquad (14.17)$$

[We leave it to Exercise 32 for the reader to verify that for Poisson input, the right-hand side of (14.17) reduces, of course, to the Erlang delay formula. This provides another illustration of the equality of the arriving customer's distribution $\{\Pi_j\}$ and the outside observer's distribution $\{P_j\}$ for systems with Poisson input.]

The calculation of the waiting-time distribution for the case of service in order of arrival closely parallels the corresponding calculation for the Erlang delay model, as done in Section 3.4. In particular, Equations (4.19)–(4.21) of Chapter 3 all remain valid for the $GI/M/s$ queue.

Corresponding to (4.22) of Chapter 3 we have

$$P\{Q=j, W>0\} = \Pi_{s+j} = A\omega^j, \qquad (14.18)$$

where the second equality in (14.18) follows from (14.10). Substitution of (14.18) and (14.17) into (4.21) of Chapter 3 yields the analogue of (4.23) of Chapter 3, namely,

$$P\{Q=j|W>0\} = (1-\omega)\omega^j \qquad (j=0,1,\ldots). \qquad (14.19)$$

Since (14.19) is the same as (4.23) of Chapter 3 with ρ replaced by ω, it follows that (4.24) of Chapter 3 will be valid for the $GI/M/s$ queue when ρ is replaced by ω; we conclude that

$$P\{W>t|W>0\} = e^{-(1-\omega)s\mu t}. \qquad (14.20)$$

Thus, for the equilibrium $GI/M/s$ queue with service in order of arrival, the conditional (given that the customer is not served immediately) waiting-time distribution function is the negative exponential, with mean value

$$E(W|W>0) = \frac{1}{(1-\omega)s\mu}. \qquad (14.21)$$

It follows from (14.17) and (14.20) that the (unconditional) waiting times are described by

$$P\{W>t\} = \frac{A}{1-\omega} e^{-(1-\omega)s\mu t}, \qquad (14.22)$$

and, from (14.17) and (14.21),

$$E(W) = \frac{A}{(1-\omega)^2 s\mu}. \qquad (14.23)$$

Of course, (14.21) and (14.23) are valid not only for the case of service in order of arrival, but for any nonbiased queue discipline. Calculations with (14.20) are facilitated by the use of the graph of Figure A.5 of the appendix.

Finally, let us look at the relationship between the outside observer's distribution $\{P_j\}$ and the arriving customer's distribution $\{\Pi_j\}$. Takács [1962a] proves that

$$P_j = \frac{\lambda}{\mu(j)} \Pi_{j-1} \qquad (j=1,2,\ldots), \qquad (14.24)$$

where λ^{-1} is the mean interarrival time, and $\mu(j) = j\mu$ when $j<s$ and

The $GI/M/s$ Queue

$\mu(j) = s\mu$ when $j \geq s$. If we write (14.24) as

$$\lambda \Pi_{j-1} = \mu(j) P_j \qquad (j = 1, 2, \ldots), \tag{14.25}$$

then we can make the following interpretation. The left-hand side of (14.25) equals the product of the arrival rate and the proportion of arrivals who find $j-1$ other customers present; hence it equals the rate at which the system moves from state E_{j-1} to E_j. The right-hand side of (14.25) is the product of the service completion rate when there are j customers present (with exponential service times) and the proportion of time that exactly j customers are present; hence, it equals the rate at which the system moves from state E_j to E_{j-1} (just as we have argued previously for birth-and-death processes). Thus, Equation (14.25) states that "rate up = rate down." [For an interesting derivation of (14.25) using Little's theorem, see Heyman and Stidham [1980].]

The development of this section follows that of Takács [1962a], Chapter 2. The reader is referred to Chapter 5 of the same book for a treatment of the corresponding model (renewal input, exponential service times) with no waiting positions.

Exercises

32. Verify that for Poisson input, Equation (14.17) reduces to

$$P\{W > 0\} = C(s, a).$$

33. Prove that in a $GI/M/s$ queue with service in order of arrival, the equilibrium conditional probability that a blocked customer will still be waiting in the queue when the next customer arrives is equal to the generalized occupancy ω. (See Exercise 29 of Chapter 3.)

34. Verify Equations (14.7) and (14.8).

35. *Derivation of* (14.14) *and the probabilities* $\Pi_0, \Pi_1, \ldots, \Pi_{s-2}$ (Takács [1962a]).
    a. Define the generating function

$$U(z) = \sum_{j=0}^{s-1} \Pi_j z^j, \tag{1}$$

and show, by using (14.3), that

$$U(z) = \int_0^\infty (1 - e^{-\mu x} + ze^{-\mu x}) U(1 - e^{-\mu x} + ze^{-\mu x}) \, dG(x)$$
$$+ A \int_0^\infty \left[\int_0^\infty e^{s\mu \omega y} (e^{-\mu y} - e^{-\mu x} + ze^{-\mu x})^s s\mu \, dy \right] dG(x)$$
$$- Az^s. \tag{2}$$

b. Show that

$$U(1) = 1 - \frac{A}{1-\omega}. \tag{3}$$

c. Define

$$U_j = \frac{1}{j!} \left(\frac{d^j U(z)}{dz^j} \right)_{z=1} \quad (j = 0, 1, \ldots, s-1), \tag{4}$$

and show that

$$U_0 = 1 - \frac{A}{1-\omega} \tag{5}$$

and

$$U_j = U_j \gamma_j + U_{j-1} \gamma_j - A \binom{s}{j} \frac{s(1-\gamma_j) - j}{s(1-\omega) - j} \quad (j = 1, 2, \ldots, s-1);$$

that is,

$$U_j = \frac{\gamma_j}{1-\gamma_j} U_{j-1} - \frac{A}{1-\gamma_j} \binom{s}{j} \frac{s(1-\gamma_j) - j}{s(1-\omega) - j} \quad (j = 1, 2, \ldots, s-1), \tag{6}$$

where γ_j is defined by (14.15).

d. To solve the difference equations (6), show that

$$\frac{U_j}{C_j} = \frac{U_{j-1}}{C_{j-1}} - \frac{A}{C_j(1-\gamma_j)} \binom{s}{j} \frac{s(1-\gamma_j) - j}{s(1-\omega) - j}, \tag{7}$$

where $C_0 = 1$ and C_j $(j \geq 1)$ is defined by (14.16). Add these equations for $j = i+1, \ldots, s-1$, and, using the fact that $U_{s-1} = \Pi_{s-1} = A\omega^{-1}$, obtain

$$\frac{U_i}{C_i} = A \sum_{j=i+1}^{s} \frac{1}{C_j(1-\gamma_j)} \binom{s}{j} \frac{s(1-\gamma_j) - j}{s(1-\omega) - j} \quad (i = 0, 1, \ldots, s-1). \tag{8}$$

Now set $i = 0$ in (8), and show that (8) and (5) together imply (14.14).

e. To obtain the unknown probabilities $\Pi_0, \Pi_1, \ldots, \Pi_{s-2}$, show that

$$U(z) = \sum_{j=0}^{s-1} U_j (z-1)^j; \tag{9}$$

and use the fact that

$$\Pi_j = \frac{1}{j!} \left(\frac{d^j U(z)}{dz^j} \right)_{z=0} \tag{10}$$

to deduce the final result, namely,

$$\Pi_j = \sum_{i=j}^{s-1} (-1)^{i-j} \binom{i}{j} U_i \quad (j=0,1,\ldots,s-2), \tag{11}$$

where U_i is defined by (8).

5.15. The $GI/M/s$ Queue with Service in Random Order

The $GI/M/s$ random-service queue has been studied by LeGall [1962], Takács [1962b], Burke [1967, unpublished], and Carter and Cooper [1972]. LeGall gives an expression for the characteristic function of the waiting-time distribution function; Takács gives a similarly complicated expression for the corresponding Laplace-Stieltjes transform. Their results are quite formidable, and reduction to practice (through either mathematical or numerical inversion) does not appear simple.

As we did in Section 5.12 for the $M/G/1$ queue with service in random order, we shall here derive algorithms (the Maclaurin-series method and the additional-conditioning-variable method) for the numerical computation of the complementary conditional waiting-time distribution function $P\{W>t|W>0\}$ for the $GI/M/s$ random-service queue. Then the unconditional waiting-time distribution can be found from the relation

$$P\{W>t\} = P\{W>t|W>0\}P\{W>0\}, \tag{15.1}$$

where $P\{W>0\}$ is given, for any nonbiased queue discipline, by (14.17).

Let N be the number of customers present when an arbitrary customer (the test customer) arrives, and define

$$W_j(t) = P\{W>t|N=s+j\} \quad (j=0,1,\ldots). \tag{15.2}$$

The probability that the test customer's waiting time W will exceed t can be written

$$P\{W>t\} = \sum_{i=0}^{\infty} P\{W>t|N=i\}P\{N=i\};$$

that is,

$$P\{W>t\} = \sum_{j=0}^{\infty} W_j(t)P\{N=s+j\}, \tag{15.3}$$

where $P\{N=s+j\} = P\{Q=j, W>0\} = A\omega^j$, as given by (14.18). Hence, it follows from (15.1) and (15.3), together with (14.17) and (14.18), that

$$P\{W>t|W>0\} = (1-\omega)\sum_{j=0}^{\infty} \omega^j W_j(t). \tag{15.4}$$

[Note that (15.4), which is valid for any nonbiased queue discipline, is the analogue of the result of Exercise 30 of Chapter 3. In particular, when service is in order of arrival, then

$$W_j(t) = \sum_{k=0}^{j} \frac{(s\mu t)^k}{k!} e^{-s\mu t},$$

and (15.4) yields (14.20).] It remains to calculate $W_j(t)$ for service in random order.

We shall first give a heuristic derivation of the Maclaurin series expansion of the conditional probability $P\{W>t|W>0\}$, and show that our results include Equation (6) of Exercise 32 of Chapter 3 (for the $M/M/s$ random-service queue) as a special case. This method, which we call the Maclaurin-series method, results in an algorithm that may be computationally useful in some cases.

We begin with the *Maclaurin-series method* (Carter and Cooper [1972]). Consider a test customer who, upon arrival at time T_c, say, finds all s servers busy and $j > 0$ other customers waiting for service. Then one of the following two events must occur: (1) the next customer arrives after time $T_c + t$, or (2) the next customer arrives prior to time $T_c + t$.

In case (1), the test customer will wait more than t for service to begin if and only if he is one of the i ($1 \leq i \leq j+1$) customers still waiting for service at time $T_c + t$. (If $i = 0$, the test customer will necessarily have begun service.) Since service times are assumed to be independently exponentially distributed with common mean μ^{-1}, the probability $p_k(x)$ that exactly k customers complete service in an interval of length x, given that all s servers are continuously busy throughout this interval, is the Poisson probability

$$p_k(x) = \frac{(s\mu x)^k}{k!} e^{-s\mu x} \quad (k=0,1,\dots). \tag{15.5}$$

Now N is the number of customers present when the test customer arrives at T_c; let $T_{c'}$ be the next arrival epoch after T_c. If $T_{c'} - T_c > t$, then the (conditional) probability that the test customer waits in excess of t for service to commence is

$$P\{W>t|N=s+j, T_{c'}-T_c>t\} = \sum_{i=1}^{j+1} \frac{i}{j+1} p_{j+1-i}(t),$$

where the factor $i/(j+1)$ is the probability that the test customer is not among the $j+1-i$ customers selected (according to the random selection procedure) to begin service during the interval $(T_c, T_c + t)$. Since the interarrival-time distribution function is $G(x)$, and since T_c is an arrival epoch, event (1) occurs with probability $1 - G(t)$.

Now consider event (2); that is, suppose that the next customer arrives at time $T_{c'} = T_c + \xi$, where $\xi \leq t$. The probability that the test customer will be among the remaining i waiting customers is

$$\frac{i}{j+1} p_{j+1-i}(\xi).$$

The test customer will now experience a total wait in excess of t for service to begin if and only if he suffers an additional delay exceeding length $t - \xi$. But since waiting customers are selected for service in random order, the probability that the test customer's additional waiting time will exceed $t - \xi$ (if he has not yet begun service) is the same as the probability that a new arrival waits in excess of $t - \xi$ for service to begin. The latter probability is $W_i(t - \xi)$. Thus, if the next customer arrives at time $T_{c'} = T_c + \xi$, where $\xi \leq t$, then the test customer's waiting time will exceed t with probability

$$P\{W > t | N = s + j, T_{c'} - T_c = \xi \leq t\} = \sum_{i=1}^{j+1} \frac{i}{j+1} p_{j+1-i}(\xi) W_i(t - \xi). \quad (15.6)$$

Finally, the probability that the next arrival epoch $T_{c'}$ will occur in an infinitesimal interval about the point $T_c + \xi$ is $dG(\xi)$.

Therefore, combining events (1) and (2), we have the following recurrence for the conditional probability $W_j(t)$ that the test customer waits in excess of t for service to begin, given that on arrival he finds all s servers busy and $j \geq 0$ other customers waiting for service:

$$W_j(t) = [1 - G(t)] \sum_{i=1}^{j+1} \frac{i}{j+1} p_{j+1-i}(t)$$

$$+ \sum_{i=1}^{j+1} \frac{i}{j+1} \int_0^t p_{j+1-i}(\xi) W_i(t - \xi) dG(\xi) \quad (j = 0, 1, \ldots), \quad (15.7)$$

where $G(x)$ is the interarrival-time distribution function and $p_k(x)$ is given by Equation (15.5).

We now assume that $W_j(t)$ for the $GI/M/s$ queue has the Maclaurin-series representation

$$W_j(t) = \sum_{\nu=0}^{\infty} \frac{t^\nu}{\nu!} W_j^{(\nu)} \quad (j = 0, 1, \ldots; \quad W_j^{(0)} = 1). \quad (15.8)$$

Equation (15.8) is the same as Equation (4) of Exercise 32 of Chapter 3, except that (15.8) refers to the $GI/M/s$ queue instead of the $M/M/s$

queue. In analogy with Equation (5) of that exercise for the $M/M/s$ queue, we have the corresponding equation for the $GI/M/s$ queue:

$$P\{W>t|W>0\} = 1+(1-\omega)\sum_{\nu=1}^{\infty}\frac{t^{\nu}}{\nu!}\sum_{j=0}^{\infty}\omega^{j}W_{j}^{(\nu)}. \qquad (15.9)$$

It remains to determine the derivatives $\{W_j^{(\nu)}\}$ appearing on the right-hand side of Equation (15.9) from the basic recurrence (15.7).

Differentiating ν times on both sides of Equation (15.7), we have

$$\frac{d^{\nu}}{dt^{\nu}}W_j(t) = \frac{d^{\nu}}{dt^{\nu}}\left([1-G(t)]\sum_{i=1}^{j+1}\frac{i}{j+1}p_{j+1-i}(t)\right)$$

$$+\sum_{i=1}^{j+1}\frac{i}{j+1}\int_0^t p_{j+1-i}(\xi)g(\xi)\frac{\partial^{\nu}}{\partial t^{\nu}}W_i(t-\xi)\,d\xi$$

$$+\sum_{i=1}^{j+1}\frac{i}{j+1}\sum_{k=0}^{\nu-1}\frac{d^k}{dt^k}[p_{j+1-i}(t)g(t)]W_i^{(\nu-1-k)}$$

$$(j=0,1,\ldots;\quad \nu=1,2,\ldots), \qquad (15.10)$$

where $g(\xi)$ is the interarrival-time density function,

$$g(\xi) = \frac{d}{d\xi}G(\xi). \qquad (15.11)$$

For convenience we set

$$a_{j+1}(t) = [1-G(t)]\sum_{i=1}^{j+1}\frac{i}{j+1}p_{j+1-i}(t) \qquad (15.12)$$

and

$$b_{j+1-i}(t) = g(t)p_{j+1-i}(t). \qquad (15.13)$$

Now set $t=0$ in Equation (15.10). The integral on the right-hand side vanishes and we have

$$W_j^{(\nu)} = a_{j+1}^{(\nu)} + \sum_{i=1}^{j+1}\frac{i}{j+1}\sum_{k=0}^{\nu-1}b_{j+1-i}^{(k)}W_i^{(\nu-1-k)}$$

$$(j=0,1,\ldots;\quad \nu=1,2,\ldots). \qquad (15.14)$$

The recurrence (15.14) permits evaluation of the sum on the right-hand side of Equation (15.9). The problem is solved if the series converges and if the assumed derivatives exist. If, in addition, the terms of the series are easy to calculate, the solution is also useful.

One might note the steps taken in the above development that are mathematically questionable. In particular, we assumed the existence of the Maclaurin series expansion (15.8) while ignoring possible effects on convergence of the values of t and ω, the differentiability of $W_j(t)$, and the vanishing of the integral on the right-hand side of (15.10) at $t=0$. It is because of these (not unreasonable) assumptions that we term this derivation "heuristic." The fact that the method gives agreement with an independently obtained result for the special case of Poisson input (see Exercise 36) is encouraging, but proves little. Clearly, work remains to be done.

Exercise

36. Show that for Poisson input this algorithm yields Equation (6) of Exercise 32 of Chapter 3.

An algorithm that does not require that the interarrival-time distribution function $G(x)$ possess a density function has been proposed by P. J. Burke [1967, unpublished]. This method, which is based on the introduction of an additional conditioning (or supplementary) variable, is essentially the same as the algorithm of Burke for the $M/G/1$ random-service queue that we discussed in Section 5.12. According to Burke, this method yields results that appear to be well suited for computation on a digital computer. For the particular case of Poisson input, Burke's method yields a power series representation for $P\{W>t|W>0\}$ that is easily shown to be convergent for all $s\mu t \geq 0$. It follows from the uniqueness property of power series representations that the Maclaurin series (6) of Exercise 32 of Chapter 3 exists and converges for all $s\mu t \geq 0$.

We now briefly describe Burke's *additional-conditioning-variable method* for the $GI/M/s$ random-service queue. The Maclaurin-series method was based on obtaining an expression for $W_j(t) = P\{W>t|N=s+j\}$ for $j=0,1,\ldots$, where W is the waiting time of the test customer, and N is the number of customers present in the system just prior to the test customer's arrival epoch T_c. Burke considers the additional random variable $X(t)$, defined as the number of customers who arrive in $(T_c, T_c + t]$, with $t > 0$, and $X(0) = 0$ with probability 1. Instead of calculating $W_j(t)$, Burke's basic calculation is that of $W_{j,k}(t)$, which we define as

$$W_{j,k}(t) = P\{W>t|N=s+j, X(t)=k\}$$
$$(j=0,1,\ldots;\ k=0,1,\ldots). \quad (15.15)$$

We have that

$$P\{X(t)=k\} = G^{*k}(t) - G^{*(k+1)}(t)$$
$$[k=0,1,\ldots;\ G^{*0}(t)=1;\ t\geq 0], \quad (15.16)$$

where $G^{*k}(t)$ is the k-fold convolution of the interarrival-time distribution function $G(t)$ with itself. Also, from the theorem of total probability,

$$W_j(t) = \sum_{k=0}^{\infty} P\{X(t)=k\} W_{j,k}(t). \tag{15.17}$$

Since the conditional waiting-time distribution function is determined by Equation (15.4), it follows from Equation (15.17) that it now remains only to determine the conditional probabilities $W_{j,k}(t)$.

First note that $X(t)=0$ if and only if $T_{c'} - T_c > t$. It follows from Equation (15.5) and the definition (15.15) of $W_{j,k}(t)$ that

$$W_{j,0}(t) = \sum_{i=1}^{j+1} \frac{i}{j+1} p_{j+1-i}(t) \qquad (j=0,1,\ldots). \tag{15.18}$$

For $k>0$, we define

$$G(\xi|k,t) = P\{T_{c'} - T_c \leq \xi | X(t)=k\}, \tag{15.19}$$

which is the conditional distribution function of the elapsed time $T_{c'} - T_c$ between the arrival epoch T_c of the test customer and the next arrival epoch $T_{c'}$, given that k arrivals occur in $(T_c, T_c+t]$. It follows from the definition of a conditional probability density function that

$$d_\xi G(\xi|k,t) = \frac{P\{X(t-\xi)=k-1\} dG(\xi)}{P\{X(t)=k\}} \qquad (k=1,2,\ldots). \tag{15.20}$$

Then, reasoning in a manner similar to that leading to Equation (15.7), we have the following recurrence for $k \geq 1$:

$$W_{j,k}(t) = \sum_{i=1}^{j+1} \frac{i}{j+1} \int_0^t p_{j+1-i}(\xi) W_{i,k-1}(t-\xi) d_\xi G(\xi|k,t)$$

$$(j=0,1,\ldots; \quad k=1,2,\ldots). \tag{15.21}$$

Thus, using Equations (15.18) and (15.21) one can compute $P\{W>t|W>0\}$ to any desired degree of accuracy.

Complicated as these algorithms for the $GI/M/s$ random-service queue are, they are clearly less complicated than the corresponding algorithms for the $M/G/1$ random-service queue that are given in Section 5.12. This is a consequence of the fact that we analyzed the $GI/M/s$ queue directly from the viewpoint of the arriving customer, whereas our analysis of the $M/G/1$ queue required a "translation" from the viewpoint of the departing customer to that of the arriving customer.

[6]
Simulation of Queueing Models

6.1. Introduction

One can define a *model* as a description that eliminates nonessential detail and captures the essence of the system being described, and *simulation* as a methodology whereby a system is studied by observing the response of its model to artificially generated input. Our interest is in mathematical models of queueing systems, in which the model is a logical, not physical, representation of the system under study. Although the concept of simulation does not depend on the existence of a computer, it is the advent of the high-speed digital computer that has made simulation a practical tool of analysis. Simulation is a subject in its own right, but its application to the analysis of queueing models is of sufficient importance and closeness to queueing theory to warrant a chapter in this text. This chapter gives an introduction to the subject, much more in the nature of a survey than the other chapters of this book. Our intention is to raise the important questions, suggest some answers, and provide references for the reader who wishes to pursue the subject further.

The preceding chapters of this book have all been concerned with the mathematical analysis of queueing models. These models were relatively simple, and presumably could provide adequate descriptions of simple real systems, or simple parts of complicated real systems. The construction of a model to describe a particular system is an art, not a science, and requires consideration of the tradeoff between the amount of detail about the real system incorporated in the model and the simplicity of the model. Roughly speaking, the inclusion of more detail will permit the analyst to draw more accurate inferences about the real system from its model if the analyst can solve the model, whereas the elimination of more detail will result in a greater likelihood that the analyst will, in fact, be able to solve the model, thereby obtaining information about the model from which inference about the real system will be drawn. In other words, the analyst wants to construct a model that strikes a balance between accuracy and tractability. It is impossible to state in general which information about a real system should be included in its model and which should not. Most readers would probably agree, for example, that a model whose purpose is to predict the proportion of patients who will be denied immediate service in a hospital emergency room should not contain information on the positions of the stars (although the arrival rate may well depend on the phase of the moon). Less obvious to the naive observer is the assertion that the type of treatment required is relevant only through the distribution of the lengths of the treatment times. A more sophisticated observer would know that if the arrivals can be said to follow a Poisson process, and if the blocked patients do not wait but instead overflow to another facility, then only the mean of this distribution is relevant to the system's performance. Clearly, the type of model chosen to describe any given real system will depend on the knowledge, insight, and wisdom of the analyst.

Digital-computer simulation permits, in principle, the analysis of arbitrarily complicated models of real systems. Conceptually, it is no more difficult to simulate a complicated model than a simple one. This fact reflects both the strength and the weakness of the methodology. For, on the one hand, simulation provides a method of analysis that is applicable when mathematical analysis is too hard, and it allows the analyst to model a real system in a way that is more complicated (and therefore, presumably, more accurate) than mathematical analysis would permit. On the other hand, all other things being equal, simulation does not yield anywhere near the insight and information that a mathematical model will yield if it can be solved. Because of the generality of simulation methodology (analysis of queueing models is only one of many kinds of applications), and because of the difficulty of obtaining the specialized skills required for the construction and solution of mathematical models, there is a tendency for applied workers automatically to adopt simulation rather than mathematical analysis, regardless of the merits of the different approaches with respect to the particular problem at hand. Of course, the optimal situation is when the analyst has competence in both simulation and mathematics, and can use either one or a combination of the two.

Historically, the earliest use of the concept of simulation is usually credited to Count de Buffon, a French naturalist, who in 1777 stated what is now known as *Buffon's needle problem*: A needle of length a is thrown at random on a plane covered with parallel lines that are all a distance d apart ($a \leqslant d$). What is the probability p that the needle will intersect one of the lines? The answer is $p = 2a/\pi d$. Buffon is supposed to have performed the experiment and, by setting the observed frequency equal to the probability p, to have calculated the value of π. This remarkable idea, the

Introduction

calculation of a deterministic quantity from the data of a random experiment, became a practical methodology with the advent of digital computers to carry out logical operations with artificially generated data at high speed. This technique was first used during the development of the atomic bomb, for the numerical evaluation of integrals, where it was called the *Monte Carlo* method. It is not widely known, however, and therefore is worth mentioning here, that the first application of the concept of simulation for the analysis of systems subject to stochastic demands, such as we are discussing here, was done much earlier. In 1907 J. C. T. Baldwin described, in an engineering report for the American Telephone and Telegraph Company, what is probably the first modern simulation study. Baldwin's study, entitled "Report of Engineering Studies for Semi-Mechanical Switchboards," used artificial data to simulate telephone traffic and determine the loads that could be handled by operators providing service with a given average delay. Baldwin called his technique the " 'Throw Down' Method." Although it did not, of course, use a computer, it embodied the essential concepts of a modern simulation study. The *throwdown* became a standard tool of teletraffic engineers. With the advent of computers, the concept of the throwdown became widely applicable, and the term *simulation* has come to be the accepted one to describe this methodology.

Our objective in this chapter is to provide an overview of the use of simulation to construct and analyze models of real queueing systems. Hence, our discussion will be restricted to *discrete-event digital simulation*, which is concerned with models of systems whose states make only discrete changes as time passes. We shall try to illustrate the relative strengths and weaknesses of simulation as opposed to mathematical analysis of queueing models. We assume that the reader is familiar with the basic concepts of computers and computer programming.

A simulation model is a description of the structure of the system under study. When the simulation program is run, the computer generates arrival times, service times, and any other stochastic variables that correspond to the demand on the system. In other words, each simulation run is an experiment in which artificial, computer-generated input data are fed through a logical structure whose response to the input data "simulates" the response that the real system would make to the same input data. Hence, the computer must have the capability of generating input data with any prescribed statistical characteristics, which are supposed to mimic the input to the real system. That is, digital-computer simulation requires algorithms for generating realizations of customer arrival processes and service times, such as Poisson input and exponential service times.

Of course, before a simulation run can be made, the simulation program must be written. The details of the programming depend on the type of computer, its operating system, and the programming language used. In particular, the analyst may have a choice between a general-purpose programming language, such as FORTRAN, and a simulation programming language, such as GPSS, SIMSCRIPT, or GASP.

Since a simulation run is an experiment in which the response (*output*) to (artificial) data (*input*) is observed, this output will itself be statistical in nature, and will therefore require some statistical analysis for its interpretation. This is the most critical part in the design and implementation of a simulation model, because it is often difficult to draw general conclusions and gain real insight (as opposed to merely noting what happened in a particular case) from output data.

We shall discuss each of these issues in turn, emphasizing those points most relevant to simulation of queueing models, and referring the reader to other sources for further information on simulation in general.

6.2. Generation of Stochastic Variables

Suppose we wish to generate independent realizations of a random variable X whose distribution function is $F_X(x)$; that is, we seek a computer algorithm to select independent samples from a given distribution. As we shall see, it is sufficient to be able to generate values of a random variable U that is uniformly distributed on $(0, 1)$.

First, observe that $F_X(x)$ is a nondecreasing function of x, on a finite or infinite interval. For $0 \leq u \leq 1$ the "inverse" function $F_X^{-1}(u)$ is defined as follows: If $F_X(x)$ has a discontinuity at $x = x_0$ and $F_X(x_0-) < u \leq F_X(x_0)$, then define $F_X^{-1}(u) = x_0$; otherwise, let $F_X^{-1}(u)$ be the realizable value of x for which $F_X(x) = u$. Now define X according to the transformation

$$X = F_X^{-1}(U), \qquad (2.1)$$

where U is a random variable uniformly distributed on $(0, 1)$. We now show that if the realization of X is generated by first realizing the random variable U and then calculating the value of X from (2.1), it will follow that for all x,

$$P\{X \leq x\} = F_X(x); \qquad (2.2)$$

that is, the values of X calculated according to this *inverse-transformation method* will be samples from a population whose distribution function is $F_X(x)$. To prove (2.2), note that

$$P\{X \leq x\} = P\{F_X^{-1}(U) \leq x\} = P\{U \leq F_X(x)\}, \qquad (2.3)$$

where the first equality in (2.3) follows from (2.1). Now recall from Equation (5.9) of Chapter 2 that if U is uniform on $(0, 1)$, then

$$P\{U \leq u\} = F_U(u) = u \qquad (0 \leq u \leq 1). \qquad (2.4)$$

Generation of Stochastic Variables

It follows from (2.4) that the right-hand side of (2.3) equals $F_X(x)$; hence (2.2) has been proved.

For example, suppose it is required to generate a value of a random variable X that is exponentially distributed with mean value τ. Then

$$F_X(x) = 1 - e^{-x/\tau} \qquad (0 \leqslant x < \infty), \tag{2.5}$$

and therefore we take

$$U = F_X(X) = 1 - e^{-X/\tau};$$

solving this for X, as in (2.1), we get

$$X = -\tau \ln(1 - U). \tag{2.6}$$

We leave it to Exercise 1 for the reader to show that if U is uniform on $(0,1)$, then so is $1 - U$. Therefore, we can replace (2.6) by the simpler calculation

$$X = -\tau \ln U. \tag{2.7}$$

Hence, we conclude from (2.7) that to obtain a realization x of a random variable X that is exponentially distributed with mean τ as in (2.5), it suffices to obtain a realization u of a random variable U that is uniform on $(0, 1)$, and then calculate x from the formula

$$x = -\tau \ln u. \tag{2.8}$$

In the simulation of queueing models, exponentially distributed random variables play a central role. In particular, a Poisson arrival stream can be generated by generating the (independent, identical) exponentially distributed interarrival times. Similarly, various other distributions can be produced by the method of phases (see Section 4.8), according to which the random variable in question is assumed to be composed of a sum of independent exponential phases. (An example of this procedure will be discussed in Section 6.5.)

Exercises

1. Show that if U is uniform on $(0, 1)$, then so is $1 - U$.

2. Let X have the Erlangian distribution function of order n,

$$F_X(x) = 1 - \sum_{j=0}^{n-1} \frac{(\lambda x)^j}{j!} e^{-\lambda x}.$$

If $U_1, U_2, \ldots, U_n$ are n independent samples, each uniformly distributed on $(0,1)$, show that

$$-(1/\lambda)\ln(U_1 \cdots U_n)$$

has the same distribution as X.

It is convenient to consider separately the special case when X is a discrete random variable, with probability distribution $P\{X = x_i\} = p_i$ ($i = 1, 2, \ldots, n$). To generate a sample from this distribution, again simply generate a value u of a random variable U that is uniform on $(0, 1)$, and take $X = x_i$ if

$$\sum_{j=1}^{i-1} p_j < u < \sum_{j=1}^{i} p_j \qquad (i = 1, 2, \ldots, n), \tag{2.9}$$

where the empty sum is taken to be zero. For example, suppose we wish to determine whether the next arrival occurs before or after the completion of service of a customer presently being served. If the customers arrive according to a Poisson process with rate λ, and the service times are exponentially distributed with mean μ^{-1}, then the probability that the arrival occurs before the service completion is $\lambda/(\lambda+\mu)$. Therefore, we say that the arrival occurs first if the realization u of the uniform random variable U satisfies $0 < u < \lambda/(\lambda+\mu)$, and the service completion occurs first if $\lambda/(\lambda+\mu) < u < 1$.

We have shown that if we can generate values of a random variable U that is uniform on $(0, 1)$, then we can calculate from these values another set of values that represent samples from any distribution we want. Therefore, we need concern ourselves only with the question of how to generate the values of U. Fortunately, there are in existence easily accessible subroutines that generate values that are uniformly distributed on $(0, 1)$. These subroutines, which are called *random-number generators*, are generally included in scientific subroutine packages and simulation programming languages. Thus, they ought to be tightly written and well tested. (However, reliance on this supposition can be costly.) It is interesting to note that most random-number generators use algorithms that produce a deterministic string of numbers with the property that they appear to be random, in that they can pass statistical tests for randomness and independence. Because these "random" numbers are in fact a deterministic sequence, they are often referred to as *pseudorandom numbers*.

Although the inverse-transformation method is, in principle, always applicable, it is not necessarily the most practical way to generate sample values from a particular given distribution. For more information on random-number generation and its use in the inverse-transformation method, and on other methods of stochastic-variable generation, see Cheng and Feast [1980], Fishman [1973, 1978a], Knuth [1969], Lewis [1975], Lewis and Shedler [1978], Schmeiser [1977], and Yakowitz [1977].

6.3. Simulation Programming Languages

Simulation programming languages are high-level problem-oriented languages that have been designed to facilitate the writing of simulation programs. They contain language statements and subroutines for certain procedures that are common to simulation programs, like random-variable generation and data collection, that are not as readily available in general-purpose languages such as FORTRAN, PASCAL, and PL/1. General-purpose languages are more flexible and can produce programs that run more efficiently than those written in simulation languages. On the other hand, it is much simpler to write a simulation in a simulation language. The choice of the type of programming language, and the particular language within that type, will depend on several factors, such as the availability and the programmer's knowledge of the various languages, as well as the characteristics of the model to be simulated.

The most widely used simulation language is probably GPSS (*G*eneral *P*urpose *S*imulation *S*ystem), a language that is particularly well suited for queueing problems. SIMSCRIPT is a widely available simulation language that is particularly comprehensive. GPSS compares favorably with SIMSCRIPT in that it is simpler; on the other hand, it is less comprehensive. Neither is uniformly better than the other. GASP (*G*eneral *A*ctivity *S*imulation *P*rogram), which is a package of FORTRAN subroutines specially written for simulation, has the advantage that it provides some of the benefits of a simulation language without requiring a programmer who knows FORTRAN to learn a completely new language. (Also, GASP IV has the capability to handle models whose states change continuously, not just discretely, in time.)

The subject of simulation languages is a substantial one, far removed from queueing theory; the best we can do here is alert the reader to the existence of this body of knowledge and identify some of the major references. Fishman [1973, 1978a] discusses simulation programming languages in general and gives detailed descriptions of GPSS, SIMSCRIPT, and SIMULA. GPSS is described in Schriber [1974], in Bobillier, Kahan, and Probst [1976], and, by its originator, in Gordon [1975], and again, along with SIMSCRIPT, in Gordon [1978]. SIMSCRIPT is described by its developers in Kiviat, Villanueva, and Markowitz [1973]. GASP IV is described by its originator in Pritsker [1974], and Pritsker and Pegden [1979] describe SLAM, which extends GASP IV to include queueing network models. Mention should be made of RESQ, which is a software package, consisting of an analytic part and a simulation part, that is designed specifically for the analysis of queueing networks of the type that arises in modeling for computer performance evaluation. RESQ is described in Sauer and MacNair [1977, 1978] and Reiser and Sauer [1978], but the program itself is presently available only within the IBM company. A good survey paper on simulation with some discussion of programming languages is Shannon [1975].

6.4. Statistical Questions

A simulation run is a statistical experiment in which the response of a logical model to artificial input data is observed. It differs from an ordinary statistical experiment, such as observing the response of a real system to real input, in that the simulation experimenter has more control over the input data and the model. That is, in a simulation experiment, the analyst can generate the input to have any specified statistical characteristics; likewise, the model can (and should) be reduced to its essential structure, without the extraneous factors that complicate understanding but have no real effect. The first problem, of course, is for the analyst to constuct a model that is optimally "clean" and yet captures the essence of the real system under study. Assuming now that the appropriate model has been constructed, the problem is to draw inference from the simulation output about the response of the model to the input data. The design and interpretation of any statistical experiment is a challenging task. In the case of a simulation experiment, the fact that the analyst has a cleaner model and greater control over the input would seem to promise that there should be a corresponding increase in the quality of the inference that can be drawn from the output data. One would expect, therefore, that simulation experiments will be more efficient and informative than real experiments with respect to drawing inference (about the model) from statistical analysis of the output data. It is probably fair to say that the general attempt to realize the potential of highly efficient and informative simulation output promised by the high degree of control over the simulation model and input has not yet been realized. This is an area of active research, but one that has not yet been especially productive in terms of useful procedures for practitioners. In this section we shall discuss some of these statistical questions (but, unfortunately, not much in the way of answers).

In order to illuminate the kinds of statistical questions that arise, let us consider a structurally simple model about which we already know, through mathematical analysis, (almost) everything, namely, the Erlang loss model. That is, for the sake of example, we consider the model characterized by Poisson arrivals, s servers, and blocked customers cleared. It is a simple matter to write a program, even in FORTRAN, to generate Poisson arrivals with any given rate and service times from any given distribution, and to keep records of the number of customers blocked and the proportion of simulated time that all servers are busy.

Since we must make an assumption about the distribution of service times, let us make the simplest assumption, namely, constant service times; and suppose that we are interested in ascertaining only the equilibrium load-loss relationship, that is, the fraction of arriving customers who find all servers busy, as a function of the offered load, after statistical equilibrium is attained.

The first question is obvious: How long will it take for statistical equilibrium to be reached? In theory, of course, statistical equilibrium will never be "reached," only approached asymptotically (unless the system is in equilibrium at time zero, which is impossible in the context of a simulation that must begin in some specified state, such as all servers idle). Unfortunately, little more can be said without the help of mathematical analysis (whose availability would probably preclude the necessity for the simulation). The practitioner's solution is simple: Ignore the first 100 (or 1,000 or 10,000) arrivals in the final calculations.

Now suppose, for the sake of argument, that we have determined how many arrivals are necessary to bring the simulation model to equilibrium. How many additional arrivals should be simulated to ensure that the observed fraction of blocked customers will be close to the "true" probability of blocking? In other words, how large should the sample size be? And is it better to make one long run or several short runs?

Let's suppose that we have answered these questions (as they must be answered, whether rationally or not, in every actual simulation). What have we got? If we call the service time unity, and generate the arrivals according to a Poisson process with rate λ [using (2.7) with $\tau = \lambda^{-1}$ to generate the interarrival times, for example], can we say that the observed fraction of blocked customers is an estimate of $B(s,\lambda)$? Or is it an estimate of $B(s,\hat{\lambda})$, where $\hat{\lambda}$ is the realized value of the arrival rate, that is, the ratio of the total number of arrivals generated to the total duration of simulated time? (Although λ and $\hat{\lambda}$ should be close, it is unlikely that they will be exactly equal.)

Thus, we see that the numerical value generated by the simulation confounds several types of error, including (1) error caused by waiting only a finite length of time before declaring the system to be in statistical equilibrium (because no real system can ever attain statistical equilibrium), (2) sampling error caused by generating only a finite number of arrivals after equilibrium has allegedly been attained, and (3) error caused by any departure from randomness of the pseudorandom-number generator. If we adopt the pragmatic point of view that we have run our simulation long enough and have a pseudorandom-number generator that is good enough so that these errors are all negligible, then we will have obtained a single point on the load-loss curve. It is an easy matter, but perhaps quite expensive, to make enough runs to obtain enough points to construct the load-loss curve—for one particular value of s. To obtain a set of curves, such as those given in the appendix, would require repetition of this process for each value of s. (Actually, in the present example of the Erlang loss system, one could alleviate this problem somewhat by recognizing that each realization for s servers contains another realization for $s-1$ servers. However, the results for each value of s would depend on the same stream of pseudorandom numbers, so that any error in this regard would have multiple effects.)

But suppose now that instead of wanting to obtain the load-loss curves for the case of constant service times only, we also wanted to consider exponential service times. One could then use the same simulation program as before, except that now the service times would be generated according to (2.7) instead of assigning each service time the same (constant) value. This would increase the variability of the response and thus, it seems reasonable to suppose, would require a larger sample size to produce results of equal accuracy. But, no matter how great the care and expense of the simulation, it is highly unlikely that the load-loss curves for exponential service times would be exactly the same as those for constant service times. Simulation could never tell us with absolute certainty that, in fact, the two sets of curves should be exactly the same and moreover, that the load-loss curves should be identical no matter what kind of service-time distribution we have.

In a similar vein, a simulation analysis would almost certainly give different values for the proportion of customers who are blocked and the proportion of time all servers are busy, even though these two probabilities are equal. These examples illustrate the crudity of simulation analysis as compared to mathematical analysis; simulation should be used only when mathematical analysis is not feasible.

One promising technique for increasing the efficiency of simulation is called the *regenerative method*. The idea is to identify an infinitely recurring state with the property that the future evolution of the system from this state is independent of the system's past history. The evolution of the system can then be viewed as consisting of a sequence of statistically independent and identical cycles, each cycle bounded by a return to this regenerative state. This method, when applicable, provides insight and some of the answers to the questions raised above. The power of the method lies in the fact that it creates a sequence of independent, identically distributed random variables from correlated random variables, and thereby permits application of the *central limit theorem* (see any textbook in statistics) and its variants.

According to the simplest version of the central limit theorem, if $R_1, R_2, \ldots$ are independent, identically distributed random variables, with common mean $E(R)$ and variance $V(R)$, both finite, and if $\overline{R}(n) = (R_1 + \cdots + R_n)/n$ is the average value of a sample of size n, then

$$P\left\{ \frac{\overline{R}(n) - E(R)}{\sqrt{V(R)/n}} \leq t \right\} \to \Phi(t) \tag{4.1}$$

as $n \to \infty$, where $\Phi(t)$ is the *standard normal* distribution function,

$$\Phi(t) = \int_{-\infty}^{t} \frac{1}{\sqrt{2\pi}} e^{-x^2/2} dx. \tag{4.2}$$

The standard normal distribution function $\Phi(t)$ is widely tabulated, and therefore, if the sample size n is sufficiently large so that the arrow in (4.1) can be replaced by an equals sign, useful information can be obtained about the relationship between the size of a sample and the accuracy of some inference made from the sample values.

We shall illustrate the use of the regenerative method and the application of the central limit theorem by considering a simulation analysis of the simple single-server Erlang loss model. Our objective is to estimate the probability Π that an arriving customer will find the server busy. Let $\hat{\Pi}(n)$ be the observed ratio of the number of arrivals who are blocked to the total number of arrivals in n cycles. Clearly, the state in which there are no customers present is a regenerative state. Let M_i be the observed number of customers blocked during the ith cycle; that is, M_i is the number of customers who arrive during the ith service time. Then, by definition,

$$\hat{\Pi}(n) = \frac{M_1 + \cdots + M_n}{n + M_1 + \cdots + M_n}; \qquad (4.3)$$

that is,

$$\hat{\Pi}(n) = \frac{\overline{M}(n)}{1 + \overline{M}(n)}, \qquad (4.4)$$

where $\overline{M}(n) = (M_1 + \cdots + M_n)/n$ is the average (over n cycles) number of customers blocked per cycle. If we denote the distribution function of $\hat{\Pi}(n)$ by $F_{\hat{\Pi}(n)}(t)$,

$$F_{\hat{\Pi}(n)}(t) = P\{\hat{\Pi}(n) \leqslant t\}, \qquad (4.5)$$

then it follows from (4.4) and (4.5) that

$$F_{\hat{\Pi}(n)}(t) = P\left\{\overline{M}(n) \leqslant \frac{t}{1-t}\right\}. \qquad (4.6)$$

Now, according to the central limit theorem, the random variable

$$\frac{\overline{M}(n) - E(M)}{\sqrt{V(M)/n}}$$

has the standard normal distribution when n is sufficiently large. It follows from Exercise 21 of Chapter 2 that

$$E(M) = \lambda \tau \qquad (4.7)$$

and
$$V(M) = \lambda\tau + \lambda^2\sigma^2, \tag{4.8}$$

where λ is the rate of the Poisson input process, and τ and σ^2 are the mean and variance, respectively, of the service-time distribution function. Hence, we can write

$$P\left\{ \frac{\overline{M}(n) - a}{\sqrt{\frac{1}{n}(a + \lambda^2\sigma^2)}} \leqslant x \right\} = \Phi(x), \tag{4.9}$$

where $a = \lambda\tau$ is the offered load; that is,

$$P\left\{ \overline{M}(n) \leqslant a + x\sqrt{\frac{1}{n}(a + \lambda^2\sigma^2)} \right\} = \Phi(x). \tag{4.10}$$

Finally, comparison of (4.6) and (4.10) yields

$$F_{\hat{\Pi}(n)}(t) = \Phi\left[\frac{\frac{t}{1-t} - a}{\sqrt{\frac{1}{n}(a + \lambda^2\sigma^2)}} \right]. \tag{4.11}$$

Equation (4.11) gives some interesting information about how the sample size n or the service-time variance σ^2 affects the accuracy of the inferences that can be drawn from the simulation output. For example, suppose we wish to compare the sample sizes required to estimate Π by $\hat{\Pi}(n)$ with the same degree of accuracy in the cases of two different service-time distribution functions, with variances σ_1^2 and σ_2^2, say. According to (4.11), the statistical properties of the estimator $\hat{\Pi}(n)$ will be the same in both cases if the argument of the standard normal distribution function Φ, on the right-hand side of (4.11), is the same for all t. In that case, it follows that

$$\frac{1}{n_1}(a + \lambda^2\sigma_1^2) = \frac{1}{n_2}(a + \lambda^2\sigma_2^2), \tag{4.12}$$

where n_1 and n_2 are the corresponding sample sizes. In particular, if we compare constant service times ($\sigma_1^2 = 0$) with exponential service times ($\sigma_2^2 = \tau^2$), then (4.12) implies that

$$n_2 = (1 + a)n_1. \tag{4.13}$$

According to (4.13), if $a = 1$ erlang, for example, then it requires exactly twice as many cycles to simulate the exponential-service-time model as it does to simulate the constant-service-time model with the same degree of accuracy (even though the true value of Π is the same in both cases).

If the analyst knew that the value of Π is the same for every service-time distribution function, but did not know how to find Π except through simulation, then he could conclude from (4.12) that he should restrict his attention to the case of constant service times, because it results in the minimum variance of the estimator $\hat{\Pi}(n)$. (This idea, choosing an estimator that has a smaller variance than another estimator for the same parameter, is called a *variance-reduction* technique.)

We can use (4.11) to investigate how accurately the statistic $\hat{\Pi}(n)$ estimates the true probability Π. In particular, if we consider the interval of half-length δ centered at Π, then the probability that the estimator $\hat{\Pi}(n)$ lies within this interval can be calculated from (4.11):

$$P\{\Pi-\delta < \hat{\Pi}(n) < \Pi+\delta\}$$

$$= \Phi\left[\frac{\frac{\Pi+\delta}{1-(\Pi+\delta)}-a}{\sqrt{\frac{1}{n}(a+\lambda^2\sigma^2)}}\right] - \Phi\left[\frac{\frac{\Pi-\delta}{1-(\Pi-\delta)}-a}{\sqrt{\frac{1}{n}(a+\lambda^2\sigma^2)}}\right]. \quad (4.14)$$

In our example we know that

$$\Pi = B(1,a) = \frac{a}{1+a}; \quad (4.15)$$

it is instructive to use (4.14) and (4.15) to investigate numerically how the sample size n and the service-time variance σ^2 affect the accuracy of the estimator $\hat{\Pi}(n)$. (See, for example, Exercise 4.)

Another interesting question is whether it is better to estimate Π by $\hat{\Pi}(n)$ or $\hat{P}(n)$, where the latter is defined as the observed proportion of simulated time that the server is busy:

$$\hat{P}(n) = \frac{X_1 + \cdots + X_n}{Y_1 + \cdots + Y_n + X_1 + \cdots + X_n}, \quad (4.16)$$

where X_i and Y_i are the observed values of the service time and the idle time, respectively, of the ith cycle. If we define the average values $\bar{X}(n) = (X_1 + \cdots + X_n)/n$ and $\bar{Y}(n) = (Y_1 + \cdots + Y_n)/n$, then (4.16) becomes

$$\hat{P}(n) = \frac{\bar{X}(n)}{\bar{Y}(n) + \bar{X}(n)}. \quad (4.17)$$

We leave it to Exercise 3 for the reader to show that the statistic $\hat{P}(n)$ has distribution function

$$F_{\hat{P}(n)}(t) = \Phi\left[\frac{\dfrac{t}{1-t} - a}{\sqrt{\dfrac{1}{n}\left[\left(\dfrac{t}{1-t}\right)^2 + \lambda^2\sigma^2\right]}}\right]. \qquad (4.18)$$

Equation (4.18), which is the analogue of (4.11), leads to the analogue of (4.14):

$$P\{\Pi - \delta < \hat{P}(n) < \Pi + \delta\}$$

$$= \Phi\left[\frac{\dfrac{\Pi+\delta}{1-(\Pi+\delta)} - a}{\sqrt{\dfrac{1}{n}\left[\left(\dfrac{\Pi+\delta}{1-(\Pi+\delta)}\right)^2 + \lambda^2\sigma^2\right]}}\right]$$

$$-\Phi\left[\frac{\dfrac{\Pi-\delta}{1-(\Pi-\delta)} - a}{\sqrt{\dfrac{1}{n}\left[\left(\dfrac{\Pi-\delta}{1-(\Pi-\delta)}\right)^2 + \lambda^2\sigma^2\right]}}\right]. \qquad (4.19)$$

Numerical investigation of (4.14) and (4.19) shows that neither of the statistics $\hat{\Pi}(n)$ and $\hat{P}(n)$ is uniformly superior to the other. For example (see Exercise 4), when $n = 100$ cycles and the service times are constant, it turns out that

$$P\{\Pi - 0.05 < \hat{\Pi}(100) < \Pi + 0.05\} = P\{\Pi - 0.05 < \hat{P}(100) < \Pi + 0.05\} = 95\%$$

when $a = 1$; when $a < 1$, then

$$P\{\Pi - 0.05 < \hat{\Pi}(100) < \Pi + 0.05\} < P\{\Pi - 0.05 < \hat{P}(100) < \Pi + 0.05\};$$

and when $a > 1$, then

$$P\{\Pi - 0.05 < \hat{\Pi}(100) < \Pi + 0.05\} > P\{\Pi - 0.05 < \hat{P}(100) < \Pi + 0.05\}.$$

In other words, when $a = 1$, then the estimators $\hat{\Pi}(100)$ and $\hat{P}(100)$ are equally likely (about 95%) to lie within $\delta = 0.05$ of the true value of Π;

whereas Π is more accurately estimated by $\hat{P}(100)$ than by $\hat{\Pi}(100)$ when $a < 1$, and vice versa when $a > 1$ (at least, when $\sigma^2 = 0$ and $\delta = 0.05$).

The statistical problems associated with the simulation of queueing models are quite similar to those associated with traffic measurement and performance measurement of real systems. For example, suppose that it is required to determine by measurement the loss probability of a real system. Is it better to estimate this loss probability by measuring the proportion of arrivals that are lost or the proportion of time that all servers are busy? If $s = 1$ and we are willing to agree that the arrivals follow a Poisson process and the service times are constant, then the above results imply that we should prefer $\hat{\Pi}$ to $\hat{P}$ if it appears that $a > 1$, and vice versa if it appears that $a < 1$. Here, even though we may know that (4.15) describes this system, we presumably don't know the value of the offered load a; and rather than measure a and then calculate Π from (4.15), we measure Π directly. This question of whether $\hat{\Pi}$ or $\hat{P}$ is a better estimator of Π is discussed from the viewpoint of traffic measurement by Descloux [1965] for the case where the system being measured can be described by an s-server Erlang loss model with exponential service times. It is interesting that Descloux used simulation to verify an approximate formula he derived for the variance of the statistic $\hat{\Pi}$. Some related papers on traffic measurement that might be of interest from the viewpoint of simulation are Beneš [1961], Descloux [1973, 1975, 1976], Kuczura and Neal [1972], and Neal and Kuczura [1973].

A more conventional use of the regenerative method is in assessing the accuracy of a parameter's *point estimate* obtained from simulation output. Suppose, for example, that a simulation has been written that calculates a statistic whose value will be taken as an estimate of the mean waiting time in a certain queueing model. The analyst would like to know how close this point estimate is to the "true" mean value $E(W)$ it is supposed to estimate. One way is to construct a *confidence interval*, which is an interval whose endpoints are calculated from the simulation output, and which will contain the true value of the parameter being estimated [in this example, $E(W)$] with a prespecified probability α.

Let N_k be the number of customers who are served during the kth cycle, and let S_k be the sum of the waiting times of those customers, and define $R_k = S_k - N_k E(W)$. Let $\bar{N}(n), \bar{S}(n)$, and $\bar{R}(n)$ be the average values (over n cycles) of these random variables. Then $\bar{R}(n) = \bar{S}(n) - \bar{N}(n)E(W)$, and according to (4.1),

$$P\left\{ \frac{\bar{S}(n) - \bar{N}(n)E(W) - E(S_k - N_k E(W))}{\sqrt{\frac{1}{n} V(S_k - N_k E(W))}} \leq t \right\} \to \Phi(t) \quad (4.20)$$

as $n\to\infty$. It can be shown (using the argument of Section 2.1 for example) that

$$E(W) = \frac{E(S_k)}{E(N_k)}; \qquad (4.21)$$

thus, $E(S_k - N_k E(W)) = 0$ and (4.20) becomes

$$P\left\{\frac{\bar{S}(n) - \bar{N}(n)E(W)}{\sqrt{\frac{1}{n}V(S_k - N_k E(W))}} \leq t\right\} \to \Phi(t). \qquad (4.22)$$

Expanding the variance term in (4.22), we have

$$V(S_k - N_k E(W)) = V(S_k) + E^2(W)V(N_k) - 2E(W)\text{Cov}(S_k, N_k).$$

If we define

$$V(n) = \frac{1}{n-1}\sum_{k=1}^{n}(S_k - \bar{S}(n))^2 + E^2(W)\frac{1}{n-1}\sum_{k=1}^{n}(N_k - \bar{N}(n))^2$$

$$- 2\frac{\bar{S}(n)}{\bar{N}(n)}\frac{1}{n-1}\sum_{k=1}^{n}(S_k - \bar{S}(n))(N_k - \bar{N}(n)), \qquad (4.23)$$

then $V(n)$ is a *consistent estimator* (see any text in statistics) of $V(S_k - N_k E(W))$, and (4.22) remains true when $V(S_k - N_k E(W))$ is replaced by $V(n)$. Thus we can say that, for sufficiently large n,

$$P\left\{\frac{\bar{S}(n) - \bar{N}(n)E(W)}{\sqrt{V(n)/n}} \leq t\right\} = \Phi(t). \qquad (4.24)$$

It follows from (4.24) that

$$P\left\{\frac{\bar{S}(n)}{\bar{N}(n)} - \frac{t}{\bar{N}(n)}\sqrt{\frac{V(n)}{n}} \leq E(W) \leq \frac{\bar{S}(n)}{\bar{N}(n)} + \frac{t}{\bar{N}(n)}\sqrt{\frac{V(n)}{n}}\right\}$$

$$= 2\Phi(t) - 1. \qquad (4.25)$$

Therefore, if t_α is the value of t for which the right-hand side of (4.25) equals α, that is

$$2\Phi(t_\alpha) - 1 = \alpha,$$

then

$$\left(\frac{\bar{S}(n)}{\bar{N}(n)} - \delta(n,\alpha), \frac{\bar{S}(n)}{\bar{N}(n)} + \delta(n,\alpha)\right)$$

is a $100\alpha\%$ confidence interval for $E(W)$, where

$$\delta(n,\alpha) = \frac{\Phi^{-1}\left(\frac{1+\alpha}{2}\right)}{\bar{N}(n)}\sqrt{\frac{V(n)}{n}}. \tag{4.26}$$

Thus, in an n-cycle simulation of a regenerative queue, the mean waiting time $E(W)$ is estimated as the ratio $\bar{S}(n)/\bar{N}(n)$; and for any specified α, the interval defined by

$$\frac{\bar{S}(n)}{\bar{N}(n)} \pm \delta(n,\alpha)$$

will, with probability α, contain the true value of the mean waiting time (if the number n of cycles is sufficiently large).

In summary, it is clear that the regenerative method, when applicable, gives much insight and provides answers to several of the questions posed earlier in this section. (In particular, note that with the regenerative method the question of how many arrivals must be generated before statistical equilibrium is attained is moot.) Unfortunately, as our examples indicate, the method is often difficult to apply to those models that are not simple enough for mathematical analysis. Some interesting progress in this area has been made by Lavenberg and Sauer [1977], who use the regenerative method to estimate confidence intervals for response variables for relatively complicated queueing models, and give sequential stopping rules that terminate a simulation run when the relative width of an estimated confidence interval reaches a prespecified value. This is an area of active and promising research. The interested reader should begin by consulting the aforementioned paper, the following papers, and some of the other papers referenced therein: Carson and Law [1980], Crane and Iglehart [1974a, 1974b, 1975], Crane and Lemoine [1977], Gunther and Wolff [1980], Heidelberger [1977, 1978a, 1978b], Heidelberger and Iglehart [1979], Heidelberger and Lewis [1980], Heidelberger and Welch [1980], Iglehart [1978], Iglehart and Shedler [1978], Lavenberg [1978], Lavenberg and Slutz [1975], and Schruben [1980]. The regenerative method is not the only method for predetermining run length and/or calculating confidence intervals from simulation output. For another approach (based on the method of *batch means*), and also a list of references that discuss other methods, see Fishman [1978b] and Law and Carson [1979].

Exercises

3. Derive Equation (4.18).

4. Consider simulation of the single-server Erlang loss model with constant service times.
   a. Show that if the offered load is $a = 1$ erlang, then for all values of δ, asymptotically,

   $$P\{\Pi - \delta < \hat{\Pi}(n) < \Pi + \delta\} = P\{\Pi - \delta < \hat{P}(n) < \Pi + \delta\},$$

   where Π is the loss probability, and $\hat{\Pi}(n)$ and $\hat{P}(n)$ are the estimators of Π defined by Equations (4.4) and (4.17).
   b. Tabulate the values of $P\{\Pi - 0.05 < \hat{\Pi}(100) < \Pi + 0.05\}$ and $P\{\Pi - 0.05 < \hat{P}(100) < \Pi + 0.05\}$ for $\Pi = 0.05, 0.10, 0.15, 0.20, \ldots, 0.95$. On the same set of axes plot the graph of each of these probabilities versus Π.

6.5. Examples

In this section we discuss two examples of simulation studies of queueing models. These particular examples were chosen because, despite their structural simplicity, they illustrate some important points in simulation methodology, and illuminate the characteristic strengths and weaknesses of simulation analysis.

Waiting Times in a Queue with Quasirandom Input and Constant Service Times

As the reader is by now well aware, queueing theory contains many results that are, at first glance, puzzling and even absolutely counterintuitive. It is, of course, a tautology that no truth is counterintuitive to one who understands it. Nevertheless, it is probably safe to assume that even the reader who has advanced this far in this book retains some naiveté about queueing theory, and therefore it is worthwhile to discuss a case in which simulation analysis uncovered a very puzzling anomaly.

The model under consideration is characterized by n sources generating quasirandom input with rate γ per idle source, s servers with constant service times, blocked customers delayed, and service in order of arrival. This model has applications in teletraffic engineering. Of particular interest in these applications is knowledge of the equilibrium waiting-time distribution as a function of the server occupancy. (The server occupancy, of course, is a strictly increasing function of γ.) C. J. Durnan and J. H. Weber [1961, unpublished] used simulation to study the waiting times in this model. (Their results are discussed also in Durnan [1967].) We will briefly

Examples

describe the logic by which this model can be simulated, the validation of the simulation program, and the results obtained by Durnan and Weber and their consequences.

We begin with the programming logic. Let us suppose that an event, either an arrival or a service completion, has just occurred in the simulation and all associated bookkeeping has been done. Let $n-j$ ($0<j<n$) be the number of idle sources, and let y be the time remaining until the next service completion is scheduled to occur. Now the simulation determines the type of event and the elapsed time to the next event. To do this, the simulation generates a value x from the exponential distribution with mean $[(n-j)\gamma]^{-1}$ (which can be implemented according to (2.8), with $\tau = [(n-j)\gamma]^{-1}$). If $x<y$, the event is an arrival: The simulation records are updated to reflect this fact; the simulation clock is advanced by the amount x; and a new value is drawn from the exponential distribution, now with the mean $[(n-j-1)\gamma]^{-1}$. If $x>y$, the event is a service completion: The simulation records are updated; the clock is advanced by the amount y; and a new value x is drawn from the exponential distribution, now with mean $[(n-j+1)\gamma]^{-1}$. This procedure is repeated (with obvious modifications when $j = 0$ or $j = n$) until the simulation is deemed to have run long enough.

Durnan and Weber wrote such a simulation, with similar or equivalent logic, to obtain information on waiting times when service is in order of arrival. In order to validate their simulation, they ran it for some special cases for which mathematical results were available, such as the case of $s=1$ server (Harrison [1959]). Also, the equilibrium distribution of the number of customers in service when $n=s$ is known [and is given by the binomial distribution that results from setting $s=n$ in (7.5) of Chapter 3]. Furthermore, it would be easy (and wise) to write the simulation so that the service times can be assigned values drawn from an exponential distribution instead of being assigned a constant value. Then the simulation results can be checked against the formulas derived in Section 3.8. In a similar fashion, Durnan and Weber checked their simulation results against mathematical formulas for various models similar to the one they were studying. Their results seemed to agree with theory in every case they tried. They concluded that their model was valid (that is, there were no logical bugs and the statistical procedures were satisfactory) and proceeded to make simulation runs to study the original model.

In particular, they used their simulation output to plot curves of the proportion (probability) of requests whose delay exceeded t service times versus the observed server occupancy. It is intuitively obvious to even the most casual observer that this probability must be a strictly increasing function of the server occupancy. Nevertheless, their results indicated that there were values of t, n, and s for which this probability *decreases* when the server occupancy increases within a certain range.

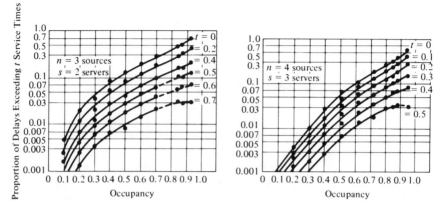

Figure 6.1.

Figure 6.1 reproduces some of the curves obtained by Durnan and Weber. The curve corresponding to $t=0.7$ for the model with $n=3$ sources and $s=2$ servers gives a hint of nonmonotonicity, while the curve for $t=0.5$, $n=4$, and $s=3$ shows the anomaly clearly.

The *Durnan-Weber dip*, as it came to be called by its friends, caused much consternation and gnashing of teeth. It was assumed, of course, that the dip was a manifestation of an error. The question was, what kind of error? The first reaction was to increase the run length to minimize the effect of sampling error. But larger sample sizes only strengthened the likelihood that the dip was really there, and not a spurious effect of statistical noise. Then the search turned to the random-number generator and to the programming logic. But the random-number generator seemed to perform acceptably in every other case, and the model was so simple that it seemed unlikely that a logical bug could escape the careful scrutiny to which the program was subjected. It is a credit to the skill and integrity of the investigators that they did not simply overlook the dip, or fudge it away, or present the curves only in the ranges where the dip was not found, or ascribe the dip to "experimental error" and ignore it. Instead, they withheld publication until they could understand what was going on. Eventually, mathematical verification of the dip was forthcoming, as reported in Durnan [1967]: "Although no intuitive justification has been advanced to support this phenomenon, its existence was subsequently verified by P. J. Burke who used numerical methods to obtain the delay distributions for the special case of three sources on two servers."

Although Burke's numerical work verified the existence of the dip, it did not explain it, and the problem remained in the air. A description of the numerical calculations for the case when $n=3$ and $s=2$ was later given in Burke [1979], as well as an intuitive justification for the existence of the dip and the following conjecture:

$$\lim_{\gamma \to \infty} F(\gamma, t) = \begin{cases} 1 & \text{when } t \leq 0.5, \\ 0 & \text{when } t > 0.5, \end{cases}$$

where $F(\gamma, t)$ is the equilibrium probability that an arbitrary request waits more than t service times for service to commence when the request rate per idle source is γ. Halfin [1979] proves that "Burke's conjecture is right (except perhaps for $t = 0.5$), even under weaker assumptions about the arrival process." Halfin's interesting paper is perhaps best summarized by its abstract:

> The behavior of queueing systems with two servers and one waiting room is investigated. It is shown that if the service time is constant, then the difference between the times to service completion of the two servers (phase difference) tends to a constant, for increasing input intensities. This phenomenon holds for a wide class of arrival processes, but not when the service time has even a small variability. These results imply that the delay is not stochastically monotone in the input intensity. In general, we conjecture that the behavior of the phase differences, and delays, depends on whether the size of the waiting room is even or odd.

In summary, this example illustrates a case where an entirely unsuspected and mathematically interesting result was uncovered through a simulation study of a simple model, conducted as part of an engineering study of a real system. In this case simulation and mathematical analysis worked together to yield results of both practical and theoretical interest. One can only speculate on the number of counterintuitive facts that have remained buried in simulation studies of more complicated models, or in simulation studies performed by less competent investigators.

Response Times in a Processor-Sharing Operating System

The methodology of simulation is usually used to study (models of) real systems that are deemed too complicated for mathematical analysis. That is, a choice is made between what are assumed to be mutually exclusive methodologies, with simulation being the methodology of choice to study complicated, real systems, and mathematics being the methodology of choice to study simple, abstract models. Of course, a certain amount of mathematics is necessarily embodied in a simulation analysis, but it is unusual for simulation methodology to be used for the stated purpose of answering "mathematical" questions. In our discussion of the queue with constant service times and quasirandom input, we saw that a simulation study stimulated mathematical analysis of a particular model by uncovering a counterintuitive fact, but this was inadvertent. We now turn to an example in which simulation was used to investigate an essentially mathematical question, namely, how the *response times* (sojourn times) reflect the variability of the *processing times* (service times) in a model of a computer operating system. An interesting point about this use of simulation is that the model in question, which provides an upper bound on the performance of a real computer operating system, is obtained as a mathematical limit of a sequence of models, each of which can, in principle, be simulated; but the limiting model itself is not physically realizable, and therefore cannot be simulated directly. We begin with a description of the *round-robin* scheduling algorithm, which is one of the best known and most widely used operating-system scheduling algorithms for time-shared computer systems. According to this algorithm, the processor allocates to each job a fixed amount of time, called a *quantum* or *time slice*. If a job's service time (the total time required from the processor) is completed in less than the quantum, it leaves; otherwise, it feeds back to the end of the queue of waiting jobs, waits its turn to receive another quantum of service, and continues in this fashion until its total service time has been obtained from the processor.

This model, which has the pleasant property that it tends to yield a shorter response time (the total time required for the job to receive its full service time from the processor) for a job with a shorter service time, was first analyzed by Kleinrock [1964]. An important advance was made when Kleinrock [1967] studied the same model in the limiting case where the quantum length goes to zero. (He called this model *processor sharing*, because it is equivalent to the assumption that all k jobs in the mix are being processed simultaneously at $1/k$ times the rate at which a single job would be processed if it were the only job in the mix.) This greatly simplified the mathematics, and led to the remarkable conclusion that in the processor-sharing model, the mean response time for all jobs whose total required service time is t is proportional to t. More precisely, Kleinrock's 1967 paper showed that under the assumptions that (1) the jobs arrive in the mix according to a Poisson process, (2) the service times are exponentially distributed, and (3) there is one processor, then the mean value $R_1(t)$ of the equilibrium response time for a job whose service time is t is given by

$$R_1(t) = \frac{1}{1-a} t, \qquad (5.1)$$

where a is the product of the arrival rate and the mean service time. The range of applicability of (5.1) was considerably widened when Sakata, Noguchi, and Oizumi [1971] showed that assumption (2) is superfluous; that is, (5.1) is valid for *any* distribution of service times whatever. (For further discussion see Kleinrock [1976], and also Coffman and Denning [1973].)

Examples

The above considerations raise the question about further generalizations of (5.1). In particular, what is the mean response time for a job whose service time is t in a processor-sharing model with an arbitrary distribution of service times and an arbitrary number s of processors? It is reasonable to conjecture that this quantity—call it $R_s(t)$—does not depend on the form of the service-time distribution function (because it has been proved so for $s = 1$, and is clearly true for $s = \infty$; it would be surprising if such a dependence would appear only for the intermediate values $1 < s < \infty$). It is also reasonable to conjecture (and we shall explain why shortly) that $R_s(t)$ is given by the following formula:

$$R_s(t) = \left(1 + \frac{C(s,a)}{s-a}\right)t, \qquad (5.2)$$

where $C(s,a)$ is the Erlang delay formula, given by Equation (4.8) of Chapter 3. Thus, we conjecture that in an s-server queueing model characterized by Poisson input and processor-sharing queue discipline, the equilibrium mean response time for a job that requires a total of t units of service time is given by (5.2), where a is the offered load.

To see why the right-hand side of (5.2) is a reasonable guess for $R_s(t)$, observe that the processor-sharing queue discipline is what we have previously termed "nonbiased," and therefore, when the service times are exponential, the mean response time R_s, say, is the same as the sum of the mean service time and the mean waiting time in the corresponding Erlang delay system [see Equation (4.7) of Chapter 3]:

$$R_s = \frac{1}{\mu}\left(1 + \frac{C(s,a)}{s-a}\right). \qquad (5.3)$$

Since

$$R_s = \int_0^\infty R_s(t)\,dH(t) \qquad (5.4)$$

for any service-time distribution function $H(t)$, it follows [from substitution of (5.2) into (5.4)] that our hypothesis (5.2) is consistent with (5.3), at least for the case of exponential service times. Equation (5.2) retains the property of linearity in t exhibited by (5.1), and reduces to (5.1) when $s = 1$. These observations all support the hypothesis that (5.2) is, in fact, the generalization of (5.1) for the case of arbitrary s and arbitrary $H(t)$.

In order to further study this hypothesis, a simulation program that would measure the response times and compare them with the values predicted by (5.2) was written as a project in a course in simulation (and reported at a conference in New York; an abstract is given in Cooper and Marmon [1978]). The service times were modeled by the method of phases;

that is, if X is the random variable that represents the total required service time of an arbitrary job, then X has distribution function $P\{X \leq x\} = H_n(x)$ given by the Erlangian distribution function of order n:

$$H_n(x) = 1 - \sum_{j=0}^{n-1} \frac{(\mu x)^j}{j!} e^{-\mu x}, \qquad (5.5)$$

where n is the number of phases, each with mean μ^{-1} and variance μ^{-2}. Hence, if we take $\mu = n$ we have

$$E(X) = 1 \qquad (5.6)$$

and

$$V(X) = \frac{1}{n}; \qquad (5.7)$$

that is, as the parameter n is assigned the values $1, 2, \ldots$, the mean service time remains constant at one, while the variance of service times runs from one to an arbitrarily small number. This corresponds to a spectrum of service-time distributions, ranging from high variability ($n = 1$, exponential service times) to no variability ($n = \infty$, constant service times), all with a mean value of unity. This model does not include all possible service-time distributions, of course, but it does provide a wide range of variability of service times. One might reasonably expect that if the mean response time does indeed depend on the form of the service-time distribution function, then the simulated response times will show a dependence on the value of the parameter n used in that particular run. Besides providing an easy way to describe a family of service-time distributions with different variances, the use of the method of phases to describe the service times has another important property in this application: It permits the adoption of a particularly simple simulation algorithm, based on the properties of the birth-and-death process. We emphasize that this algorithm gives an exact representation of the processor-sharing model, not an approximation for a "negligible" value of the time slice. This is consistent with the objective of using simulation to test a mathematical formula that is conjectured to describe exactly a mathematical model that is itself a physically unrealizable idealization of the round-robin operating system.

We now describe the simulation algorithm. Suppose there are k jobs present in the system. Then each job is receiving service at rate μ if $k \leq s$ and rate $s\mu / k$ if $k > s$, where μ^{-1} is the mean duration of an (exponentially distributed) phase of service. If we define

$$r(k) = \begin{cases} k\mu & (k = 0, 1, \ldots, s), \\ s\mu & (k = s+1, s+2, \ldots), \end{cases} \qquad (5.8)$$

then the probability $p(k)$ that the next event is a phase completion is

$$p(k) = \frac{r(k)}{\lambda + r(k)} \quad (k=0,1,\ldots), \tag{5.9}$$

and the probability $q(k)$ that the next event is an arrival of a new job in the mix is

$$q(k) = 1 - p(k) \quad (k=0,1,\ldots), \tag{5.10}$$

where λ is the rate of the Poisson arrival process. If the event is a phase completion, then the probability that it corresponds to any given job is $1/k$. Random numbers are generated and decisions as to the kind of event generated are made according to these probabilities. Records are kept as to the number of jobs present and the number of phases of service completed by each job in the system. The time between events is calculated by drawing a sample value $x = x(k)$ from an exponential distribution with mean $[\lambda + r(k)]^{-1}$, where k is the number of jobs in the mix during the time between events. Thus, whenever an arrival or phase completion occurs, the cumulative response time for each job present is increased by the amount $x(k)$. If $k \leq s$, each job's cumulative service time is also increased by the same amount x; but if $k > s$, each job's cumulative service time is increased by an amount $x_1 = (s/k)x$.

Thus, whenever a job leaves the system, it has two times associated with it—the actual response time experienced by the job and the actual amount of service time t that the job received. From these data the simulation calculates, for a range of values of t, s, and a, the average value $\overline{R}_s(t)$ of the response time for a job that received (required) t units of service.

Finally, the values of $\overline{R}_s(t)$ produced by the simulation are compared with the hypothesized corresponding theoretical values of $R_s(t)$ given by (5.2). Some examples of these comparisons are given in Figures 6.2, 6.3, and 6.4, where $\overline{R}_s(t)$ is plotted against $R_s(t)$ for $n = 1, 2$, and 10 and $s = 1, 2$, and 3. If the hypothesis (5.2) is correct, each of these graphs should be the

Figure 6.2.

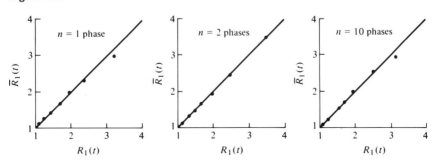

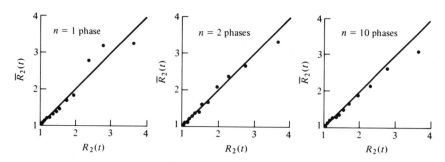

Figure 6.3.

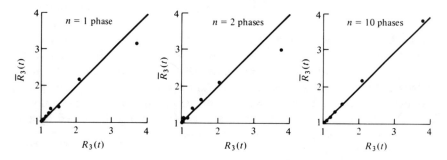

Figure 6.4.

straight line with unit slope, with any deviation attributable solely to statistical sampling error and roundoff error. Figure 6.2 corresponds to the case $s=1$, for which case, according to Sakata, Noguchi, and Oizumi [1971], Equation (5.2) is correct. Inspection of Figures 6.3 and 6.4 shows the same quality of agreement between simulation and theory that Figure 6.2 shows. This indicates that the deviations from the straight line with unit slope in Figures 6.3 and 6.4 are manifestations of the same source of error that appears in Figure 6.2, namely, statistical and roundoff error. In summary, the simulation results support the conjecture that (5.2) is correct.

In conclusion, this simulation study not only provides the usual kind of engineering information about the performance of a system as a function of the relevant parameters, but, more interestingly, illustrates the use of simulation as a complement to mathematical analysis in the theoretical study of mathematical models.

As an interesting postscript, we observe that even stronger evidence for the hypothesis (5.2) is available: In a relatively obscure paper, presented at a conference in Hawaii, Sakata, Noguchi, and Oizumi [1969] proved Equation (5.2). (Surprisingly, these authors did not reference their 1969 paper in their 1971 paper, which is on the same subject and is much more accessible.) In addition, J. W. Cohen presented a paper at a 1977 conference in the Netherlands that, independently, also arrived at (5.2) (an abstract is given in Cohen [1978]). Finally, R. Schassberger has pointed out to this author that formula (5.2) can be obtained after much calculation as a special case of formula (12) of his unpublished paper, Schassberger [1978c]. Needless to say, each author's work was done with no knowledge of the work of any of the others.

[7]
Annotated Bibliography

To the casual observer, the very idea that so simple-minded a concept as that of a queue should form the basis of a mathematical theory to which otherwise rational people devote their professional lives may seem ludicrous. Perhaps. But, as the reader will have recognized by now, queueing theory is an unending subject to which this book provides only an introduction. It is the purpose of this chapter to provide a guide to the textbooks, reference works, tables, and research journals on queueing theory.

We begin with a list of textbooks on queueing theory. The list includes, to the best of the author's knowledge, every English-language hardcover textbook devoted primarily to queues, as well as some other books that fall outside this category. Each book is referenced by author, date, and title; additional publishing information is given in the References.

Bear [1976], *Principles of Telecommunication-Traffic Engineering*. The application of probability (and queueing) theory in the engineering and administration of telephone systems.

Beckmann [1968], *Introduction to Elementary Queuing Theory and Telephone Traffic*. Short outline of basic formulas and applications to telephone traffic engineering.

Conolly [1975], *Lecture Notes on Queueing Systems*. Lecture notes from a graduate-level course.

Cox and Smith [1961], *Queues*. Short and elegant overview of the theory, with an eye toward application. Especially valuable is the carefully selected list of references in an appendix.

Fry [1928, 1965], *Probability and Its Engineering Uses*. Probably the first reference and/or textbook to treat queueing theory (mainly in the context of telephone traffic theory). Remarkable in 1928, but the second edition, 1965, adds little.

Giffin [1978], *Queueing: Basic Theory and Applications*. A "primer on queues...for practitioners rather than theoretical mathematicians."

Gross and Harris [1974], *Fundamentals of Queueing Theory*. Designed for a graduate-level course in operations research.

Hillier and Stidham [1981, in preparation], *Queueing Theory: Describing and Optimizing Queueing Systems*.

Kleinrock [1975], *Queueing Systems, Volume I: Theory*. Designed for a graduate-level course. Prepares the reader for *Volume II: Computer Applications*.

Kosten [1973], *Stochastic Theory of Service Systems*. Treats some of the standard models and variants, some of which have a telecommunications flavor.

Lee [1966], *Applied Queueing Theory*. Short exposition of theory followed by illustrative case studies.

Mina [1974], *Introduction to Teletraffic Engineering*. A manual of telephone traffic engineering practice, with examples, graphs, and tables.

Morse [1958], *Queues, Inventories and Maintenance*. Probably the first widely used and influential book on the subject. Clear explanation of mathematical concepts and applications. Still useful as a textbook.

Murdoch [1978], *Queueing Theory: Worked Examples and Problems*. "It is designed to meet the needs not only of management science training programmes, but also of mangement (*sic*) teaching programmes."

Page [1972], *Queueing Theory in OR*. Outlines the derivation of formulas for some classical queueing models, gives numerical examples and problems.

Panico [1969], *Queuing Theory: A Study of Waiting Lines for Business, Economics, and Science*. Totally confused and confusing.

Ruiź-Palá, Ávila-Beloso, and Hines [1967], *Waiting-Line Models*.

White, Schmidt, and Bennett [1975], *Analysis of Queueing Systems*. Well-written textbook for applications-oriented students with an engineering background.

The following are important reference works and monographs.

Beneš [1963], *General Stochastic Processes in the Theory of Queues*. An attempt to present the theory of single-server queues simultaneously to mathematicians and engineers.

Beneš [1965], *Mathematical Theory of Connecting Networks and Telephone Traffic*. Reprints of papers originally published in *The Bell System Technical Journal*.

Borovkov [1976], *Stochastic Processes in Queueing Theory*. "... an attempt to set forth the contemporary state (as it appears to us) of the mathematical theory of queues from the most unified and general standpoint possible."

Brockmeyer, Halstrøm, and Jensen [1948], *The Life and Works of A. K. Erlang*. An especially interesting and important historical work, which should be on the shelf of every afficianado.

Cohen [1969], *The Single Server Queue*. A massive, rigorous mathematical treatment, relying heavily on complex-variable theory, of the classical single-server models and some of their variants. Contains substantial preliminary material on stochastic processes and an extensive bibliography.

Cohen [1976], *On Regenerative Processes in Queueing Theory*. Views the classical models in the light of recent results in the theory of regenerative processes.

Conway, Maxwell, and Miller [1967], *Theory of Scheduling*. About one-third of this book concerns queueing theory.

Franken, König, Arndt, and Schmidt [1980], *Queues and Point Processes*.

Gnedenko and Kovalenko [1968], *Introduction to Queueing Theory*. Emphasis on theory. References many Russian-language papers.

Jaiswal [1968], *Priority Queues*. "... addressed primarily to research students...."

Kelly [1979], *Reversibility and Stochastic Networks*. Discusses the concept of reversibility, including its application to the analysis of networks of queues.

Khintchine [1969], *Mathematical Methods in the Theory of Queueing*. A short, rigorous, mathematical study of some classical models and the properties of the underlying stochastic processes.

LeGall [1962], *Les systèmes avec ou sans attente et les processes stochastiques*, Tome 1.

Neuts [1981, in preparation], *Matrix-Geometric Solutions in Stochastic Models—An Algorithmic Approach*.

Pollaczek [1957], *Problèmes stochastiques posés par le phénomène de formation d'une queue d'attente à un guichet et par des phénomènes apparentés*.

Pollaczek [1961], *Théorie analytique des problèmes stochastiques relatifs à un group de lignes téléphoniques avec dispositif d'attente*.

Prabhu [1965], *Queues and Inventories*. "The purpose of this book is to give a unified treatment of [their basic stochastic] properties."

Riordan [1962], *Stochastic Service Systems*. Queueing theory with an emphasis on topics relating to teletraffic theory.

Saaty [1961], *Elements of Queueing Theory*. A text and summary of scattered papers and monographs, with an extensive bibliography.

Schassberger [1973], *Warteschlangen*.

Annotated Bibliography

Syski [1960], *Introduction to Congestion Theory in Telephone Systems.* Encylopedic. Treats standard queueing models and background probability theory; contains summaries of a large number of papers relating to teletraffic theory, extensive bibliography.

Takács [1962a], *Introduction to the Theory of Queues.* A rigorous mathematical treatment of some classical models. Contains exercises and their solutions.

There are many textbooks on methods of operations research and management science, most of which devote at least a chapter to queueing theory and its applications. These chapters might provide an overview, but in most cases this is poorly done. Two exceptions are

Hillier and Lieberman [1980], *Introduction to Operations Research,* 3rd ed.

Wagner [1975], *Principles of Operations Research,* 2nd ed.

Some books concerned with applications of queueing theory in computer science are

Allen [1978], *Probability, Statistics, and Queueing Theory, with Computer Science Applications.* "... a junior–senior level textbook on applied probability and statistics with computer science applications."

Coffman and Denning [1973], *Operating Systems Theory.* Chapter 4 treats queueing theory.

Courtois [1977], *Decomposability: Queueing and Computer System Applications.* A monograph on queueing networks and computer system performance evaluation.

Ferrari [1978], *Computer Systems Performance Evaluation.* Chapter 4 discusses queueing models.

Kleinrock [1964b], *Communication Nets: Stochastic Message Flow and Delay.* Probably the first application of queueing theory to the analysis of store-and-forward communication networks.

Kleinrock [1976], *Queueing Systems, Volume II: Computer Applications.* Queueing theory applied to computer performance evaluation and to the design and analysis of computer-communication networks.

Kobayashi [1978], *Modeling and Analysis: An Introduction to System Performance Evaluation Methodology.* Contains material on queueing theory and its application to computer performance evaluation.

Sauer and Chandy [1981, in preparation], *Computer Systems Performance Modeling: A Primer.*

Queueing Network Models of Computer System Performance, a special issue of *ACM Computing Surveys* (Vol. 10, No. 3, September 1978) devoted entirely to performance evaluation of computer systems, with focus on queueing network models.

Queueing theory with an eye toward application in transportation engineering is presented in

Newell [1971], *Applications of Queueing Theory*.
Newell [1973], *Approximate Stochastic Behavior of n-Server Service Systems with Large n*.
Newell [1979], *Approximate Behavior of Tandem Queues*.

Two books that are not nominally about queueing theory, but contain substantial amounts of related material are

Takács [1960a], *Stochastic Processes: Problems and Solutions*. A collection of challenging problems imbedded in a small amount of text, with complete and detailed solutions.
Takács [1967], *Combinatorial Methods in the Theory of Stochastic Processes*. An exposition of the author's original research in which he shows that "for a wide class of random variables and for a wide class of stochastic processes we can obtain explicit results in a simple and elementary way using a generalization of the classical ballot theorem."

There are many textbooks on probability and/or stochastic processes that present queueing theory in the context of illuminating and illustrating a more general theory. Some notable ones are

Bhat [1972], *Elements of Applied Stochastic Processes*.
Çinlar [1975], *Introduction to Stochastic Processes*.
Cox and Miller [1965], *The Theory of Stochastic Processes*.
Feller [1971], *An Introduction to Probability Theory and Its Applications*, Volume II, 2nd ed.
Heyman and Sobel [1981, in preparation], *Stochastic Models in Operations Research* (two volumes).
Karlin [1968], *A First Course in Stochastic Processes*.
Ross [1970], *Applied Probability Models with Optimization Applications*.

Tables of queueing formulas are given in

Descloux [1962], *Delay Tables for Finite- and Infinite-Source Systems*. Tables of formulas for the Erlang C model and its quasirandom-input counterpart, prefaced by a concise derivation of these formulas.
Dietrich, Kruse, Michel, Ondra, Peter, and Wagner [1966], *Teletraffic Engineering Manual*.
Frankel [1976], *Tables for Traffic Management and Design: Book 1— Trunking*. A manual for teletraffic engineering.
Hillier and Lo [1971], *Tables for Multiple-Server Queueing Systems Involving Erlang Distributions*. Extensive tabulation of formulas relating to the $E_m/E_k/s$ model, with a comprehensive introduction describing the use of the tables and the numerical analysis underlying their construction.

Hillier and Yu [1981], *Queueing Tables and Graphs.*
Peck and Hazelwood [1958], *Finite Queuing Tables.*

The number of papers on queueing theory that have been published since the seminal work of Erlang is enormous. One would think that there would be little left to say, but it seems that new papers are appearing at an ever-increasing rate. Following is a list of journals that regularly carry papers on queueing theory or its application:

Advances in Applied Probability. Contains review and expository papers, as well as mathematical and scientific papers of interest to probabilists. Each issue invariably contains at least one paper on queueing theory.
The Bell System Technical Journal. Occasionally contains papers on queueing theory, usually relating to teletraffic theory or engineering.
IBM Journal of Research and Development. Occasionally contains papers on queueing theory and its applications in computer science.
IEEE Transactions on Communication. Regularly publishes papers on queueing theory and its application to communication system analysis.
IEEE Transactions on Software Engineering. Regularly publishes papers on queueing theory and its application to computer system analysis.
Journal of Applied Probability. A companion publication to *Advances in Applied Probability*, it contains "research papers and notes on applications of Probability Theory to the biological, physical, social and technological sciences." Each issue invariably contains several papers on queueing theory. These papers are "applied" only in the sense that the whole subject of queueing theory is an application of probability theory; they are not case studies.
Journal of the Association for Computing Machinery. Often contains papers on queueing theory and its applications in computer science.
Journal of the Operations Research Society of Japan. Often contains papers on queueing theory.
Management Science. Emphasizes the methodological and practical aspects of the development and implementation of management science models. Averages about one paper per issue related to queueing theory.
Mathematics of Operations Research. Averages slightly less than one paper per issue on the mathematics of queueing theory.
Naval Research Logistics Quarterly. Occasionally contains papers on queueing theory.
Operations Research. Emphasizes practice, operational science and methodologies. Averages about one paper per issue related to queueing theory.
Opsearch. The journal of The Operational Research Society of India. Often contains papers on queueing theory.

Proceedings of the International Teletraffic Congress. Held every third year beginning in 1955. Many papers on the application of probability and queueing theory to teletraffic theory and engineering. (Available from several libraries throughout the world, including the Engineering Societies Library, 345 East 47 Street, New York, New York 10017, U.S.A.)

Studies in Congestion Theory. A continuing series of tables and reports published by the Institute of Switching and Data Technics at the University of Stuttgart.

Appendix
Engineering Curves for Systems with Poisson Input

In the following figures, the number of servers is denoted by s, and the magnitude in erlangs of the offered load is denoted by a ($a = \lambda/\mu$, where λ is the customer arrival rate, and μ^{-1} is the mean service time).

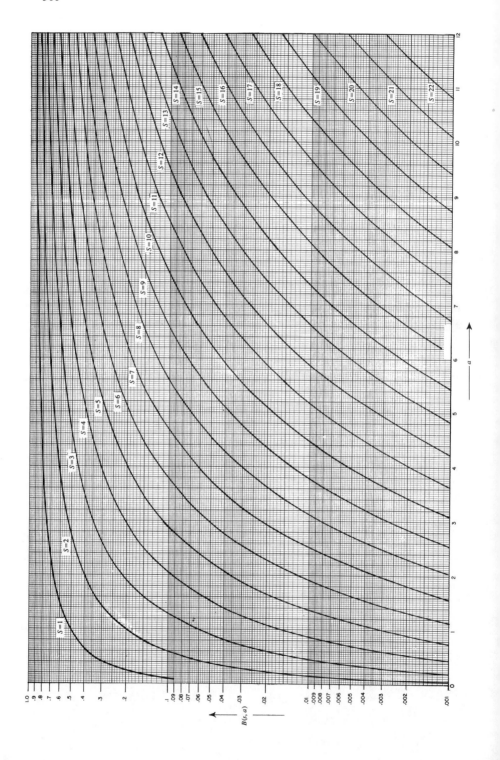

Figure A.1. Erlang loss formula $B(s,a)$ plotted against offered load a in erlangs for different values of the number s of servers. For systems with Poisson input in which blocked customers are cleared, $B(s,a)$ is the proportion of customers who find all servers busy and are consequently cleared from the system without receiving service. The service-time distribution need not be exponential.

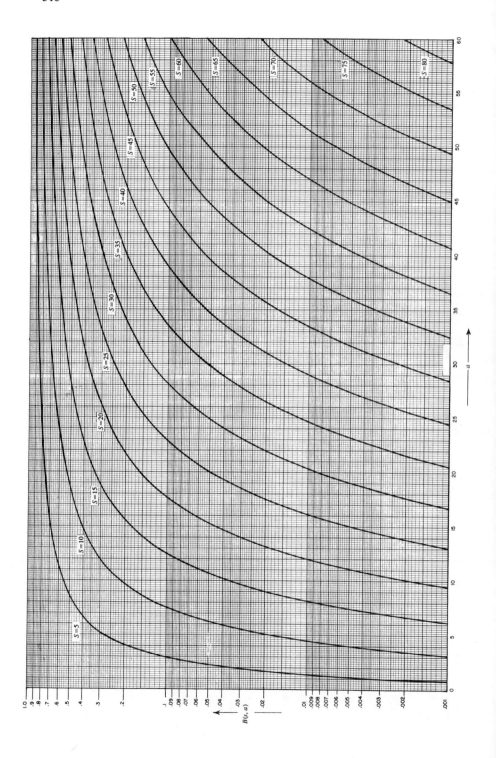

Figure A.2. Erlang loss formula $B(s,a)$ plotted against offered load a in erlangs for different values of the number s of servers. For systems with Poisson input in which blocked customers are cleared, $B(s,a)$ is the proportion of customers who find all servers busy and are consequently cleared from the system without receiving service. The service-time distribution need not be exponential.

Figure A.3. Erlang delay formula $C(s,a)$ plotted against offered load a in erlangs for different values of the number s of servers. For systems with Poisson input and exponential service times in which blocked customers are delayed (blocked customers wait until served), $C(s,a)$ is the proportion of customers who find all servers busy and consequently wait until served. The order of service of waiting customers is irrelevant.

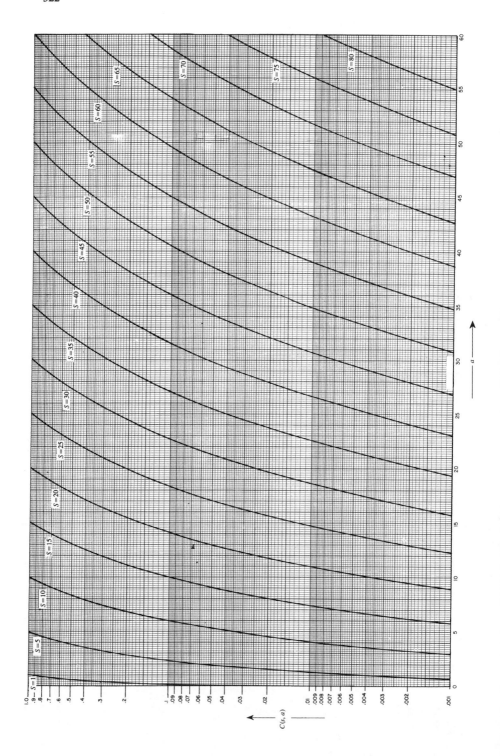

Figure A.4. Erlang delay formula $C(s,a)$ plotted against offered load a in erlangs for different values of the number s of servers. For systems with Poisson input and exponential service times in which blocked customers are delayed (blocked customers wait until served), $C(s,a)$ is the proportion of customers who find all servers busy and consequently wait until served. The order of service of waiting customers is irrelevant.

324

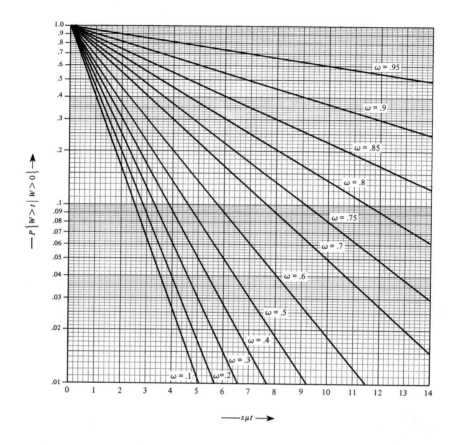

Figure A.5. Conditional probability $P\{W>t|W>0\}=e^{-(1-\omega)s\mu t}$ that a blocked customer waits beyond t for service to commence in a delay system (blocked customers wait until served) with exponential service times and service in order of arrival, plotted against $s\mu t$, for different values of the generalized occupancy ω. For Erlang delay systems (Poisson input, exponential service times), $\omega=\rho=a/s$. To calculate the unconditional probability $P\{W>t\}$ that a customer waits beyond t for service to commence in an Erlang delay system with service in order of arrival, use the formula $P\{W>t\}=P\{W>t|W>0\}C(s,a)$, where $C(s,a)$ is determined from Figure A.3 or A.4.

References

ALLEN, A. O. [1978]. *Probability, Statistics, and Queueing Theory, with Computer Science Applications*. New York: Academic Press.

ASHTON, W. D. [1966]. *The Theory of Road Traffic Flow*. New York: John Wiley and Sons.

BALACHANDRAN, K. R. [1973]. Control Policies for a Single Server System. *Management Science* 19, No. 9 (May), 1013–1018.

――――, and H. TIJMS. [1975]. On the D-Policy for the $M/G/1$ Queue. *Management Science* 21, No. 9 (May), 1073–1076.

BARBOUR, A. D. [1976]. Networks of Queues and the Method of Stages. *Advances in Applied Probability* 8, No. 3 (September), 584–591.

BASKETT, F., K. M. CHANDY, R. R. MUNTZ, and F. G. PALACIOS. [1975]. Open, Closed and Mixed Networks of Queues with Different Classes of Customers. *Journal of the Association for Computing Machinery* 22, No. 2 (April), 248–260.

BEAR, D. [1976]. *Principles of Telecommunication-Traffic Engineering*. England: Peter Peregrinus Ltd.

BECKMANN, P. [1968]. *Introduction to Elementary Queuing Theory and Telephone Traffic*. Boulder, CO: The Golem Press.

BENEŠ, V. E. [1957]. On Queues with Poisson Arrivals. *The Annals of Mathematical Statistics* 28, 670–677.

――――. [1959]. On Trunks with Negative Exponential Holding Times Serving a Renewal Process. *The Bell System Technical Journal* 38, No. 1 (January), 211–258.

――――. [1961]. The Covariance Function of a Simple Trunk Group, with Applications to Traffic Measurement. *The Bell System Technical Journal* 40, No. 1 (January), 117–148.

――――. [1963]. *General Stochastic Processes in the Theory of Queues*. Reading, MA: Addison-Wesley.

――――. [1965]. *Mathematical Theory of Connecting Networks and Telephone Traffic*. New York: Academic Press.

BEUTLER, F. J., and B. MELAMED. [1978]. Decomposition and Customer Streams of Feedback Networks of Queues in Equilibrium. *Operations Research* 26, No. 6 (November–December), 1059–1072.

BEUTLER, F. J., B. MELAMED, and B. P. ZIEGLER. [1977]. Equilibrium Properties of Arbitrarily Interconnected Queueing Networks. In *Proceedings of the Fourth International Symposium on Multivariate Analysis* (P. R. Krishniaih, ed.). Amsterdam: North-Holland, pp. 351–370.

BHAT, U. N. [1972]. *Elements of Applied Stochastic Processes*. New York: John Wiley and Sons.

———. [1973]. Some Problems in Finite Queues. In (A. B. Clarke, ed.) *Mathematical Methods in Queueing Theory* (Proceedings of a Conference at Western Michigan University, May 10–12, 1973). Berlin and New York: Springer-Verlag Lecture Notes in Economics and Mathematical Systems, Vol. 98, pp. 139–156. 1974.

BOBILLIER, P. A., B. C. KAHAN, and A. R. PROBST. [1976]. *Simulation with GPSS and GPSS V*. Englewood Cliffs, NJ: Prentice-Hall.

BOREL, E. [1942]. Sur l'emploi du théorème de Bernoulli pour faciliter le calcul d'un infinité de coefficients. Application au problème de l'attente à un guichet. *Comptes Rendus Académie des Sciences*, Paris 214, 452–456.

BOROVKOV, A. A. [1976]. *Stochastic Processes in Queueing Theory*. New York: Springer-Verlag.

BOXMA, O. J. [1979a]. On a Tandem Queueing Model with Identical Service Times at Both Counters, I. *Advances in Applied Probability* 11, No. 3 (September), 616–643.

———. [1979b]. On a Tandem Queueing Model with Identical Service Times at Both Counters, II. *Advances in Applied Probability* 11, No. 3 (September), 644–659.

BRANDWAJN, A. [1979]. An Iterative Solution of Two-Dimensional Birth and Death Processes. *Operations Research* 27, No. 3 (May–June), 595–605.

BRETSCHNEIDER, G. [1956]. Die Berechnung von Leitungsgruppen für überfliessenden Verkehr in Fernsprechwahlanlagen. *Nachrichtentechnische Zeitschrift* 9, 533–540.

BROCKMEYER, E. [1954]. The Simple Overflow Problem in the Theory of Telephone Traffic. *Teleteknik* 5 (December), 361–374.

———, H. A. HALSTRØM, and A. JENSEN. [1948]. *The Life and Works of A. K. Erlang*. Trans. 2. Copenhagen: Danish Academy of Technical Sciences, pp. 1–277.

BRUMELLE, S. L. [1971]. On the Relation Between Customer and Time Averages in Queues. *Journal of Applied Probability* 8, No. 3 (September), 508–520.

———. [1978]. A Generalization of Erlang's Loss System to State Dependent Arrival and Service Rates. *Mathematics of Operations Research* 3, No. 1 (February), 10–16.

BUCHNER, M. M., JR., and S. R. NEAL. [1971]. Inherent Load-Balancing in Step-by-Step Switching Systems. *The Bell System Technical Journal* 50, No. 1 (January), 135–165.

References

Burke, P. J. [1956]. The Output of a Queuing System. *Operations Research* 4, No. 6 (December), 699–704.

———. [1959]. Equilibrium Delay Distribution for One Channel with Constant Holding Time, Poisson Input and Random Service, *The Bell System Technical Journal* 38 (July), 1021–1031.

———. [1968]. The Output Process of a Stationary $M/M/s$ Queueing System. *The Annals of Mathematical Statistics* 39, No. 4, 1144–1152.

———. [1971]. The Overflow Distribution for Constant Holding Time. *The Bell System Technical Journal* 50, No. 10 (December), 3195–3210.

———. [1972]. Output Processes and Tandem Queues. *Proceedings of the Symposium on Computer-Communications Networks and Teletraffic*. New York: Polytechnic Institute of Brooklyn Press, pp. 419–428.

———. [1975]. Delays in Single-Server Queues with Batch Input. *Operations Research* 23, No. 4 (July–August), 830–833.

———. [1976]. Proof of a Conjecture on the Interarrival-Time Distribution in an $M/M/1$ Queue with Feedback. *IEEE Transactions on Communications* (May), pp. 575–576.

———. [1979]. A Queuing-Theoretic Anomaly. *Journal of Applied Probability* 16, No. 2 (June), 373–383.

Buzen, J. [1973]. Computational Algorithms for Closed Queueing Networks with Exponential Servers. *Communications of the Association for Computing Machinery* 16, No. 9 (September), 527–531.

Carson, J. S., and A. M. Law. [1980]. Conservation Equations and Variance Reduction in Queueing Simulations. *Operations Research* 28, No. 3, Part I (May–June), 535–546.

Carter, G. M., and R. B. Cooper. [1972]. Queues with Service in Random Order. *Operations Research* 20, No. 2 (March–April), 389–405.

Chaiken, J. M. [1971]. The Number of Emergency Units Busy at Alarms Which Require Multiple Servers. R-531-NYC/HUD. Santa Monica: The Rand Corporation. See also Chapter 7 of *Fire Department Deployment Analysis* (W. E. Walker, J. M. Chaiken, and E. J. Ignall, eds.). New York: North Holland, 1979.

Chandy, K. M., J. H. Howard, Jr., and D. F. Towsley. [1977]. Product Form and Local Balance in Queueing Networks. *Journal of the Association for Computing Machinery* 24, No. 2 (April), 250–263.

Cheng, R. C. H., and G. M. Feast. [1980]. Gamma Variate Generators with Increased Shape Parameter Range. *Communications of the Association for Computing Machinery* 23, No. 7 (July), 389–394.

Çinlar, E. [1975]. *Introduction to Stochastic Processes*. Englewood Cliffs, NJ: Prentice-Hall.

———, and R. L. Disney. [1967]. Stream of Overflows from a Finite Queue. *Operations Research* 15, No. 1 (January–February), 131–134.

Clarke, A. B., and R. L. Disney. [1970]. *Probability and Random Processes for Engineers and Scientists*. New York: John Wiley and Sons.

Coffman, E. G., Jr., and P. J. Denning. [1973]. *Operating Systems Theory*. Englewood Cliffs, NJ: Prentice-Hall.

COHEN, J. W. [1957]. The Generalized Engset Formulae. *Philips Telecommunication Review* 18, No. 4 (November), 158–170.

———. [1969]. *The Single Server Queue*. Amsterdam: North-Holland.

———. [1976]. *On Regenerative Processes in Queueing Theory*. Berlin and New York: Springer-Verlag Lecture Notes in Economics and Mathematical Systems, Vol. 121.

———. [1978]. On a General Class of Processor Sharing Service Facilities. Abstract in *Advances in Applied Probability* 10, No. 2 (June), 334–335.

CONOLLY, B. W. [1975]. *Lecture Notes on Queueing Systems*. New York: Halsted (a division of John Wiley and Sons).

———, and J. CHAN. [1977]. Generalised Birth and Death Queueing Processes: Recent Results. *Advances in Applied Probability* 9, No. 1 (March), 125–140.

CONWAY, R. W., W. L. MAXWELL, and L. W. MILLER. [1967]. *Theory of Scheduling*. Reading, MA: Addison-Wesley.

COOPER, R. B. [1970]. Queues Served in Cyclic Order: Waiting Times. *The Bell System Technical Journal* 49, No. 3 (March), 399–413.

———. [1976]. Queues with Ordered Servers That Work at Different Rates. *Opsearch* 13, No. 2 (June), 69–78.

———, and R. I. MARMON. [1978]. A Conjecture on a Generalization of a Theorem Concerning Response Times in a Processor-Sharing Operating System, and Its Investigation by Simulation. Abstract in the *TIMS/ORSA Bulletin*, No. 5, p. 211 (Joint National TIMS/ORSA Meeting, New York, May 1–3).

———, and G. MURRAY. [1969]. Queues Served in Cyclic Order. *The Bell System Technical Journal* 48, No. 3 (March), 675–689.

———, and B. TILT. [1976]. On the Relationship Between the Distribution of Maximal Queue Length in the $M/G/1$ Queue and the Mean Busy Period in the $M/G/1/n$ Queue. *Journal of Applied Probability* 13, No. 1 (March), 195–199.

COURTOIS, P. J. [1977]. *Decomposability: Queueing and Computer System Applications*. New York: Academic Press.

COX, D. R. [1955]. A Use of Complex Probabilities in the Theory of Stochastic Processes. *Proceedings of the Cambridge Philosophical Society*, 51, 313–319.

———, and H. D. MILLER. [1965]. *The Theory of Stochastic Processes*. London: Methuen and Co.

———, and W. L. SMITH. [1961]. *Queues*. London: Methuen and Co.

CRABILL, T. B., D. GROSS, and M. J. MAGAZINE. [1977]. A Classified Bibliography of Research on Optimal Design and Control of Queues. *Operations Research* 25, No. 2 (March–April), 219–232.

CRAMÉR, H. [1930]. On the Mathematical Theory of Risk. *Skandia Jubilee Volume*, Stockholm.

CRANE, M. A., and D. L. IGLEHART. [1974a]. Simulating Stable Stochastic Systems, I: General Multiserver Queues. *Journal of the Association for Computing Machinery* 21, No. 1 (January), 103–113.

———. [1974b]. Simulating Stable Stochastic Systems, II: Markov Chains. *Journal of the Association for Computing Machinery* 21, No. 1 (January), 114–123.

———. [1975]. Simulating Stable Stochastic Systems, III: Regenerative Processes and Discrete-Event Simulations. *Operations Research*, 23, No. 1 (January–February), 33–45.

CRANE, M. A., and A. J. LEMOINE. [1977]. *An Introduction to the Regenerative Method for Simulation Analysis*. New York: Springer-Verlag.

DALEY, D. J. [1976]. Queueing Output Processes. *Advances in Applied Probability* 8, No. 2 (June), 395–415.

DESCLOUX, A. [1962]. *Delay Tables for Finite- and Infinite-Source Systems*. New York: McGraw-Hill.

———. [1963]. On Overflow Processes of Trunk Groups with Poisson Inputs and Exponential Service Times. *The Bell System Technical Journal* 42, No. 2 (March), 383–398.

———. [1965]. On the Accuracy of Loss Estimates. *The Bell System Technical Journal* 44, No. 6 (July–August), 1139–1164.

———. [1967]. On the Validity of the Particular Subscriber's Point of View. Abstract in *Proceedings of the Fifth International Teletraffic Congress, New York, June 14–20*.

———. [1970]. On Markovian Servers with Recurrent Input. *Proceedings of the Sixth International Teletraffic Congress, Munich, September 9–15*.

———. [1973]. Traffic Measurement Biases Induced by Partial Sampling. *The Bell System Technical Journal* 52, No. 8 (October), 1375–1402.

———. [1975]. Variance of Load Measurements in Markovian Service Systems. *The Bell System Technical Journal* 54, No. 7 (September), 1277–1300.

———. [1976]. Some Properties of the Variance of the Switch-Count Load. *The Bell System Technical Journal* 55, No. 1 (January), 59–88.

DE SMIT, J. H. A. [1973]. Some General Results for Many Server Queues. *Advances in Applied Probability* 5, No. 1 (April), 153–169.

DIETRICH, G., W. KRUSE, R. MICHEL, F. ONDRA, E. PETER, and H. WAGNER. [1966]. *Teletraffic Engineering Manual*. Stuttgart: Standard Elektrik Lorenz AG.

DISNEY, R. L. [1975]. Random Flow in Queueing Networks: A Review and a Critique. *AIIE Transactions* 7, No. 3 (September), 268–288.

———, R. L. FARRELL, and P. R. DE MORAIS. [1973]. A Characterization of $M/G/1$ Queues with Renewal Departure Processes. *Management Science* 19, No. 11 (July), 1222–1228.

———, D. C. MCNICKLE, and B. SIMON [1980]. The $M/G/1$ Queue with Instantaneous Bernoulli Feedback. *Naval Research Logistics Quarterly* (in press).

DURNAN, C. J. [1967]. Estimates of the Delay Distributions for Queueing Systems with Constant Server Holding Time and Random Order of Service. *Proceedings of the Fifth International Teletraffic Congress, New York, June 14–20*, pp. 96–106..

EISENBERG, M. [1972]. Queues with Periodic Service and Changeover Time. *Operations Research* 20, No. 2 (March–April), 440–451.

FELLER, W. [1968]. *An Introduction to Probability Theory and Its Applications*, Vol. I, 3rd ed. New York: John Wiley and Sons.

———. [1971]. *An Introduction to Probability Theory and Its Applications*, Vol. II, 2nd ed. New York: John Wiley and Sons.

FERRARI, D. [1978]. *Computer Systems Performance Evaluation*. Englewood Cliffs, NJ: Prentice-Hall.

FINCH, P. D. [1959]. Cyclic Queues with Feedback. *Journal of the Royal Statistical Society* B21, No. 1, 153–157.

FISHMAN, G. S. [1973]. *Concepts and Methods in Discrete Event Digital Simulation*. New York: John Wiley and Sons.

———. [1978a]. *Principles of Discrete Event Simulation*. New York: John Wiley and Sons.

———. [1978b]. Grouping Observations in Digital Simulation. *Management Science* 24, No. 5 (January), 510–521.

FISZ, M. [1963]. *Probability Theory and Mathematical Statistics*, 3rd ed. New York: John Wiley and Sons.

FOLEY, R. D. [1979]. The $M/G/1$ Queue with Delayed Feedback. Dissertation (Ph.D., Industrial and Operations Engineering), The University of Michigan, Ann Arbor.

FORTET, R., and C. GRANDJEAN. [1963]. Study of the Congestion in a Loss System. Paper presented at the Fourth International Teletraffic Congress, London, July 15–21, 1964. Laboratoire Central de Télécommunications, 46 Avenue de Breteuil, Paris (VIIe), as internal memorandum, February 12, 1963.

FRANKEL, T. [1976]. *Tables for Traffic Management and Design: Book 1—Trunking*. Geneva, IL: Lee's abc of the Telephone.

FRANKEN, P., D. KÖNIG, U. ARNDT, and V. SCHMIDT. [1980]. *Queues and Point Processes*. Berlin: Akadomie-Verlag.

FREDERICKS, A. A. [1980]. Congestion in Blocking Systems—A Simple Approximation Technique. *The Bell System Technical Journal* 59, No. 6 (July–August), 805–827.

FRY, T. C. [1965]. *Probability and Its Engineering Uses*, 2nd ed. New York: Van Nostrand. First edition published in 1928.

GARABEDIAN, P. R. [1964]. *Partial Differential Equations*. New York: John Wiley and Sons. Chapter 2, Section 1.

GAVER, D. P. [1954]. The Influence of Servicing Times in Queuing Processes. *Operations Research* 2, 139–149.

GIFFIN, W. C. [1978]. *Queueing: Basic Theory and Applications*. Columbus, OH: Grid Publishing.

GNEDENKO, B. V., and I. N. KOVALENKO. [1968]. *Introduction to Queueing Theory*. Jerusalem: Israel Program for Scientific Translations.

GORDON, G. [1975]. *The Application of GPSS V to Discrete System Simulation*. Englewood Cliffs, NJ: Prentice-Hall.

———. [1978]. *System Simulation*, 2nd ed. Englewood Cliffs, NJ: Prentice-Hall.

GORDON, W. J., and G. F. NEWELL. [1967]. Closed Queuing Systems with Exponential Servers. *Operations Research* 15, No. 2 (March–April), 254–265.

GROSS, D., and C. M. HARRIS. [1974]. *Fundamentals of Queueing Theory*. New York: John Wiley and Sons.

GUMBEL, H. [1960]. Waiting Lines with Heterogeneous Servers. *Operations Research* 8, No. 4, 504–511.

GUNTHER, F. L., and R. W. WOLFF [1980]. The Almost Regenerative Method for Stochastic System Simulations. *Operations Research* 28, No. 2 (March–April), 375–386.

HAIGHT, F. A. [1963]. *Mathematical Theories of Traffic Flow*. New York: Academic Press.

HALFIN, S. [1975]. An Approximate Method for Calculating Delays for a Family of Cyclic-Type Queues. *The Bell System Technical Journal* 54, No. 10 (December), 1733–1754.

———. [1979]. A Singular Property of Multiserver Systems with Constant Service Time. *Journal of Applied Probability* 16, No. 2 (June), 362–372.

HALMSTAD, D. G. [1974]. Actuarial Techniques and Their Relations to Noninsurance Models. *Operations Research* 22, No. 5 (September–October), 942–953.

HARRISON, G. [1959]. Stationary Single-Server Queuing Processes with a Finite Number of Sources. *Operations Research* 7, No. 4 (July–August), 458–467.

HASOFER, A. M. [1963]. On the Integrability, Continuity and Differentiability of a Family of Functions Introduced by L. Takács. *The Annals of Mathematical Statistics* 34, 1045–1049.

HEIDELBERGER, P. [1977]. Variance Reduction Techniques for the Simulation of Markov Processes, I: Multiple Estimates. *Technical Report 42*, Stanford, CA: Stanford University, Department of Operations Research, October.

———. [1978a]. Variance Reduction Techniques for the Simulation of Markov Processes, II: Matrix Iterative Methods. *Technical Report 44*. Stanford, CA: Stanford University, Department of Operations Research, January.

———. [1978b]. Variance Reduction Techniques for the Simulation of Markov Processes, III: Increasing the Frequency of Regenerations. *Technical Report 45*. Stanford, CA: Stanford University, Department of Operations Research, January.

———, and D. L. IGLEHART [1979]. Comparing Stochastic Systems Using Regenerative Simulation with Common Random Numbers. *Advances in Applied Probability* 11, No. 4 (December), 804–819.

———, and P. A. W. LEWIS. [1980]. Regression-Adjusted Estimates for Regenerative Simulations, with Graphics. To appear as *IBM Research Report*. Yorktown Heights, NY: IBM Thomas J. Watson Research Center.

———, and P. D. WELCH. [1980]. A Spectral Method for Simulation Confidence Interval Generation and Run Length Control. *IBM Research Report RC 8264 (#35826)*, April 28. Yorktown Heights, NY: IBM Thomas J. Watson Research Center.

HEYMAN, D. P. [1968]. Optimal Operating Policies for $M/G/1$ Queuing Systems. *Operations Research* 16, No. 2 (March–April), 362–382.

———. [1977]. The T-Policy for the $M/G/1$ Queue. *Management Science* 23, No. (March), 775–778.

———, and M. J. SOBEL [1981]. *Stochastic Models in Operations Research* (two volumes). New York: McGraw-Hill. (In preparation.)

———, and S. STIDHAM, JR. [1980]. The Relation Between Customer and Time Averages in Queues. *Operations Research* 28, No. 4 (July–August).

HILLIER, F. S., and G. J. LIEBERMAN. [1980]. *Introduction to Operations Research*, 3rd ed. San Francisco: Holden-Day.

_____, and F. D. Lo. [1971]. *Tables for Multiple-Server Queueing Systems Involving Erlang Distributions*. Technical Report No. 31, Dept. of Operations Research, Stanford University.

_____, and S. STIDHAM, JR. [1981]. *Queueing Theory: Describing and Optimizing Queueing Systems*. Englewood Cliffs, NJ: Prentice-Hall. (In preparation.)

_____, and O. S. YU. [1981]. *Queueing Tables and Graphs*. New York: North Holland.

HOKSTAD, P. [1977]. Asymptotic Behavior of the $E_k/G/1$ Queue with Finite Waiting Room. *Journal of Applied Probability* 14, No. 2 (June), 358–366.

HOLTZMAN, J. M. [1973]. The Accuracy of the Equivalent Random Method with Renewal Inputs. *The Bell System Technical Journal* 52, No. 9 (November), 1673–1679.

IGLEHART, D. L. [1978]. The Regenerative Method for Simulation Analysis. In *Current Trends in Programming Methodology, Volume III: Software Modeling* (K. M. Chandy and R. T. Yeh, eds.). Englewood Cliffs, NJ: Prentice-Hall, pp. 52–71.

_____, and G. S. SHEDLER. [1978]. Regenerative Simulation of Response Times in Networks of Queues. *Journal of the Association for Computing Machinery* 25, No. 3 (July), 449–460.

JACKSON, J. R. [1957]. Networks of Waiting Lines. *Operations Research* 5, 518–521.

JACKSON, R. R. P. [1954]. Queueing Systems with Phase Type Service. *Operational Research Quarterly* 5 (December), 109–120.

_____. [1956]. Queueing Processes with Phase-type Service. *Journal of the Royal Statistical Society* B18, No. 1, 129–132.

JAGERMAN, D. L. [1974]. Some Properties of the Erlang Loss Function. *The Bell System Technical Journal* 53, No. 3 (March), 525–551.

_____. [1975]. Nonstationary Blocking in Telephone Traffic. *The Bell System Technical Journal* 54, No. 3 (March), 625–661.

_____. [1978]. An Inversion Technique for the Laplace Transform with Application to Approximation. *The Bell System Technical Journal* 57, No. 3 (March), 669–710.

JAISWAL, N. K. [1968]. *Priority Queues*. New York: Academic Press.

KARLIN, S. [1968]. *A First Course in Stochastic Processes*. New York: Academic Press.

KATZ, S. S. [1967]. Statistical Performance Analysis of a Switched Communications Network. *Proceedings of the Fifth International Teletraffic Congress, Rockefeller University, New York, June 14–20*, pp. 566–575.

KEILSON, J. [1965]. A Review of Transient Behavior in Regular Diffusion and Birth-Death Processes. Part II. *Journal of Applied Probability*, 2, 405–428.

_____. [1979]. *Markov Chain Models—Rarity and Exponentiality*. New York: Springer-Verlag.

KELLY, F. P. [1976]. Networks of Queues. *Advances in Applied Probability* 8, No. 2 (June), 416–432.

_____. [1979]. *Reversibility and Stochastic Networks*. Chichester: John Wiley and Sons Ltd.

KENDALL, D. G. [1951]. Some Problems in the Theory of Queues. *Journal of the Royal Statistical Society* B13, 151–185.

_____. [1953]. Stochastic Processes Occurring in the Theory of Queues and Their Analysis by Means of the Imbedded Markov Chain. *The Annals of Mathematical Statistics* 24, 338–354.

KHINTCHINE, A. Y. [1932]. Mathematical Theory of a Stationary Queue. (In Russian.) *Matematicheskii Sbornik* 39, No. 4, 73–84.

_____. [1969]. *Mathematical Methods in the Theory of Queueing*, 2nd ed. New York: Hafner.

KINGMAN, J. F. C. [1962]. On Queues in Which Customers Are Served in Random Order. *Proceedings of the Cambridge Philosophical Society* 58, 79–91.

KIVIAT, P. J., R. VILLANUEVA, and H. M. MARKOWITZ. [1973]. In *SIMSCRIPT II.5 Programming Language* (E. C. Russell, ed.). Los Angeles: Consolidated Analysis Centers.

KLEINROCK, L. [1964a]. Analysis of a Time-Shared Processor. *Naval Research Logistics Quarterly* 11, No. 10 (March), 59–73.

_____. [1964b]. *Communication Nets: Stochastic Message Flow and Delay*. New York: McGraw-Hill. Reprinted by Dover Publications, New York.

_____. [1967]. Time-Shared Systems: A Theoretical Treatment. *Journal of the Association for Computing Machinery* 14, No. 2 (April), 242–261.

_____. [1975]. *Queueing Systems, Volume I: Theory*. New York: John Wiley and Sons.

_____. [1976]. *Queueing Systems, Volume II: Computer Applications*. New York: John Wiley and Sons.

KNUTH, D. E. [1969]. *The Art of Computer Programming: Seminumerical Algorithms*, Vol. 2. Reading, MA: Addison-Wesley.

KOBAYASHI, H. [1978]. *Modeling and Analysis: An Introduction to System Performance Evaluation Methodology*. Reading, MA: Addison-Wesley.

KOENIGSBERG, E. [1958]. Cyclic Queues. *Operational Research Quarterly* 9, No. 1, 22–35.

KOSTEN, L. [1937]. Über Sperrungswahrscheinlichkeiten bei Staffelschaltungen, *Elektro Nachrichten–Technik* 14, 5–12.

_____. [1973]. *Stochastic Theory of Service Systems*. New York: Pergamon Press.

KUCZURA, A. [1973]. The Interrupted Poisson Process as an Overflow Process. *The Bell System Technical Journal* 52, No. 3 (March), 437–448.

_____, and S. R. NEAL. [1972]. The Accuracy of Call-Congestion Measurements for Loss Systems with Renewal Input. *The Bell System Technical Journal* 51, No. 10 (December), 2197–2208.

KUEHN, P. J. [1979]. Multiqueue Systems with Nonexhaustive Cyclic Service. *The Bell System Technical Journal* 58, No. 3 (March), 671–698.

LAVENBERG, S. S. [1978]. Regenerative Simulation of Queueing Networks. *IBM*

Research Report RC 7087 (#30391), April 24. Yorktown Heights, NY: IBM Thomas J. Watson Research Center.

_____, and C. H. SAUER. [1977]. Sequential Stopping Rules for the Regenerative Method of Simulation. *IBM Journal of Research and Development* 21, No. 6 (November), 545–558.

_____, and D. R. SLUTZ. [1975]. Introduction to Regenerative Simulation. *IBM Journal of Research and Development* 19, No. 5 (September), 458–463.

LAW, A. M., and J. S. CARSON. [1979]. A Sequential Procedure for Determining the Length of a Steady-State Simulation. *Operations Research* 27, No. 5 (September–October), 1011–1025.

LEE, A. M. [1966]. *Applied Queueing Theory*. New York: St. Martin's Press.

LEGALL, P. [1962]. *Les systèmes avec ou sans attente et les processes stochastiques*, Tome I. Paris: Dunod.

LEIBOWITZ, M. A. [1968]. Queues. *Scientific American* 219, No. 2 (August), 96–103.

LEMOINE, A. J. [1977]. Networks of Queues—A Survey of Equilibrium Analysis. *Management Science* 24, No. 4 (December), 464–481.

_____. [1978]. Networks of Queues—A Survey of Weak Convergence Results. *Management Science* 24, No. 11 (July), 1175–1193.

_____. [1979]. On Total Sojourn Time in Networks of Queues. *Management Science* 25, No. 10 (October), 1034–1035.

LEVY, Y., and U. YECHIALI. [1975]. Utilization of Idle Time in an $M/G/1$ Queueing System. *Management Science* 22, No. 2 (October), 202–211.

LEWIS, P. A. W., and G. S. SHEDLER. [1978]. Simulation of Nonhomogeneous Poisson Processes by Thinning. *Technical Report NPS-55-78-014*, Monterey, CA: Naval Postgraduate School.

LEWIS, T. G. [1975]. *Distribution Sampling for Computer Simulation*. Lexington, MA: D. C. Heath and Company.

LITTLE, J. D. C. [1961]. A Proof for the Queuing Formula: $L=\lambda W$. *Operations Research* 9, No. 3, 383–387.

LOTZE, A. [1964]. A Traffic Variance Method for Gradings of Arbitrary Type. *Proceedings of the Fourth International Teletraffic Congress, London, July 15–21*, Volume III.

MELAMED, B. [1979a]. On Poisson Traffic Processes in Discrete-State Markovian Systems with Applications to Queueing Theory. *Advances in Applied Probability* 11, No. 1 (March), 218–239.

_____. [1979b]. Characterizations of Poisson Traffic Streams in Jackson Queueing Networks. *Advances in Applied Probability* 11, No. 2 (June), 422–438.

MESSERLI, E. J. [1972]. Proof of a Convexity Property of the Erlang B Formula. *The Bell System Technical Journal* 51, No. 4 (April), 951–953.

MINA, R. R. [1974]. *Introduction to Teletraffic Engineering*. Chicago: Telephony Publishing Corporation.

MITRANI, I. [1979]. A Critical Note on a Result by Lemoine. *Management Science* 25, No. 10 (October), 1026–1027.

MIYAZAWA, M. [1977]. Time and Customer Processes in Queues with Stationary Inputs. *Journal of Applied Probability* 14, No. 2 (June), 349–357.

MORSE, P. [1958]. *Queues, Inventories and Maintenance*. New York: John Wiley and Sons.

MURDOCH, J. [1978]. *Queueing Theory: Worked Examples and Problems*. London: The Macmillan Free Press Ltd.

NEAL, S. R. [1971]. Combining Correlated Streams of Nonrandom Traffic. *The Bell System Technical Journal* 50, No. 6 (July–August), 2015–2037.

———, and A. KUCZURA. [1973]. A Theory of Traffic-Measurement Errors for Loss Systems with Renewal Input. *The Bell System Technical Journal* 52, No. 6 (July–August), 967–990.

NEUTS, M. F. [1973]. *Probability*. Boston: Allyn and Bacon.

———. [1975a]. Probability Distributions of Phase Type, in *Liber Amicorum Professor Emeritus H. Florin*, Louvain, Belgium: University of Louvain, Department of Mathematics, pp. 173–206.

———. [1975b]. Computational Uses of the Method of Phases in the Theory of Queues. *Computers and Mathematics with Applications* I, 151–166.

———. [1977]. Algorithms for the Waiting Time Distributions Under Various Queue Disciplines in the $M/G/1$ Queue with Service Time Distributions of Phase Type. In *Algorithmic Methods in Probability* (M. F. Neuts, ed.). Amsterdam: North-Holland/TIMS Studies in the Management Sciences 7, pp. 177–197.

———. [1978a]. Markov Chains with Applications in Queueing Theory, Which Have a Matrix-Geometric Invariant Probability Vector. *Advances in Applied Probability* 10, No. 1 (March), 185–212.

———. [1978b]. Renewal Processes of Phase Type. *Naval Research Logistics Quarterly* 25, No. 3 (September), 445–454.

———. [1980]. Explicit Steady-State Solutions to Some Elementary Queueing Models. *Applied Mathematics Institute*, Technical Report No. 53B, March. Newark, DE: University of Delaware.

———. [1981]. *Matrix-Geometric Solutions in Stochastic Models—An Algorithmic Approach*. Baltimore, MD: The Johns Hopkins University Press. (In preparation.)

NEWELL, G. F. [1971]. *Applications of Queueing Theory*. London: Chapman and Hall.

———. [1973]. *Approximate Stochastic Behavior of n-Server Service Systems with Large n*. Berlin and New York: Springer-Verlag Lecture Notes in Economics and Mathematical Systems, Vol. 87.

———. [1979]. *Approximate Behavior of Tandem Queues*. Berlin and New York: Springer-Verlag Lecture Notes in Economics and Mathematical Systems, Vol. 171.

NOETZEL, A. S. [1979]. A Generalized Queueing Discipline for Product Form Network Solutions. *Journal of the Association for Computing Machinery* 26, No. 4 (October), 779–793.

OAKES, D. [1976]. Random Overlapping Intervals—A Generalisation of Erlang's Loss Formula. *Annals of Probability* 4, No. 6, 940–946].

PAGE, E. [1972]. *Queueing Theory in OR*. New York: Crane Russak and Company.

PALM, C. [1943]. Intensitätsschwankungen im Fernsprechverkehr. *Ericsson Technics* 44, 1–189.

Panico, J. A. [1969]. *Queuing Theory: A Study of Waiting Lines for Business, Economics, and Science.* Englewood Cliffs, NJ: Prentice-Hall.

Peck, L. G., and R. N. Hazelwood. [1958]. *Finite Queuing Tables.* New York: John Wiley and Sons.

Pittel, B. [1979]. Closed Exponential Networks of Queues with Saturation: The Jackson-Type Stationary Distribution and Its Asymptotic Analysis. *Mathematics of Operations Research* 4, No. 4 (November), 357–378.

Pollaczek, F. [1930]. Über eine Aufgabe der Wahrscheinlichkeitstheorie, I–II. *Mathem. Zeitschrift* 32, 64–100, 729–750.

———. [1946]. La Loi d'attente des appels téléphoniques. *Comptes Rendus Académie des Sciences, Paris* 222, 353–355.

———. [1957]. *Problèmes stochastiques posés par le phénomène de formation d'une queue d'attente à un guichet et par des phénomènes apparentés.* (Mémorial de Sciences Mathématiques No. 136.) Paris: Gauthier–Villars.

———. [1961]. *Théorie analytique des problèmes stochastiques relatifs à un groupe de lignes téléphoniques avec dispositif d'attente.* (Mémorial des Sciences Mathématiques, No. 150.) Paris: Gauthier–Villars.

Prabhu, N. U. [1965]. *Queues and Inventories.* New York: John Wiley and Sons.

Pritsker, A. A. B. [1974]. *GASP IV Simulation Language.* New York: John Wiley and Sons.

———, and C. D. Pegden. [1979]. *Introduction to Simulation and SLAM.* New York: Halsted (a division of John Wiley and Sons).

Rapp, Y. [1964]. Planning of Junction Network in a Multiexchange Area. *Ericsson Technics* 20, No. 1, 77–130.

Reich, E. [1957]. Waiting Times When Queues Are in Tandem. *The Annals of Mathematical Statistics* 28, 768–773.

Reiser, M., and C. H. Sauer. [1978]. Queueing Network Models: Methods of Solution and Their Program Implementation. In *Current Trends in Programming Methodology, Volume III: Software Modeling* (K. M. Chandy and R. T. Yeh, eds.). Englewood Cliffs, NJ: Prentice-Hall, pp. 115–167.

Riordan, J. [1953]. Delay Curves for Calls Served at Random. *The Bell System Technical Journal* 32, 100–119.

———. [1956]. Derivation of Moments of Overflow Traffic. (Appendix I of R. I. Wilkinson, Theories for Toll Traffic Engineering in the U.S.A.) *The Bell System Technical Journal* 35, No. 2 (March), 421–514.

———. [1961]. Delays for Last-Come First-Served Service and the Busy Period. *The Bell System Technical Journal* 40, No. 3 (May), 785–793.

———. [1962]. *Stochastic Service Systems.* New York: John Wiley and Sons.

———. [1968]. *Combinatorial Identities.* New York: John Wiley and Sons.

Ross, S. M. [1970]. *Applied Probability Models with Optimization Applications.* San Francisco: Holden-Day.

———. [1972]. *Introduction to Probability Models.* New York: Academic Press.

Ruiz-Palá, E., C. Ávila-Beloso, and W. W. Hines. [1967]. *Waiting-Line Models: An Introduction to Their Theory and Application.* New York: Reinhold.

References

RUNNENBURG, J. TH. [1965]. In *Proceedings of the Symposium on Congestion Theory* (W. L. Smith and W. E. Wilkinson, eds.). Chapel Hill: University of North Carolina Press, Chapter 13, On the Use of the Method of Collective Marks in Queueing Theory, pp. 399–438.

SAATY, T. L. [1961]. *Elements of Queueing Theory*. New York: McGraw-Hill.

SAKATA, M., S. NOGUCHI, and J. OIZUMI. [1969]. Analysis of a Processor Shared Queueing Model for Time Sharing Systems. *Proceedings of the Second Hawaii International Conference on System Sciences, University of Hawaii, Honolulu, January*, pp. 625–628.

———. [1971]. An Analysis of the $M/G/1$ Queue Under Round-Robin Scheduling. *Operations Research* 19, No. 2 (March–April), 371–385.

SAUER, C. H., and K. M. CHANDY. [1981]. *Computer Systems Performance Modeling: A Primer*. Englewood Cliffs., NJ: Prentice-Hall. (In preparation.)

———, and E. A. MACNAIR. [1977]. Computer/Communication System Modeling with Extended Queueing Networks. *IBM Research Report RC 6654 (#28630)*, July 22. Yorktown Heights, NY: IBM Thomas J. Watson Research Center.

———. [1978]. Queueing Network Software for Systems Modeling. *IBM Research Report RC 7143 (#30592)*, May 24. Yorktown Heights, NY: IBM Thomas J. Watson Research Center.

SCHASSBERGER, R. [1973]. *Warteschlangen*. Vienna: Springer-Verlag.

———. [1978a]. The Insensitivity of Stationary Probabilities in Networks of Queues. *Advances in Applied Probability* 10, No. 4 (December), 906–912. See also correction in the same journal, Vol. 12, No. 2 (June, 1980), 541.

———. [1978b]. Insensitivity of Steady-State Distributions of Generalized Semi-Markov Processes with Speeds. *Advances in Applied Probability* 10, No. 4 (December), 836–851.

———. [1978c]. Mean Sojourn Times in Insensitive Generalized Semi-Markov Schemes. *Research Paper 380*, Calgary, Alberta: The University of Calgary, Department of Mathematics and Statistics, January, revised April.

SCHERR, A. L. [1967]. *An Analysis of Time-Shared Computer Systems*. Cambridge, MA: M.I.T. Press.

SCHMEISER, B. W. [1977]. Methods for Modelling and Generating Probabilistic Components in Digital Computer Simulation When the Standard Distributions Are Not Adequate: A Survey. *Proceedings of the 1977 Winter Simulation Conference*, Gaithersburg, MD, December, pp. 51–57.

SCHRIBER, T. J. [1974]. *Simulation Using GPSS*. New York: John Wiley and Sons.

SCHRUBEN, L. W. [1980]. A Coverage Function for Interval Estimators of Simulation Response. *Management Science* 26, No. 1 (January), 18–27.

SEAL, H. L. [1969]. *Stochastic Theory of a Risk Business*. New York: John Wiley and Sons.

SEVAST'YANOV, B. A. [1957]. An Ergodic Theorem for Markov Processes and Its Application to Telephone Systems with Refusals. *Theory of Probabilities and Its Application* 2, 104–112.

SHANNON, R. E. [1975]. Simulation: A Survey with Research Suggestions. *AIIE Transactions* 7, No. 3 (September), 289–301.

Simon, B., and Foley, R. D. [1979]. Some Results on Sojourn Times in Acyclic Jackson Networks. *Management Science* 25, No. 10 (October), 1027–1034.

Smith, W. L. [1953]. On the Distribution of Queueing Times, *Proceedings of the Cambridge Philosophical Society* 49, 449–461.

Sobel, M. J. [1973]. Optimal Operation of Queues. In (A. B. Clarke, ed.) *Mathematical Methods in Queueing Theory* (Proceedings of a Conference at Western Michigan University, May 10–12, 1973). Berlin and New York: Springer-Verlag Lecture Notes in Economics and Mathematical Systems, Vol. 98, pp. 231–261. 1974.

Stidham, S., Jr. [1972]. Regenerative Processes in the Theory of Queues, with Applications to the Alternating-Priority Queue. *Advances in Applied Probability* 4, No. 3 (December), 542–577.

———. [1974]. A Last Word on $L=\lambda W$. *Operations Research* 22, No. 2 (March–April), 417–421.

———. [1977]. Optimal Control of Stochastic Service Systems. Abstract in *Advances in Applied Probability* 10, No. 2 (June 1978), 277–278 (Seventh Conference on Stochastic Processes and Their Applications, Enschede, The Netherlands, August 15–19, 1977).

———, and N. U. Prabhu. [1973]. Optimal Control of Queueing Systems. In (A. B. Clarke, ed.) *Mathematical Methods in Queueing Theory* (Proceedings of a Conference at Western Michigan University, May 10–12, 1973). Berlin and New York: Springer-Verlag Lecture Notes in Economics and Mathematical Systems, Vol. 98, pp. 263–294. 1974.

Stoyan, D. [1977]. Queueing Networks—Insensitivity and a Heuristic Approximation. Abstract in *Advances in Applied Probability* 10, No. 2 (June 1978), 318 (Seventh Conference on Stochastic Processes and Their Applications, Enschede, The Netherlands, August 15–19, 1977).

Syski, R. [1960]. *Introduction to Congestion Theory in Telephone Systems*. London: Oliver and Boyd.

Takács, L. [1955]. Investigation of Waiting Time Problems by Reduction to Markov Processes. *Acta Math. Acad. Sci. Hungary* 6, 101–129.

———. [1959]. On the Limiting Distribution of the Number of Coincidences Concerning Telephone Traffic. *Annals of Mathematical Statistics* 30, 134–142.

———. [1960a]. *Stochastic Processes: Problems and Solutions*. New York: John Wiley and Sons.

———. [1960b]. Transient Behavior of Single-Server Queuing Processes with Recurrent Input and Exponentially Distributed Service Times. *Operations Research* 8, 231–245.

———. [1962a]. *Introduction to the Theory of Queues*. New York: Oxford University Press.

———. [1962b]. Delay Distributions for Simple Trunk Groups with Recurrent Input and Exponential Service Times. *The Bell System Technical Journal* 41, No. 1 (January), 311–320.

———. [1963]. Delay Distributions for One Line with Poisson Input, General Holding Times, and Various Orders of Service. *The Bell System Technical Journal* 42, No. 2 (March), 487–503.

———. [1967]. *Combinatorial Methods in the Theory of Stochastic Processes.* New York: John Wiley and Sons.

———. [1969]. On Erlang's Formula. *The Annals of Mathematical Statistics* 40, 71–78.

TANNER, J. C. [1953]. A Problem of Interference Between Two Queues. *Biometrika* 40, 58–69.

———. [1961]. A Derivation of the Borel Distribution. *Biometrika* 48, 222–224.

VANTILBORGH, H. [1978]. Exact Aggregation in Exponential Queueing Networks. *Journal of the Association for Computing Machinery* 25, No. 4 (October), 620–629.

VAULOT, E. [1925]. Application du calcul des probabilités à l'exploitation téléphonique. *Annales des Postes, Télégraphes et Téléphones* 14, No. 2, 135–156.

WAGNER, H. M. [1975]. *Principles of Operations Research*, 2nd ed. Englewood Cliffs, NJ: Prentice-Hall.

WALLACE, V. L. [1973]. Algebraic Techniques for Numerical Solution of Queueing Networks. In (A. B. Clarke, ed.) *Mathematical Methods in Queueing Theory* (Proceedings of a Conference at Western Michigan University, May 10–12, 1973). Berlin and New York: Springer-Verlag Lecture Notes in Economics and Mathematical Systems, Vol. 98, pp. 295–305. 1974.

WALLSTRÖM, B. [1966]. Congestion Studies in Telephone Systems with Overflow Facilities. *Ericsson Technics*, No. 3, pp. 190–351.

WHITE, J. A., J. W. SCHMIDT, and G. K. BENNETT. [1975]. *Analysis of Queueing Systems.* New York: Academic Press.

WIDDER, D. V. [1941]. *The Laplace Transform.* Princeton, NJ: Princeton University Press.

———. [1961]. *Advanced Calculus*, 2nd ed. Englewood Cliffs, NJ: Prentice-Hall.

———. [1971]. *An Introduction to Transform Theory.* New York: Academic Press.

WILKINSON, R. I. [1956]. Theories for Toll Traffic Engineering in the U.S.A. *The Bell System Technical Journal* 35, No. 2 (March), 421–514.

———. [1970]. *Nonrandom Traffic Curves and Tables for Engineering and Administrative Purposes.* Holmdel, NJ: Bell Telephone Laboratories, Traffic Studies Center.

WISHART, D. M. G. [1960]. Queuing Systems in Which the Discipline Is 'Last-Come, First-Served.' *Operations Research* 8, 591–599.

———. [1961]. An Application of Ergodic Theorems in the Theory of Queues. *Proceedings of the Fourth Berkeley Symposium on Mathematical Statistics and Probability, Vol. 2.* Berkeley and Los Angeles: University of California Press, 581–592.

YADIN, M., and P. NAOR. [1963]. Queueing Systems with a Removable Service Station. *Operational Research Quarterly* 14, No. 4, 393–405.

YAKOWITZ, S. J. [1977]. *Computational Probability and Simulation.* Reading, MA: Addison-Wesley.

YOUNG, D. M. [1971]. *Iterative Solution of Large Linear Systems.* New York: Academic Press.

Index

A

Abel's generalization of binomial theorem, 233
Actual waiting time, 98
Additional conditioning variable, 256, 260, 275, 279; *see also* Supplementary variable
Advances in Applied Probability, 313
Age, *see* Recurrence time, backward
Allen, A.O., 311, 325
Arndt, U., 310, 330
Arriving customer's distribution, 57, 77, 105, 185
Ashton, W.D., 71, 325
At random, 53
Ávila-Beloso, C., 309, 337
Axness, Carl, xv

B

Balachandran, K.R., 253, 325
Balance equations, 24
Baldwin, J.C.T., 283
Ballot theorems, 231, 312
Barbour, A.D., 135, 175, 325
Baskett, F., 135, 325
Batch arrivals, 241
Batch means, 297
Bayes' rule, 78
BCC, *see* Blocked customers cleared
BCD, *see* Blocked customers delayed
Bear, D., 308, 325
Beckmann, P., 308, 325
Bell System Technical Journal, The, 313
Beneš, V.E., 117, 217, 295, 309, 325, 326
Bennett, G.K., 309, 339
Bernoulli distribution, 34
Bernoulli trials, 34, 42

Bessel function, 121
Beutler, F.J., 135, 136, 326
Bhat, U.N., 72, 237, 312, 326
Binomial distribution, 17, 34, 46, 53
Binomial theorem, 30
Birth-and-death process, 14, 22, 24, 44, 74
Birth-and-death queueing models, 74
  multidimensional, 123
Birth rate, 16
Blocked customers cleared, 74, 79, 108
Blocked customers delayed, 74, 90, 111
Blocked customers held, 87
Bobillier, P.A., 287, 326
Borel, E., 231, 326
Borel distribution, 231
Borel-Tanner distribution, 231
Borovkov, A.A., 309, 326
Boxma, O.J., 135, 326
Brandwajn, A., 160, 326
Bretschneider, G., 168, 326
Brockmeyer, E., 80, 91, 140, 310, 326
Brumelle, S.L., 79, 84, 181, 326
Buchner, M.M., Jr., 131, 327
Buffon's Needle Problem, 282
Burke, Paul J., xv, 126, 135, 136, 148, 154, 180, 186, 188, 189, 242, 253, 256, 261, 275, 279, 300, 327
Burke's theorem, 135, 136, 188
Busy period, 13, 93, 102, 228, 239
Buzen, J., 138, 160, 327

C

Carson, J.S., 297, 327
Carter, Grace M., xv, 253, 258, 259, 260, 275, 276, 327; *see also* Murray, G.

Central limit theorem, 290
Chaiken, J.M., 138, 139, 327
Chan, J., 79, 328
Chandy, K.M., 135, 311, 325, 327, 337
Cheng, R.C.H., 286, 327
Çinlar, E., 9, 13, 26, 72, 138, 208, 212, 312, 327, 328
Clarke, A.B., 72, 328
Classical ruin problem, 241
Coffman, E.G., Jr., 302, 311, 328
Cohen, J.W., 14, 26, 72, 84, 111, 117, 122, 200, 212, 214, 220, 223, 237, 307, 310, 328
Collective marks, method of, 31, 220
Combinatorial methods, 231, 241
Compound distribution, 30
Compound Poisson process, 61
Computer science, *see* Computer system performance evaluation
Computer system performance evaluation, 2, 111, 115, 135, 301, 311
Confidence interval, 295
Congestion theory, 2
Conolly, B.W., 79, 308, 328
Conservation of flow, 3, 4, 24, 123
Convolution, 28, 198
Conway, R.W., 310, 328
Cooper, Robert B., 157, 223, 241, 253, 258, 259, 260, 266, 267, 275, 276, 303, 327, 328
Counting process, 50
Courtois, P.J., 135, 311, 328
Covering interval, 54, 56, 205
Cox, D.R., 72, 111, 122, 173, 308, 312, 328
Crabill, T.B., 244, 328
Cramér, H., 219, 220, 329
Crane, M.A., 297, 329
Cycle, 13
Cycle time, 206
Cyclic order, service in, 261

D

D-policy, 253
Daley, D.J., 135, 329
Dams, 2
De Buffon, Count, 282
Death rate, 16
Decomposition, of Poisson Process, 59
DeMorais, P.R., 136, 329
Denning, P.J., 302, 311, 328
Descloux, A., 91, 97, 113, 117, 125, 140, 148, 149, 153, 208, 295, 312, 329
DeSmit, J.H.A., 178, 329
Departing customer's distribution, 185
Dietrich, G., 81, 91, 97, 312, 329
Disney, Ralph L., xv, 72, 135, 136, 208, 328, 329

Do-nothing policy, 245
Durnan, C.J., 298, 299, 300, 330
Durnan-Weber dip, 300

E

$E_k M/s$, 178
Eisenberg, M., 266, 330
Emmons, Hamilton, xv
Engset formula, 108, 117
Equilibrium, *see* Statistical equilibrium
Equivalent random method, 140, 165
Ergodic property, 20
Erl, 4; *see also* Erlangs
Erlang, A.K., 4, 80, 84, 91, 173, 309
Erlang B formula, *see* Erlang loss formula
Erlang C formula, *see* Erlang delay formula
Erland delay formula, 7, 91, 320, 322
Erlang delay system, 90, 156, 320, 324
Erlang loss distribution, 80, 90
Erlang loss formula, 5, 80, 316, 318
Erlang loss system, 79, 95, 117, 154, 171, 175, 179, 288
Erlangian distribution, 64, 97, 172
Erlangs, 4, 76; *see also* Erl
Erlang's first formula, *see* Erlang loss formula
Erlang's second formula, *see* Erlang delay formula
Event, 10
Exponential distribution, *see* Negative-exponential distribution
Exponential service times, 74

F

Failure rate, 49
Farrell, R.L., 136, 329
Feast, G.M., 286, 327
Feedback, 118, 136, 302
Feller, W., 9, 14, 18, 22, 33, 41, 45, 60, 71, 72, 111, 197, 198, 200, 212, 312, 330
Ferrari, D., 311, 330
Finch, P.D., 138, 330
Finite-source input, *see* Quasirandom input
Finite-source systems, with nonidentical sources, 125
Finite waiting room, 7, 92, 235, 238
Fishman, G.S., 286, 287, 297, 330
Fisz, M., 9, 28, 72, 330
Flat-rate trunks, 88, 89, 95, 119
Foley, F.D., 135, 136, 330, 338
Fortet, R., 128, 330
FORTRAN, 284, 287
Frankel, T., 312, 330
Franken, P., 310, 330
Fredericks, A.A., 171, 330
Fry, T.C., 117, 308, 330

Index

G

Gambler's ruin problem, *see* Classical ruin problem
Gamma distribution, 66; *see also* Erlangian distribution
Gaps, in expressway traffic, 71
Garabedian, P.R., 33, 142, 330
GASP, 284, 287
GASP IV, 287
Gating, M/G/1 with, 267
Gauss-Seidel iteration, 158
Gaver, D.P., 173, 330
Generalized Engset formula, 111
Generalized occupancy, 270
Generating function, *see* Probability-generating function
Geometric distribution, 7, 36, 43, 63, 64, 96
GI/G/s, 178
G I/G/1, 222
GI/M/s, 177, 267, 275
Giffin, W.C., 309, 330
Gnedenko, B.V., 310, 330
Gordon, G., 287, 331
Gordon, W.J., 137, 331
GPSS, 284, 287
Grandjean, C., 128, 330
Gross, D., 244, 309, 328, 331
Gumbel, H., 157, 331
Gunther, F.L., 297, 331

H

Haight, F.A., 71, 331
Halfin, S., 266, 301, 331
Halmstad, D.G., 222, 331
Halstrøm, H.A., 310, 326
Harris, Carl M., xv, 309, 331
Harrison, G., 229, 331
Hasofer, A.M., 227, 331
Hazelwood, R.N., 113, 117, 313, 336
Heidelberger, Philip, xv, 297, 331
Heterogeneous servers, 131, 157
Heyman, Daniel P., xv, 181, 245, 252, 253, 273, 312, 332
Hillier, F.S., 309, 311, 312, 313, 332
Hines, W.W., 309, 337
Hokstad, P., 237, 332
Holding cost, 244
Holding times, 3; *see also* Service times
Holtzman, J.M., 171, 332
Howard, J.H., Jr., 135, 327
Hyperexponential distribution, 67, 172

I

IBM Journal of Research and Development, 313
Idle period, 13
IEEE Transactions on Communications, 313
IEEE Transactions on Software Engineering, 313

Iglehart, D.L., 297, 329, 331, 332
Infinite-source input, *see* Poisson input
Infinitely divisible process, 60
Input process, 2, 73, 188
Institute of Switching and Data Technics at the University of Stuttgart, *see. Studies in Congestion Theory*
Insurance, 2, 219, 220
Integrodifferential equation of Takács, 227
Intended offered load, 104, 110, 115
International Teletraffic Congress, Proceedings of the, 314
Inventories, 2
Inverse-transformation method, 284
Iteration matrix, 159

J

j-busy period, 228
Jackson, J.R., 136, 332
Jackson, R.R.P., 135, 332
Jagerman, D.L., 81, 117, 198, 258, 332
Jaiswal, N.K., 310, 332
Jensen, A., 310, 326
Journal of Applied Probability, 313
Journal of the Association for Computing Machinery, 313
Journal of the Operations Research Society of Japan, 313
Journals, 313

K

Kahan, B.C., 287, 326
Karlin, S., 26, 72, 212, 312, 333
Katz, S.S., 171, 333
Keilson, J., 127, 333
Kelly, F.P., 127, 135, 175, 310, 333
Kendall, D.G., 176, 177, 190, 210, 212, 219, 333
Khintchine, A.Y., 26, 140, 219, 310, 333
Kingman, J.F.C., 253, 333
Kiviat, P.J., 287, 333
Kleinrock, L., 116, 172, 302, 309, 311, 333
Knuth, D.E., 286, 333
Kobayashi, H., 116, 311, 333
Koenigsberg, E., 138, 333
Kolmogorov's criterion for reversibility, 127
König, D., 310, 330
Kosten, L., 66, 140, 146, 309, 333, 334
Kovalenko, I.N., 310, 330
Kruse, W., 312, 329
Kuczura, A., 117, 171, 295, 334, 335
Kuehn, P.J., 266, 334

L

$L=\lambda W$, *see* Little's theorem
Laplace transform, 197

Laplace-Stieltjes transform, 63, 197
Lavenberg, S.S., 297, 334
Law, A.M., 297, 327, 334
Lee, A.M., 253, 309, 334
LeGall, P., 253, 275, 310, 334
Leibowitz, M.A., 266, 334
Lemoine, A.J., 135, 297, 329, 334
Levy, Y., 223, 334
Lewis, P.A.W., 286, 297, 331, 334
Lewis, T.G., 286, 334
Lieberman, G.J., 311, 332
Liebmann iteration, 159
Little, J.D.C., 178, 334
Little's theorem, 76, 99, 116, 178, 273
Lo, F.D., 312, 332
Load
 carried, 76, 104, 110, 114, 179
 lost, 81
 offered, 4, 76, 110, 114, 179, 208
Load carried by ordered server, 89, 94
Loss-delay system, see Finite waiting room
Lotze, A., 149, 334

M

$M/D/s$, 178
$M/D/1$, 189, 230, 253, 257, 260
$M/E_k/1$, 253
$M/G/1$, 64, 177, 189, 200, 208, 235, 241, 244, 253, 267
$M/G/\infty$, 32, 86, 119
$M/M/s$, 90, 178
$M/M/1$, 120, 190
Machine interference model, 111
Maclaurin series (for waiting times), 256, 258, 275
MacNair, E.A., 287, 337
Macrostate, 150
Magazine, M.J., 244, 328
Management Science, 313
Management science, see Operations research
Markov chain, 9, 83, 138, 212
 imbedded, 176, 210
Markov process, 26, 211
Markov property, 37, 43
Markowitz, H.M., 287, 333
Marmon, R.I., 303, 328
Mass service, theory of, 2
Mathematics of Operations Research, 313
Maximum, of independent random variables, 47, 49
Maxwell, W.L., 310, 328
McNickle, D.C., 136, 329
Mean queue length, 189
Mean waiting time, 97, 114, 189, 199, 209, 272
Measured-rate trunks, 88, 89, 95, 119
Melamed, B., 135, 136, 326, 335
Merging, into expressway, 70, 71
Messerli, E.J., 89, 335
Michel, R., 312, 329
Microstate, 150

Miller, H.D., 72, 312, 328
Miller, L.W., 310, 328
Mina, R.R., 309, 335
Minimum, of independent random variables, 46, 49
Mitrani, I., 135, 335
Mixed traffic, 126
Mixture, 172, 175
Miyazawa, M., 181, 335
Molina, E.C., 141
Monte Carlo method, 283
Morse, P., 67, 172, 309, 335
Multinomial distribution, 35
Muntz, R.R., 135, 325
Murdoch, J., 309, 335
Murray, G., 266, 328; *see also* Carter, Grace M.

N

N-policy, 243, 245
Naor, P., 252, 340
Naval Research Logistics Quarterly, 313
Neal, S.R., 117, 131, 148, 171, 295, 327, 334, 335
Negative-binomial distribution, 37
Negative-exponential distribution, 42, 50
Nekrasov iteration, 159
Networks of queues, 135, 136, 175
Neuts, Marcel F., xv, 9, 26, 31, 73, 173, 200, 212, 253, 258, 310, 335
Newell, G.F., 137, 312, 331, 335, 336
Noetzel, A.S., 135, 336
Noguchi, S., 302, 306, 337
Nonbiased queue discipline, 95, 98, 180
Nonscheduled queue discipline, *see* Nonbiased queue discipline
Normal distribution, 290
Numerical solution, 158
Nyquist, H., 141

O

$o(h)$, 15
Oakes, D., 84, 336
Occupancy, *see* Server occupancy
Offered load per idle source, 104
Oizumi, J., 302, 306, 337
Ondra, F., 312, 329
Operations Research, 313
Operations research, 2, 311
Opsearch, 313
Optimal control, *see* Optimal design
Optimal design, 243
Optimal operating policy, 244
Order statistics
 from uniform distribution, 39
 from exponential distribution, 46
Ordered hunt, 88, 90, 93, 153, 207
Output process, 188
Outside observer's distribution, 57, 77, 105

Index

Overflow distribution
  apportioning moments, 148
  mean, 140, 169
  variance, 140, 169
Overflow stream, 208
Overflow traffic, 139, 165
Overrelaxation iteration, 158

P

P-policy, 245
Page, E., 111, 309, 336
Palacios, F.G., 135, 325
Palm, C., 140, 208, 336
Panico, J.A., 309, 336
Particle-counting device, 240
PASCAL, 287
Pascal distribution, see Negative-binomial distribution
Peakedness factor, 149, 170
Peck, L.G., 113, 117, 313, 336
Pegden, C.D., 287, 336
Peter, E., 312, 329
Phase, 65
Phase type, 173
Phases, method of, 171, 303
  extended, 173
Pittel, B., 135, 336
PL/1, 287
Point estimate, 295
Poisson distribution, 17, 50
Poisson input, 52, 73, 103, 117
  time-varying, 119
Poisson process, 5, 17, 51, 136, 198, 200
Polite customer, 233
Pollaczek, F., 101, 219, 310, 336
Pollaczek-Khintchine formula, 189, 200, 209, 216
Prabhu, N.U., 230, 231, 244, 245, 310, 336, 338
Preemptive repeat, 69
Preemptive resume, 69
Priority queue, 69, 118, 119
Priority reservation, 118, 150
Pritsker, A.A.B., 287, 336
Probability, theory of, 9
Probability-generating function, 26, 139, 197
Probst, A.R., 287, 326
Processing times, 302
Processor sharing, 301
Product solutions, 126
Pseudorandom numbers, 286
Pure birth process, 16, 17, 51
Pure death process, 16, 17, 46

Q

Quantum, 302
Quasirandom input, 73, 102, 105, 117, 130, 188, 298

Queue discipline, 2, 74
Queueing theory, 2

R

Random input, see Poisson input
Random number generator, 286
Random variable, 9
Range, of sample from exponential distribution, 49
Rapp, Y., 170, 336
Recurrence time
  backward, 54, 205
  forward, 54, 56, 202
Recurrent input, 51, 201
Regenerative method, 290
Reich, E., 135, 136, 189, 336
Reiser, M., 287, 336
Reliability theory, 48, 49
Renege, 79
Renewal function, 201
Renewal input, see Recurrent input
Renewal point, 211
Renewal process, 201
  equilibrium, 203
Renewal sequence, ordinary, 201
Renewal theorem
  elementary, 201
  key, 204
Renewal theory, 200
Repairman model, 111
Response time, 115, 301
RESQ, 287
Retrials, 82
Reversibility, 127
Riemann integral, 48
Riemann-Stieltjes integral, 48, 62, 192
Riordan, J., 82, 101, 122, 140, 148, 233, 235, 310, 336, 337
Ross, S.M., 9, 72, 200, 212, 244, 312, 337
Round-robin, 302
Ruiz-Palá, E., 309, 337
Runnenburg, J. Th., 31, 337
Running Cost, 244

S

Saaty, T.L., 310, 337
St. John, Stephen, xv
Sakata, M., 302, 306, 337
Sanford, Donald G., xv
Sauer, C.H., 287, 297, 311, 334, 336, 337
Schassberger, R., 135, 173, 175, 307, 310, 337
Scherr, A.L., 111, 337
Schmeiser, Bruce W., xv, 286, 337
Schmidt, J.W., 309, 339
Schmidt, V., 310, 330
Schriber, T.J., 287, 337
Schruben, L.W., 297, 337
Seal, H.L., 222, 337

Seidel Iteration, 159
Semi-Markov process, 83
Separation of variables, 128
Server occupancy, 76
Service mechanism, 2, 73
Service times, 5
Sevast'yanov, B.A., 84, 338
Shannon, R.E., 287, 338
Shedler, G.S., 286, 297, 332, 334
Simon, B., 135, 136, 329, 338
Simple overflow model, 140
SIMSCRIPT, 284, 287
SIMULA, 287
Simulation, 281
SLAM, 287
Slutz, D.R., 297, 334
Smith, W. L., 111, 122, 308, 328, 338
Sobel, M. J., 244, 312, 332, 338
Sojourn time, 88, 99
Stage, see Phase
Stages, method of, see Phases, method of
Stationary policy, 244
Stationary property, 20
Statistical equilibrium, 4, 19, 20
Statistical-equilibrium distribution, 22
Statistical-equilibrium state equations, 24
Statistical questions (in simulation), 288
Stephenson, Samuel S., xv
Stidham, Shaler, Jr., xv, 178, 181, 183, 244, 245, 266, 273, 309, 332, 338
Stieltjes integral, see Riemann-Stieltjes integral
Stochastic dominance, 62
Stochastic process, 9, 14, 312
Stochastic service systems, theory of, 2
Stoyan, D., 135, 338
Studies in Congestion Theory, 314
Superposition, of Poisson Process, 59
Supplementary variable, 256; see also Additional conditioning variable
Switch point, 263
Switching cost, 244
Syski, Ryszard, xv, 72, 84, 101, 111, 117, 122, 148, 311, 338

T

T-policy, 243, 245, 251
Tables, of queueing formulas, 312
Takács, L., 72, 84, 122, 186, 205, 208, 219, 220, 222, 224, 225, 227, 228, 231, 232, 234, 235, 241, 253, 270, 271, 272, 273, 275, 311, 312, 338, 339
Tandem queues, 126, 132, 189
Tanner, J.C., 231, 339
Telephone traffic engineering, see Teletraffic engineering
Teletraffic engineering, 2, 3, 80, 87, 89, 95, 98, 111, 117, 118, 119, 129, 130, 131, 138, 298

Test customer, 13
Think time, 115
Throughput, 115
Throwdown, 283
Tijms, H., 253, 325
Tilt, Børge, xiii, xv, 120, 241, 328
Time slice, 302
Total probability, theorem of, 13
Towsley, D.F., 135, 327
Traffic intensity, 208
Traffic measurement, 295
Traffic theory, 2
Transaction, 115
Transient solution, 19, 86, 119, 120
Transportation engineering, 311; see also Vehicular traffic theory
Transition probabilities, 211
Truncated binomial distribution, 109
Truncated Poisson distribution, see Erlang loss distribution

U

Uniform distribution, 38, 54, 56
Utilization factor, 76

V

Vacation times, $M/G/1$, with, 222, 247
van Dantzig, D., 31
van der Vaart, 31
Vantilborgh, H., 135, 339
Variance reduction, 293
Vaulot, E., 94, 154, 339
Vehicular traffic theory, 71; see also Transportation engineering
Villanueva, R., 287, 333
Virtual waiting time, 98, 219, 227, 228

W

Wagner, H., 312, 329
Wagner, H.M., 311, 339
Waiting times
  service in order of arrival, 95, 113, 216, 225, 227, 271, 298, 324
  service in random order, 101, 253, 275
  service in reverse order of arrival, 102, 234
Wallace, V.L., 160, 339
Wallström, B., 140, 148, 339
Weber, J.H., 298, 299, 300
Welch, P.D., 297, 331
White, J.A., 309, 339
Widder, D.V., 192, 193, 197, 198, 340
Wilkinson, R.I., 167, 170, 336, 340
Wishart, D.M.G., 224, 235, 340
Wolff, R.W., 297, 331
Wolman, Eric, xv

Y

Yadin, M., 252, 340
Yakowitz, S.J., 286, 340
Yechiali, U., 223, 334
Young, D.M., 160, 340
Yu, O.S., 313, 332

Z

Ziegler, B.P., 135, 326